HANDBOOK OF
PROTON IONIZATION HEATS

HANDBOOK OF PROTON IONIZATION HEATS

and Related Thermodynamic Quantities

JAMES J. CHRISTENSEN
PROFESSOR OF CHEMICAL ENGINEERING
DEPARTMENT OF CHEMICAL ENGINEERING SCIENCE
BRIGHAM YOUNG UNIVERSITY
PROVO, UTAH

LEE D. HANSEN
ASSOCIATE PROFESSOR OF CHEMISTRY
DEPARTMENT OF CHEMISTRY
BRIGHAM YOUNG UNIVERSITY
PROVO, UTAH

REED M. IZATT
PROFESSOR OF CHEMISTRY
DEPARTMENT OF CHEMISTRY
BRIGHAM YOUNG UNIVERSITY
PROVO, UTAH

A WILEY-INTERSCIENCE PUBLICATION
JOHN WILEY & SONS, New York • London • Sydney • Toronto

Contribution No. 85 from the Center for Thermochemical Studies, Brigham Young University, Provo, Utah

Library of Congress Cataloging in Publication Data:

Christensen, James J

Handbook of proton ionization heats and related thermodynamic quantities.

(Contribution from the Center for Thermochemical Studies, Brigham Young University, Provo, Utah; no. 85)

"A Wiley-Interscience publication."

Includes indexes.

1. Thermodynamics—Tables, calculations, etc. 2. Protons—Tables. 3. Ionization—Tables. I. Hansen, Lee D., joint author. II. Izatt, Reed McNeil, 1926- joint author. III. Title.

IV. Series: Brigham Young University, Provo, Utah. Center for Thermochemical Studies. Contribution; no. 85.

QD511.8.C48 539.7'212 76-16511

ISBN 0-471-01991-7

Printed in the United States of America

10 9 8 7 6 5 4 3 2 1

To
V.B.C., J.W.H. and H.F.I.

PREFACE

We have undertaken the compilation of heats of proton ionization in solution to aid those working in the fields of chemistry, physics, biology, medicine, engineering, etc., where these data are necessary and useful. We have found that an enormous amount of such data has been collected and in some cases partly tabulated, as in the table of "Heats of Proton Ionization and Related Thermodynamic Quantities" by R. M. Izatt and J. J. Christensen in *Handbook of Biochemistry and Selected Data for Molecular Biology,* ed. H. Sober, The Chemical Rubber Publishing Company, Cleveland Ohio, Second Edition, 1970 pp. J58-J173, but that up to now there has been no one reasonably complete source of this information.

This book is a compilation of enthalpy changes for proton ionization, ΔH, together with the related thermodynamic quantities, pK, ΔS, and ΔCp, where available. The book consists of a Table in which are summarized the published literature values through mid-1975 under the headings of the various acids. The criterion for including an acid in the Table was that a ΔH value(s) must have been reported in the published literature. Values appearing in theses, technical reports, books, or other nonrefereed sources were not included. In addition to the ΔH, pK, ΔS, and ΔCp values, the following information is included in the Table: the appropriate reaction, temperature, method, and conditions of measurement of ΔH, literature references, and pertinent supplemental information.

Three indexes–Empirical Formula, Synonym, and Reference–are included in the book. The Empirical Formula Index contains the empirical formula of each acid together with the location of the acid in the table. The Synonym Index is an alphabetical listing of the common synonyms of the acids appearing in the Table and the location of the acids in the table. The Reference Index includes a year by year alphabetical listing of all references cited in the Table.

The compilers have read the original of each paper, if possible, and data from other sources are reported only if the original paper could not be obtained through any of the means available, including interlibrary loans. All values are those reported by the original investigators except when only pK or ΔS was reported in addition to ΔH, and in such cases the third quantity, either pK or ΔS, was calculated from the other two. Values calculated in this manner are indicated by being underlined. The authors do not claim to be infallible with respect to the selection of the most correct value in these cases and are solely responsible for any errors which may have been made in either compiling or critically evaluating the data. Comments concerning errors and omitted data would be most appreciated.

This book is intended as a companion to the following published Tables.

"Handbook of Metal Ligand Heats and Related Thermodynamic Quantitites" by J. J. Christensen, D. J. Eatough, and R. M. Izatt, Marcel Dekker Publishing Company, New York, N.Y., Second Edition, 1975.

"Stability Constants of Metal-Ion Complexes" by L. G. Sillen and A. E. Martell, Special Publication No. 17, The Chemical Society, Burlington House, London, 1964.

"Stability Constants of Metal-Ion Complexes, Supplement No. 1" by L. G. Sillen and A. E. Martell, Special Publication No. 25, The Chemical Society, Burlington House, London, 1971.

We express our gratitude to Robert Reeder and Neil Izatt for editorial assistance and to Susan Belnap for technical typing of the manuscript.

JAMES J. CHRISTENSEN
LEE D. HANSEN
REED M. IZATT

Brigam Young University
Provo, Utah
April 1976

CONTENTS

USE OF TABLE AND INDEXES

A. *Acids*

1. The acids are ordered alphabetically and each is assigned a letter-number combination, i.e., A1, A2, A3, etc. In general, the nomenclature follows that in the 54th Edition of the *Handbook of Chemistry and Physics,* The Chemical Rubber Co., 1973-1974.

B. *Formula, Synonyms*

1. Empirical and structural formulas for the neutral form of each acid are given. Empirical formulas are also given in the Empirical Formula Index where they are arranged with the elements in the order C, H, O, N, Cl, Br, I, F, S, P followed by other elements in the alphabetical order of their symbols according to Beilstein's system.
2. Synonyms for each acid are given alphabetically in the Synonym Index.

C. *Reactions*

1. ΔH, pK, and ΔS values are valid for the reaction designated by the underlined number which is in parentheses immediately following the pK value. The charge on the ionizing acid is designated by the sign and magnitude of this number. These numbers have the following meanings where HA is the acid.

Number	*Reaction*
−3	$HA^{3-} = H^+ + A^{4-}$
−2	$HA^{2-} = H^+ + A^{3-}$
−1	$HA^{1-} = H^+ + A^{2-}$
0	$HA^0 = H^+ + A^{1-}$
+1	$HA^{1+} = H^+ + A^0$
+2	$HA^{2+} = H^+ + A^{1+}$
+3	$HA^{3+} = H^+ + A^{2+}$

Only the proton ionizing from the acid is shown. Other ionizable protons, if any, are included in the A part of the formula. If no number appears after the pK value, see Conditions or Remarks Column for the reaction. The reaction designation is that given by the author(s). In many cases where the reaction was not reported, the compilers have taken the liberty to assign the reaction designation.

2. When the acid is part of a complex molecule or ion consisting of a metal ion and one or more acids, the formula of the complex is given in the Conditions column. For example, if a complex ion containing Maleic Acid had the formula

$$[(NH_3)_5Cr(III)O_2CCH = CHCO_2H)^{2+}$$

the reaction would be designated by (+2) and HA would be given in the conditions column as

$$HA = [(NH_3)_5Cr(III)O_2CCH = CHCO_2H]^{2+}$$

D. ΔH

1. The first criterion for construction fo the Table was to include all reliable ΔH data. When only one ΔH value was reported for a given reaction, this value is given in the Table irrespective of its apparent validity. In those cases where values reported by different workers disagree widely with no apparent reason all values are given.

2. The order of the thermodynamic quantities is as follows. The reactions are ordered according to their pK values; going from the lowest pK value to the highest. All values for the ionization having the lowest pK value are presented before the one having the next higher pK value. For a given reaction, values in water are presented first with those for other solvents following, the solvents being arranged alphabetically. For a given solvent, values are arranged in order of increasing ionic strength. For a given ionic strength values are arranged in order of increasing temperature. This order is given below in outline form.

   a. Reaction
      (1) Ionization reactions for the simple acid starting with the ionization having the lowest pK value.
      (2) Ionization reactions for the acid in the complex molecule or ion containing a metal ion according to order given for metals in L. G. Sillen and A. E. Martell, "Stability Constants," Chem. Soc. Special Pub. No. 17, 1964.

   b. Solvent
      (1) Water
      (2) Other solvents arranged alphabetically

   c. Ionic Strength, μ
      (1) $\mu = 0$
      (2) Increasing μ values

   d. Temperature—lowest temperature first

3. The ΔH values are valid at the temperature or over the temperature range listed in the T, °C column.

4. Uncertainties of the ΔH values are not reported; however, the number of significant figures given is that reported in the original article.

5. The units on ΔH are kcal/gram mole.

E. *pK*

1. Each pK value is valid for the reaction designated by the underlined number which is in parentheses immediately following the pK value. The significance of these numbers is given in Section C. If no number appears after the pK value see Conditions or Remarks column for the reaction.

2. Each pK value is valid at the temperature listed in the T, °C column, unless otherwise specified in parentheses immediately following the pK value.

3. Uncertainties of the pK values are not reported; however, the number of significant figures given is that reported in the original article if pK is reported directly. If ΔG or K is reported in the original article then pK is usually reported with one additional significant figure.

F. ΔS

1. The ΔS values are valid at the temperatures listed in the T, °C column if the ΔH and pK values are valid at the same temperature. If the ΔH and pK values are not valid at the same temperature, no attempt has been made to designate the temperature at which ΔS is valid.

2. Uncertainties of the ΔS values are not reported: however, the number of significant figures given is that reported in the original article.

3. The units on ΔS are cal./°C-gram-mole.

G. *Values Calculated by Compilers*

1. In those cases where ΔH and only one of the other two thermodynamic quantities, pK or ΔS, are given, the compilers have taken the liberty to calculate the third quantity. This calculated value is underlined in each case.

H. *T, °C*

1. Where ΔH values were determined calorimetrically, they are valid at the stated temperatures.

2. Where ΔH values were determined by the temperature variation method, the temperature or temperature range is that given by the author(s).

I. *Method*

1. The method of ΔH determination is designated by C (calorimetric) or T (temperature variation).

J. *Conditions*

1. The experimental conditions under which ΔH was determined are given. Supporting electrolyte, when present, is identified by its composition and molarity (M). Composition of non-aqueous solvents, when present, is indicated by the symbol, S: followed by the composition of the solvent. In a solvent mixture if the second component is not stipulated it is understood to be water.

2. If the reaction is not stipulated in parentheses immediately following the pK values (see Section C), it is indicated either in the Conditions column by R: followed by the chemical reaction or in the Remarks column by an alphabetic letter.

3. Abbreviations used are defined under Section L.

4. The units of ΔCp are cal./°C-gram-mole.

K. *References*

1. The references given are those from which the data were taken.

2. All references in the Table are listed in the Reference Index according to the year of publication. Under each year the references are in order alphabetically according to the authors. In the Reference Index all journal titles are abbreviated in accordance with Chemical Abstracts usage as given in *ACCESS,* 1969 edition.

L. *Remarks*

1. When the same comment was applicable to several data entries, that comment was given an alphabetic letter. These letters have the following meanings.

   a. Indicates value for heat of ionization used to calculate ΔH, e.g., if value for heat of ionization of water was 13.5 kcal/mole then a:13.5 appears in Remarks column.

   b. ΔH values are calculated from pK values determined at only two temperatures.

   c. ΔH values are average values over the indicated temperature range, resulting in a combination of thermodynamic data valid at different temperatures.

   d. Some data in this reference appear to be unreliable (private communication from author(s) to compilers).

   e. Indicates group from which proton was reported to be ionizing from, e.g., if proton was reported to be ionizing from NH_3^+ group then a e:NH_3^+ appears in Remarks column.

   f. These data appear unreliable.

   g. ΔH value was computed from earlier work, e.g., if earlier work has a reference number of

50Ha then g:50Ha appears in the Remarks column.

h. These data should be disregarded; see Reference 62 Cb.

i. Data are not internally consistent.

j. The salt of the acid was used. This is indicated by placing the cation formula after the letter, e.g., if the sodium salt of the acid was used then j:Na appears in the Remarks column.

k. ΔH value was calculated from K values where K was in molar units.

l. The ΔH value was computed from a compilation of earlier work.

m. No correction was made for the dissociation of urea prior to calculation of ΔH.

n. The data are questionable; see Reference 62 Cb.

o. ΔH value was corrected for presence of microspecies.

p. The ΔH value is an average of two or more ΔH values reported in this reference for the same temperature and ionic strength.

q. The author has placed a large uncertainty on ΔH, e.g., if the uncertainty was reported as ±3 kcal./mole then q:±3 appears in the Remarks column.

M. *Abbreviations*

1. The following abbreviations were used in the Table primarily in the Conditions column.

HA = acid	C = concentration of acid in moles/liter
H = proton	μ = ionic strength
R = reaction	D = dielectric constant
v = volume	w = weight
M = moles/liter	Cp = heat capacity in cal./deg.-mole
m = moles/kg	

N. *Heats of Metal Ligand Interaction and Related Thermodynamic Quantities*

1. In many cases it is desirable to know related thermodynamic values for metal ligand interaction. A table containing heats of metal ligand interaction and related thermodynamic quantities for many of the acids contained in the present table has been published by the compilers in the *Handbook of Metal Ligand Heats and Related Thermodynamic Quantities,* Marcel Dekker Publishing Co., New York, N.Y., Second Edition, 1975.

HANDBOOK OF PROTON IONIZATION HEATS

ΔH, kcal /mole	pK	ΔS, cal /mole°K	T,°C	M	Conditions	Ref.	Remarks

A

A1 ACETIC ACID $C_2H_4O_2$ CH_3CO_2H

ΔH, kcal /mole	pK	ΔS, cal /mole°K	T,°C	M	Conditions	Ref.	Remarks
0.804	(0)	-----	0	C	μ=0	70Ld	
0.714	4.781(0)	-19.2	0	T	μ=0	33Ha	
0.781	4.780(0)	-18.8	0	T	μ=0	66Pa	
	(0)	-----	2.5	C	μ=0 ΔCp=-29.4	70Ld	
0.657	(0)	-----	5	C	μ=0	70Ld	
0.552	4.770(0)	-19.7	5	T	μ=0	33Ha	
0.47	4.762(0)	-20.2	10	C	μ=0	68Cd	
0.389	4.762(0)	-20.3	10	T	μ=0	33Ha	
0.410	(0)	-----	10	T	μ=0	34Hc	
	(0)	-----	10	C	μ=0 ΔCp=-38.2	70Ld	
0.275	(0)	-----	15	C	μ=0	70Ld	
0.223	4.758(0)	-20.9	15	T	μ=0	33Ha	
0.057	4.756(0)	-21.5	20	T	μ=0	33Ha	
	(0)	-----	20	C	μ=0 ΔCp=41.2	70Ld	
-0.02	4.756(0)	-21.9	25	C	μ=0	67Ca	
-0.09	4.76(0)	-22.1	25	C	μ=0	71Ac	
						66Ae	
						67Ad	
-0.137	(0)	-----	25	C	μ=0	70Ld	
-0.07	4.76(0)	-22.0	25	C	μ=0	58Ca	a:13.50
-0.096	4.76(0)	-22.2	25	C	μ=0	74Md	
-0.10	(0)	-----	25	C	μ=0	70Ab	
						71Ad	
-0.112	4.756(0)	-22.1	25	T	μ=0	33Ha	
-0.100	(0)	-----	25	T	μ=0	34Hc	
-0.105	4.7554(0)	-22.11	25	T	μ=0	52Ea	
-0.100	4.757(0)	-22.1	25	T	μ=0	57Ka	
-0.098	4.756(0)	-22.1	25	T	μ=0 ΔCp=-41.3	39Hb	1:66Pa
						66Pa	
-0.095	4.756(0)	-22.1	25	T	μ=0 ΔCp=-37	65Pa	
	(0)	-----	30	C	μ=0 ΔCp=-29.3	70Ld	
-0.282	4.757(0)	-22.6	30	T	μ=0	33Ha	
-0.430	(0)	-----	35	C	μ=0	70Ld	
-0.455	4.762(0)	-23.2	35	T	μ=0	33Ha	
	(0)	-----	40	C	μ=0 ΔCp=-24.1	70Ld	
-0.50	4.769(0)	-23.4	40	C	μ=0	68Cd	
-0.628	4.769(0)	-23.8	40	T	μ=0	33Ha	
-1.1047	4.787(0)	-25.1	40	T	μ=0	66Pa	
-0.730	(0)	-----	40	T	μ=0	34Hc	
-0.671	(0)	-----	45	C	μ=0	70Ld	
-0.804	4.777(0)	-24.3	45	T	μ=0	33Ha	
	(0)	-----	50	C	μ=0 ΔCp=-23.0	70Ld	
-0.982	4.787(0)	-24.8	50	T	μ=0	33Ha	
-0.901	(0)	-----	55	C	μ=0	70Ld	
-1.161	4.799(0)	-25.4	55	T	μ=0	33Ha	
-1.342	4.812(0)	-26.0	60	T	μ=0	33Ha	
-0.05	(0)	-----	25	C	μ=0.01M(Na_2SO_4)	71Ha	
0.280	(0)	-----	20	C	μ=0.1M($NaNO_3$)	72Sa	
0.09	(0)	-----	25	C	μ=0.1	64Nb	
0.18	(0)	-----	25	C	μ=0.2(NaCl)	62Wa	q:±0.05
0.721	(0)	-----	25	C	μ=2.0($NaClO_4$)	64Gc	
1.395	(0)	-----	25	T	μ=2.7M(NaCl)	53Ea	
0.745	5.015(0)	-20.3	25	C	μ=3.00M($NaClO_4$)	67Ga	
0.721	(0)	-----	25	C	μ=3M($NaClO_4$)	61Sa	
0.214	4.738(35°,0)	-21	0&35	T	--	39Wa	b,c
-0.090	4.756(0)	-22.1	25	C	--	67Ac	
						67Ad	
0.09	4.754(0)	-22.1	25	T	--	62Aa	
-0.100	4.752(0)	-22.1	25	T	--	39Ea	
-0.656	4.72(25°,0)	-23.8	25-90	T	--	67Ma	
0.0	4.80(0)	-21.9	90	T	--	67Ma	
4.850	5.46(0)	-8.7	90-190	T	--	67Ma	

ΔH, kcal /mole	pK	ΔS, cal /mole°K	T,°C	M	Conditions	Ref.	Remarks
Al, cont.							
0.0	4.19(0)	-19.2	190-210	T	--	67Ma	
-7.466	2.07(0)	-144.9	210-300	T	--	67Ma	
5.7	12.62(0)	-38.6	25	C	S: Acetic Acid (Anhydrous)	52Ja	
-0.28	4.86(0)	-23.2	25	C	S: 5 w % Acetone	71Ac	
-0.35	4.99(0)	-24.0	25	C	S: 10 w % Acetone	71Ac	
-0.25	5.13(0)	-24.3	25	C	S: 15 w % Acetone	71Ac	
-0.07	5.27(0)	-24.4	25	C	S: 20 w % Acetone	71Ac	
-0.10	5.44(0)	-25.2	25	C	S: 25 w % Acetone	71Ac	
-0.14	5.61(0)	-26.1	25	C	S: 30 w % Acetone	71Ac	
-0.23	5.99(0)	-28.2	25	C	S: 40 w % Acetone	71Ac	
-0.35	6.43(0)	-30.6	25	C	S: 50 w % Acetone	71Ac	
-0.46	6.99(0)	-33.5	25	C	S: 60 w % Acetone	71Ac	
-0.59	7.67(0)	-37.1	25	C	S: 70 w % Acetone	71Ac	
0.024	4.87(0)	-22.2	25	C	μ=0 S: 5 w % tert-Butanol	74Md	
0.048	4.95(0)	-22.5	25	C	μ=0 S: 10 w % tert-Butanol	74Md	
0.19	(0)	-----	25	C	μ=0 S: 10 w % tert-Butanol	71Ad 70Ab	
0.12	5.05(0)	-22.7	25	C	S: 15 w % tert-Butanol	74Md	
0.24	(0)	-----	25	C	S: 15 w % tert-Butanol	71Ad 70Ab	
0.26	5.11(0)	-22.5	25	C	μ=0 S: 17.5 w % tert-Butanol	74Md	
0.38	5.17(0)	-22.5	25	C	μ=0 S: 20 w % tert-Butanol	74Md	
0.31	(0)	-----	25	C	μ=0 S: 20 w % tert-Butanol	70Ab 70Ad	
0.36	(0)	-----	25	C	S: 22 w % tert-Butanol	71Ad 70Ab	
0.22	5.28(0)	-23.4	25	C	μ=0 S: 25 w % tert-Butanol	74Md	
0.36	(0)	-----	25	C	S: 25 w % tert-Butanol	71Ad 70Ab	
0.24	5.43(0)	-24.8	25	C	μ=0 S: 30 w % tert-Butanol	74Md	
0.41	(0)	-----	25	C	S: 30 w % tert-Butanol	71Ad 70Ab	
-0.096	5.57(0)	-25.8	25	C	μ=0 S: 35 w % tert-Butanol	74Md	
0.41	(0)	-----	25	C	S: 38 w % tert-Butanol	71Ad 70Ab	
-0.24	5.71	-27.0	25	C	μ=0 S: 40 w % tert-Butanol	74Md	
0.36	(0)	-----	25	C	S: 40 w % tert-Butanol	71Ad 70Ab	
0.2	(0)	-----	25	C	S: 48 w % tert-Butanol	71Ad 70Ab	
0.14	(0)	-----	25	C	μ=0 S: 50 w % tert-Butanol	71Ad 70Ab	
-0.10	(0)	-----	25	C	S: 60 w % tert-Butanol	71Ad 70Ab	
-0.18	13.30(0)	-61.5	15	T	S: Dimethylformamide	68Pc	
1.98	13.29(0)	-54.0	20	T	S: Dimethylformamide	68Pc	
4.19	13.25(0)	-46.6	25	T	S: Dimethylformamide	68Pc	
6.43	13.19(0)	-39.1	30	T	S: Dimethylformamide	68Pc	
8.71	13.10(0)	-31.7	35	T	S: Dimethylformamide	68Pc	
11.02	12.99(0)	-24.2	40	T	S: Dimethylformamide	68Pc	
-0.05	5.290(0)	-24.4	25	T	S: 20 w % Dioxane	62Aa	
-0.102	5.293(0)	-24.6	25	T	S: 20 w % Dioxane	45Ja	
-0.016	5.292(0)	-24.3	25	T	S: 20 w % Dioxane ΔCp=-40.8	39Hb	

Al, cont.

ΔH, kcal /mole	pK	ΔS, cal /mole°K	T, °C	M	Conditions	Ref.	Re-marks
-0.44	6.318(0)	-30.3	25	T	S: 45 w % Dioxane	62Aa	
-0.392	6.302(0)	-30.2	25	T	S: 45 w % Dioxane	45Ja	
-0.394	6.307(0)	-30.2	25	T	S: 45 w % Dioxane ΔCp=-43.3	39Hb	1
-0.61	8.318(0)	-40.1	25	T	S: 70 w % Dioxane	62Aa	
-0.615	8.276(0)	-39.9	25	T	S: 70 w % Dioxane	45Ja	
-0.594	8.321(0)	-40.1	25	T	S: 70 w % Dioxane ΔCp=-44.6	39Hb	1
-1.34	10.135(0)	-50.8	25	T	S: 82 w % Dioxane	62Aa	
-0.17	4.93(0)	-23.0	25	C	μ=0 S: 10 w % Ethanol	67Ad,69Ab	
-0.10	5.06(0)	-23.5	25	C	μ=0 S: 17 w % Ethanol	67Ad,69Ab	
-0.11	5.11(0)	-23.8	25	C	μ=0 S: 20 w % Ethanol	67Ad,69Ab	
-0.10	5.25(0)	-24.4	25	C	μ=0 S: 26 w % Ethanol	67Ad,69Ab	
-0.10	5.32(0)	-24.7	25	C	μ=0 S: 30 w % Ethanol	67Ad,69Ab	
-0.13	5.43(0)	-25.3	25	C	μ=0 S: 35 w % Ethanol	67Ad,69Ab	
-0.15	5.55(0)	-25.9	25	C	μ=0 S: 40 w % Ethanol	67Ad,69Ab	
-0.40	5.80(0)	-27.9	25	C	μ=0 S: 50 w % Ethanol	67Ad,69Ab	
-0.50	6.10(0)	-29.6	25	C	μ=0 S: 60 w % Ethanol	67Ad,69Ab	
-0.72	6.45(0)	-31.9	25	C	μ=0 S: 70 w % Ethanol	67Ad,69Ab	
-1.24	6.82(0)	-35.4	25	C	μ=0 S: 80 w % Ethanol	67Ad,69Ab	
0.05	(0)	-----	25	C	S: 10 w % Isopropanol	71Ad 70Ab	
0.05	(0)	-----	25	C	S: 15 w % Isopropanol	71Ad 70Ab	
0.05	(0)	-----	25	C	S: 20 w % Isopropanol	71Ad 70Ab	
0.05	(0)	-----	25	C	S: 25 w % Isopropanol	71Ad 70Ab	
0.05	(0)	-----	25	C	S: 30 w % Isopropanol	71Ad 70Ab	
0.05	(0)	-----	25	C	S: 35 w % Isopropanol	71Ad 70Ab	
0.05	(0)	-----	25	C	S: 40 w % Isopropanol	71Ad 70Ab	
-0.14	(0)	-----	25	C	S: 50 w % Isopropanol	71Ad 70Ab	
-0.41	(0)	-----	25	C	S: 60 w % Isopropanol	71Ad 70Ab	
-0.10	4.81(0)	-22.3	25	C	S: 5 w % Methanol	71Ac	
0.25	4.90(0)	-21.6	25	C	S: 10 w % Methanol	71Ac	
0.081	4.905(0)	-22.2	25	T	S: 10 w % Methanol ΔCp=-40.1	39Hb	1
0.35	5.07(0)	-22.0	25	C	S: 20 w % Methanol	71Ac	
0.264	5.081(0)	-22.4	25	T	S: 20 w % Methanol ΔCp=-38.9	39Hb	1
0.60	5.26(0)	-22.0	25	C	S: 30 w % Methanol	71Ac	
0.40	5.46(0)	-23.7	25	C	S: 40 w % Methanol	71Ac	
0.15	5.69(0)	-25.5	25	C	S: 50 w % Methanol	71Ac	
-0.052	5.660(0)	-26.1	25	T	μ=0 S: 50 w % Methanol ΔCp=-54	65Pa	
0.10	5.93(0)	-26.8	25	C	S: 60 w % Methanol	71Ac	
-0.20	6.27(0)	-29.4	25	C	S: 70 w % Methanol	71Ac	
-0.30	6.68(0)	-31.6	25	C	S: 80 w % Methanol	71Ac	
-2.12	(0)	-----	25	C	S: 95 v % Methanol	64Gc	
-4.450	(0)	-----	17	C	S: Methanol	27Wa	
5.13	9.625(0)	-26.8	25	T	μ=0 S: Methanol ΔCp=0	70Bf	
5.0	9.625(0)	-27	25	T	S: Methanol	69Rd	
4.984	8.245(0)	-19.8	5	T	μ=0 S: N-Methylpropionamide	72Eb	
4.872	8.177(0)	-20.2	10	T	μ=0 S: N-Methylpropionamide	72Eb	
4.757	8.111(0)	-20.6	15	T	μ=0 S: N-Methylpropionamide	72Eb	
4.641	8.050(0)	-21.0	20	T	μ=0	72Eb	

ΔH, kcal /mole	pK	ΔS, cal /mole°K	T, °C	M	Conditions	Ref.	Remarks
A1, cont.							
4.523	7.995(0)	-21.4	25	T	S: N-Methylpropionamide μ=0	72Eb	
4.403	7.940(0)	-21.8	30	T	S: N-Methylpropionamide μ=0	72Eb	
4.281	7.890(0)	-22.2	35	T	S: N-Methylpropionamide μ=0	72Eb	
4.156	7.841(0)	-22.6	40	T	S: N-Methylpropionamide μ=0	72Eb	
4.030	7.796(0)	-23.0	45	T	S: N-Methylpropionamide μ=0	72Eb	
3.902	7.754(0)	-23.4	50	T	S: N-Methylpropionamide μ=0	72Eb	
3.772	7.715(0)	-23.8	55	T	S: N-Methylpropionamide μ=0	72Eb	
0.421	(0)	-----	25	C	S: N-Methylpropionamide μ=3M($NaClO_4$) R: $HA_2^- = H^+ + 2A^-$	61Sa	q:±0.1
1.017	(0)	-----	25	C	μ=3M($NaClO_4$) R: $H_2A_2 = 2H^+ + 2A^-$	61Sa	
A2 ACETIC ACID, duterium			$C_2H_3DO_2$		CH_3CO_2D		
1.192	5.3463(0)	-20.2	0	T	μ=0 HA=CH_3COOD	65Ga	
0.275	5.3130(0)	-23.4	25	T	μ=0 HA=CH_3COOD	65Ga	
-0.092	4.756(0)	-22.1	25	T	HA=CH_3COOD	41Ha	
-0.730	5.3245(0)	-26.6	40	T	HA=CH_3COOD	65Ga	
-0.052	5.292(0)	-24.4	25	T	S: 20 w % Dioxane HA=CH_3COOD	41Ha	
-0.442	6.308(0)	-30.3	25	T	S: 45 w % Dioxane HA=CH_3COOD	41Ha	
-0.610	8.322(0)	-40.1	25	T	S: 70 w % Dioxane HA=CH_3COOD	41Ha	
-1.138	10.140(0)	-50.8	25	T	S: 82 w % Dioxane HA=CH_3COOD	41Ha	
1.174	5.360(5°,0)	-20.3	0	T	μ=0 S: D_2O HA=CH_3COOD	66Pb	
0.279	5.325(0)	-23.4	25	T	μ=0 S: D_2O HA=CH_3COOD	66Pb	
-0.695	5.336(0)	-26.6	50	T	μ=0 S: D_2O HA=CH_3COOD	66Pb	
0.735	4.795(0)	-19.2	0	T	μ=0 HA=CD_3COOH	66Pa	
0.069	4.771(0)	-22.1	25	T	μ=0 HA=CD_3COOH	66Pa	
-0.945	4.799(0)	-24.9	50	T	μ=0 HA=CD_3COOH	66Pa	
A3 ACETIC ACID, amide, N, N-dimethyl-			C_4H_9ON		$H_3CCON(CH_3)_2$		
32.0	(0)	-----	25	C	C$\sim 10^{-3}$M S: Fluorosulfuric acid	74Ab	
A4 ACETIC ACID, nitrile			C_2H_3N		CH_3CN		
13.6	(0)	-----	25	C	C$\sim 10^{-3}$M S: Fluorosulfuric acid	74Ab	
A5 ACETIC ACID, nitriolotri-			$C_6H_9O_6N$		$N(CH_2CO_2H)_3$		
2.0	1.687(0)	-2.1	0	T	μ=0	56Ha	
0.25	1.65(0)	-7.3	10	T	μ=0	56Ha	
-0.25	1.65(0)	-8.4	20	T	μ=0	56Ha	
-0.30	1.88(0)	-9.6	25	C	μ=0	65Mb	
-0.75	1.66(0)	-9.5	30	T	μ=0	56Ha	

ΔH, kcal /mole	pK	ΔS, cal /mole°K	T,°C	M	Conditions	Ref.	Remarks
A5, cont.,							
-1.75	1.686(0)	-12.3	40	T	μ=0	56Ha	
0.4	2.953(-1)	-12.1	0	T	μ=0	56Ha	
0.3	2.948(-1)	-12.5	10	T	μ=0	56Ha	
0	2.940(-1)	-13.5	20	T	μ=0	56Ha	
-0.30	2.48(-1)	-12.3	25	C	μ=0	65Mb	
-0.8	2.956(-1)	-16.1	30	T	μ=0	56Ha	
-1.1	2.978(-1)	-17.2	40	T	μ=0	56Ha	
5.20	10.594(-2)	-29.6	0	T	μ=0	56Ha	
4.80	10.454(-2)	-30.9	10	T	μ=0	56Ha	
4.40	10.334(-2)	-32.4	20	T	μ=0	56Ha	
3.70	9.685(-2)	-31.9	25	C	μ=0	65Mb	
4.05	10.230(-2)	-33.6	30	T	μ=0	56Ha	
4.566	(-2)	-----	20	C	μ=0.01	64Eb	
4.53	10.28(-2)	-31.9	25	C	μ=0.04	69Cc	
4.68	(-2)	-----	15-40	T	0.10F(KNO_3)	62Ma	
4.56	9.73(-2)	-29	20	C	μ=0.1	64Ha	
4.73	9.73(-2)	-28.4	20	C	μ=0.1(KNO_3)	64Aa	
4.90	(-2)	-----	25	C	μ=0.1M(KNO_3)	71Gd	
-0.7	1.82(25°,+2)	-10.7	6&25	T	μ=1M($NaClO_4$) HA=[Co(III)(NH_3)5-(Nitrilotriacetic acid)]$^{+2}$	75Oa	b,c
3.0	7.96(25°,+1)	-26.4	6&25	T	μ=1M($NaClO_4$) HA=[Co(III)(NH_3)5-(Nitrilotriacetic acid)]$^{+1}$	75Oa	b,c

A6 ACETIC ACID, bromo- $C_2H_3O_2Br$ $BrCH_2CO_2H$

ΔH, kcal /mole	pK	ΔS, cal /mole°K	T,°C	M	Conditions	Ref.	Remarks
-1.054	2.8877(0)	-16.8	20	T	μ=0	55Ia	
-1.10	2.902(0)	-17.0	25	C	μ=0	68Cd	
-0.620	2.902(0)	-15.4	25	C	μ=0	66Ae	
-1.200	2.901(0)	-17.2	25	T	μ=0	57Ka	
-1.239	2.9021(0)	-17.4	25	T	μ=0	55Ia	
-1.435	2.9180(0)	-18.1	30	T	μ=0	55Ia	
-1.43	2.93(0)	-19.9	40	C	μ=0	68Cd	

A7 ACETIC ACID, chloro- $C_2H_3O_2Cl$ $ClCH_2CO_2H$

ΔH, kcal /mole	pK	ΔS, cal /mole°K	T,°C	M	Conditions	Ref.	Remarks
-0.149	2.816(0)	-13.4	0	T	μ=0	34Wa	
-0.46	2.827(0)	-15.0	10	C	μ=0	68Cd	
-0.593	2.827(0)	-15.0	10	T	μ=0	34Wa	
-0.560	(0)	-----	10	T	μ=0	34Hc	
-0.914	2.833(0)	-16.1	18	T	μ=0	34Wa	
-0.897	2.8556(0)	-16.1	20	T	μ=0	55Ia	
-0.99	2.858(0)	-16.4	25	C	μ=0	68Cd	
-0.730	2.869(0)	-15.6	25	C	μ=0	66Ae	
-1.123	2.8699(0)	-16.9	25	T	μ=0	55Ia	
-1.236	2.861(0)	-13.9	25	T	μ=0 ΔCp=-49.0	39Hb	1
-1.107	2.859(0)	-7.25	25	T	μ=0	45Ja	
-1.170	2.858(0)	-17.0	25	T	μ=0	34Wa	
-1.080	2.867(0)	-16.8	25	T	μ=0	57Ka	
-1.240	(0)	-----	25	T	μ=0	34Hc	
-1.362	2.8894(0)	-18.7	30	T	μ=0	55Ia	
-1.402	2.883(0)	-17.9	32	T	μ=0	34Wa	
-1.53	2.910(0)	-18.2	40	C	μ=0	68Cd	
-1.639	2.910(0)	-18.5	40	T	μ=0	34Wa	
-2.030	(0)	-----	40	T	μ=0	34Hc	
-1.09	-2.77(0)	9.12	25	C	μ=0.05	65Pc	
-1.09	2.860(0)	-16.7	25	C	μ=0.05M	61Ba	
2.37	10.13(0)	-38.3	20	T	S: Dimethylformamide	68Pc	
3.97	10.10(0)	-32.9	25	T	S: Dimethylformamide	68Pc	

ΔH, kcal /mole	pK	ΔS, cal /mole°K	T,°C	M	Conditions	Ref.	Remarks
A7, cont.							
5.58	10.04(0)	-27.5	30	T	S: Dimethylformamide	68Pc	
7.23	9.96(0)	-22.1	35	T	S: Dimethylformamide	68Pc	
8.90	9.87(0)	-16.7	40	T	S: Dimethylformamide	68Pc	
A8 ACETIC ACID, chloro-, amide, N,N-dimethyl-			C_4H_8ONCl		$ClH_2CCON(CH_3)_2$		
27.4	(0)	-----	25	C	C∿10^{-3}M S: Fluorosulfuric acid	74Ab	
A9 ACETIC ACID, chlorodifluoro-			$C_2HO_2F_2Cl$		ClF_2CCO_2H		
-1.4	0.46(0)	-7	25	T	μ=0	69Kb	
A10 ACETIC ACID, cyclohexyl-			$C_8H_{14}O_2$		$\underline{CH_2CH_2CH_2CH_2CH_2CH}CH_2CO_2H$		
-0.56	4.45(0)	-22.4	10	C	μ=0	72Cb	
-1.39	4.51(0)	-25.3	25	C	μ=0	72Cb	
-2.04	4.55(0)	-27.3	40	C	μ=0	72Cb	
-----	(0)	-----	10-40	C	μ=0 ΔCp=-49	72Cb	
A11 ACETIC ACID, dibromo-			$C_2H_2O_2Br_2$		Br_2CHCO_2H		
-0.5	1.39(0)	-8	25	-	--	69La	
A12 ACETIC ACID, dichloro-			$C_2H_2O_2Cl_2$		Cl_2CHCO_2H		
0.17	1.30(0)	-6.5	25	C	μ=0	66Ae	
-1.9	1.36(0)	-12	25	T	μ=0	69Kb	
-0.1	1.25(0)	-6	25	-	--	69La	
A13 ACETIC ACID, difluoro-			$C_2H_2O_2F_2$		F_2CHCO_2H		
-2.0	1.34(0)	-13	25	T	μ=0	69Kb	
0	1.32(0)	-6	25	-	--	69La	
A14 ACETIC ACID, diphenylhydroxy-			$C_{14}H_{12}O_3$		$(C_6H_5)_2COHCO_2H$		
-0.168	3.049(35°,0)	-14	0&35	T	--	39Wa	b,c
A15 ACETIC ACID, fluoro-			$C_2H_3O_2F$		FCH_2CO_2H		
-1.232	2.5711(0)	-16.0	20	T	μ=0	55Ia	
-0.680	2.586(0)	-14.1	25	C	μ=0	66Ae	
-1.350	2.585(0)	-16.3	25	T	μ=0	57Ka	
-1.390	2.5864(0)	-16.5	25	G	μ=0	55Ia	
-1.558	2.6038(0)	-17.1	30	T	μ=0	55Ia	
A16 ACETIC ACID, hydroxy-			$C_2H_4O_3$		$HOCH_2CO_2H$		
1.020	3.875(0)	-14.0	0.3	T	μ=0	36Na	
0.53	3.848(0)	-15.8	10	C	μ=0	68Cd	
0.660	3.844(0)	-15.29	12.5	T	μ=0	36Na	
0.11	3.831(0)	-17.2	25	C	μ=0	67Ca	
0.15	3.831(0)	-17.1	25	C	μ=0	68Cd	

ΔH, kcal /mole	pK	ΔS, cal /mole°K	T,°C	M	Conditions	Ref.	Remarks
A16, cont.							
0.211	3.831(0)	-16.8	25	T	μ=0	39Hb	1:39Hb
					ΔCp=-39.2	36Na	
0.180	3.832(0)	-16.9	25	T	μ=0	57Ka	
-0.320	3.833(0)	-18.59	37.5	T	μ=0	36Na	
-0.39	3.835(0)	-18.8	40	C	μ=0	68Cd	
-0.920	3.849(0)	-20.55	50	T	μ=0	36Na	
0.56	(0)	-----	25	C	μ=2.0($NaClO_4$)	64Gc	
A17 ACETIC ACID, hydroxy-(phenyl)-		$C_8H_8O_3$			$C_6H_5CHOHCO_2H$		
-0.067	3.372(35°,0)	-16	0&35	T	--	39Wa	b,c
A18 ACETIC ACID, iminodi-		$C_4H_7O_4N$			$HN(CH_2CO_2H)_2$		
1.00	-2.33(+1)	14	25	C	μ=1.0M($NaClO_4$)	72Gd	
-0.95	2.58(25°,0)	-15.0	25-90	T	C_L=5x10^{-3}	71Bd	c,q: ±0.10
1.6	2.52(0)	-6.2	25	C	μ=0.1(KNO_3)	70Kb	
0.81	-1.40(0)	9.1	25	C	μ=1.0M($NaClO_4$)	72Gd	
6.98	9.89(-1)	-21.8	25	C	μ=0.02	69Cc	
8.15	9.44(-1)	-15.4	20	C	0.1M(KNO_3)	64Ab	
7.9	9.30(-1)	-16	25	C	μ=0.1(KNO_3)	70Kb	
7.90	(-1)	-----	25	C	μ=0.1M(KNO_3)	71Gd	
8.52	3.17(-1)	-14	25	C	μ=1.0M($NaClO_4$)	72Gd	
A19 ACETIC ACID, iminodi, N-(2-carboxyethyl)-		$C_7H_{11}O_6N$			$HO_2CCH_2CH_2N(CH_2CO_2H)_2$		
5.22	9.61(-2)	-26.4	25	C	μ=0.1(KNO_3)	69Ua 70Ma	
A20 ACETIC ACID, iminodi, N-(2-carboxyphenyl)-		$C_{11}H_{11}O_6N$			$(HO_2CC_6H_4)N(CH_2CO_2H)_2$		
1.70	7.77(-2)	-29.9	25	C	μ=0.1(KNO_3)	69Ua 70Ma	
A21 ACETIC ACID, iminodi, N-(2,5-dicarboxyphenyl)-		$C_{12}H_{11}O_8N$			$[(HO_2C)_2C_6H_3]N(CH_2CO_2H)_2$		
0.4	7.54(-3)	-33.1	25	C	μ=0.1M(KNO_3)	70Mb	
A22 ACETIC ACID, iminodi, N(2-hydroxyethyl)-		$C_6H_{11}O_5N$			$HOCH_2CH_2N(CH_2CO_2H)_2$		
5.4	8.65(25°,-1)	-22	20-40	T	μ=0.1M(KNO_3)	69Sb	c
A23 ACETIC ACID, iminodi, N-methyl-		$C_5H_9O_4N$			$CH_3N(CH_2CO_2H)_2$		
0	2.150(0)	-9.5	30	T	μ=0	56Oa	
6.8	9.920(0)	-23	30	T	μ=0	56Oa	
6.94	9.65(-1)	-20.5	20	C	0.1M(KNO_3)	65Aa	
6.3	9.41(25°,-1)	-22	20-40	T	μ=0.1M(KNO_3)	69Sb	
A24 ACETIC ACID, iodo-		$C_2H_3O_2I$			ICH_2CO_2H		
-0.75	3.124(0)	-16.9	10	C	μ=0	68Cd	

ΔH, kcal /mole	pK	ΔS, cal /mole°K	T, °C	M	Conditions	Ref.	Remarks
A24, cont.							
-1.256	3.1582(0)	-18.7	20	T	μ=0	55Ia	
-1.21	3.175(0)	-18.6	25	C	μ=0	68Cd	
-1.16	3.175(0)	-18.4	25	C	μ=0	66Ae	
-1.416	3.1752(0)	-19.3	25	T	μ=0	55Ia	
-1.370	3.174(0)	-19.1	25	T	μ=0	56Ka	
-1.585	3.1934(0)	-19.8	30	T	μ=0	55Ia	
-1.51	3.211(0)	-19.5	40	C	μ=0	68Cd	
A25 ACETIC ACID, mercapto-			$C_2H_4O_2S$		$HSCH_2CO_2H$		
6.2	10.56(-1)	-27.6	25	C	μ=0	64Wb	
6.28	10.68(-1)	-27.8	25	C	μ=0.05	65Ia	
A26 ACETIC ACID, mercapto-, S-phenyl-			$C_8H_8O_2S$		$C_6H_5SCH_2CO_2H$		
-0.80	3.39(0)	-18.2	25	C	μ=0.10M	69Ba	
-0.600	3.56(25°,0)	-18.3	0-25	T	--	34Ca	b,c
A27 ACETIC ACID, methoxy-			$C_3H_6O_3$		$CH_3OCH_2CO_2H$		
-0.35	3.544(0)	-17.4	10	C	μ=0	68Cd	
-0.94	3.570(0)	-19.4	25	C	μ=0	68Cd	
-0.960	3.5704(0)	-19.6	25	T	μ=0	60Kb	
-1.148	3.613(0)	-22.3	40	C	μ=0	68Cd	
A28 ACETIC ACID, oxo-			$C_2H_2O_3$		$HCOCO_2H$		
0.53	3.46(0)	-14	25	C	μ=0	67Ob	
0.641	3.46(0)	-13.68	25	C	μ=0.05	67Ob	
A29 ACETIC ACID, oxydi-			$C_4H_6O_5$		$O(CH_2CO_2H)_2$		
0.988	(0)	-----	5	C	μ=1.0M($NaClO_4$)	72Ge	
0.420	(0)	-----	20	C	μ=1.0M($NaClO_4$)	72Ge	
	(0)	-----	25	C	μ=1.0M($NaClO_4$) ΔCp=-29.5	72Ge	
0.027	(0)	-----	35	C	μ=1.0M($NaClO_4$)	72Ge	
-0.574	(0)	-----	50	C	μ=1.0M($NaClO_4$)	72Ge	
-0.143	(-1)	-----	5	C	μ=1.0M($NaClO_4$)	72Ge	
-0.644	(-1)	-----	20	C	μ=1.0M($NaClO_4$)	72Ge	
	(-1)	-----	25	C	μ=1.0M($NaClO_4$) ΔCp=-29.1	72Ge	
-1.089	(-1)	0	35	C	μ=1.0M($NaClO_4$)	72Ge	
-1.641	(-1)	0	35	C	μ=1.0M($NaClO_4$)	72Ge	
A30 ACETIC ACID, phenoxy-			$C_8H_8O_3$		$C_6H_5OCH_2CO_2H$		
-1.12	3.16(0)	-18.2	25	C	μ=0.10M	69Ba	
A31 ACETIC ACID, phenyl-			$C_8H_8O_2$		$C_6H_5CH_2CO_2H$		
-0.311	4.296(35°,0)	-21	0&35	T	--	39Wa	b,c

ΔH, kcal /mole	pK	ΔS, cal /mole°K	T, °C	M	Conditions	Ref.	Remarks
A32 ACETIC ACID, phenylsulfonyl-			$C_8H_8O_4S$		$C_6H_5SO_2CH_2CO_2H$		
-1.300	2.44(25°,0)	-15.5	0&25	T	--	34Ca	b,c
A33 ACETIC ACID, phenylsulphenyl			$C_8H_8O_3S$		$C_6H_5SOCH_2CO_2H$		
-1.000	2.66(25°,0)	-15.5	0&25	T	34Ca	34Ca	b,c
A34 ACETIC ACID, phenylseleno-			$C_8H_8O_2Se$		$C_6H_5SeCH_2CO_2H$		
-0.22	3.75(0)	-17.9	25	C	μ=0.10M	69Ba	
A35 ACETIC ACID, thiolo-			C_2H_4OS		CH_3COSH		
0.56	3.62(0)	-14.7	25	C	μ=0.015	64Ia	
A36 ACETIC ACID, tribromo-			$C_2HO_2Br_3$		Br_3CCO_2H		
-0.8	-0.147(0)	-2	25	-	--	69La	
A37 ACETIC ACID, trichloro-			$C_2HO_2Cl_3$		Cl_3CCO_2H		
0.28	0.515(0)	-1.4	25	T	μ=0 ΔCp=-42	69Kb	
1.5	1.54(0)	-2	25	-	--	69La	
A38 ACETIC ACID, trifluoro-			$C_2HO_2F_3$		CF_3CO_2H		
0.4	0.50(0)	-1	25	T	μ=0	69Kb	
0	0.22(0)	-1	25	-	--	69La	
11.88	6.56(30°,0)	45.9	25-35	T	μ=0 S: 1,2-Dichloroethane	72Bf	c
A39 ACETIC ACID, trifluoro-, amide, N,N-dimethyl-			$C_4H_6ONF_3$		$F_3CCON(CH_3)_2$		
20.9	(0)	-----	25	C	C∿10^{-3}M S: Fluorosulfuric acid	74Ab	
A40 ACETOPHENONE			C_8H_8O		$CH_3COC_6H_5$		
15.6	5.66(40°,0)	23.5	25-60	T	S: Aqueous Solution of Sulfuric Acid	73Ke	c,q: ±1.6
A41 ACETOPHENONE, 2-hydroxy			$C_8H_8O_2$		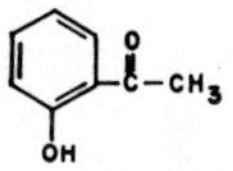		
8.6	10.8(0)	-21	25	T	μ=3.0	56Aa	

ΔH, kcal /mole	pK	ΔS, cal /mole°K	T,°C	M	Conditions	Ref.	Remarks
A42 ACRIDINE			$C_{13}H_{14}N$		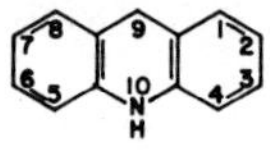		
8.32	5.65(0)	-3.01	15	T	--	58Pa	
3.480	4.07(0)	-6.95	25	C	S: 50 w % Ethanol	64Sc	
A43 ACRIDINE, 1-hydroxy-			$C_{13}H_{14}ON$		see A42		
6.71	5.72(0)	-0.56	15	T	--	58Pa	
A44 ACRIDINE, 2-hydroxy-			$C_{13}H_{14}ON$		see A42		
7.69	5.62(0)	0.54	15	T	--	58Pa	
A45 ACRIDINE, 3-hydroxy-			$C_{13}H_{14}ON$		see A42		
5.10	5.30(0)	-1.64	15	T	--	58Pa	
A46 ACTIN							
4.5	8.10(25°,0)	-22	12-40	T	2 X 10^{-3} M Tris - 50 X 10^{-3} M KCl Dissociation from Mercaptopurine Moiety 6-Mercaptopurine complex of double-stranded helical, fibrous actin	71Mc	c
5.6	8.37(25°,0)	-20	12-40	T	2 X 10^{-3} M Tris Dissociation from Mercaptopurine Moiety 6-Mercaptopurine complex of monomeric globular actin	71Mc	c
5.1	8.37(25°,0)	-21	12-40	T	2 X 10^{-3} M Tris Dissociation from Mercaptopurine Moiety Actin with adenine nucleotide in ATP replaced by a 6-mercaptopurine nucleotide	71Mc	c
6.9	7.99(25°,0)	-13	12-40	T	2 X 10^{-3} M Tris - 50 X 10^{-3} M KCl Dissociation from Mercaptopurine Moiety Actin with adenine nucleotide in ATP replaced by a 6-mercaptopurine nucleotide	71Mc	c

ΔH, kcal /mole	pK	ΔS, cal /mole$°K$	T,$°C$	M	Conditions	Ref.	Re-marks

A47   ADENOSINE $C_{10}H_{13}O_4N_5$

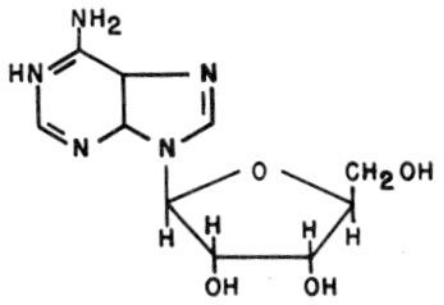

3.1	3.5(+1)	−5.7	25	C	μ=0	62Cc	
3.91	3.50(+1)	−2.92	25	C	μ=0	70Cb	
3.4	3.51(+1)	−5	25	T	μ=0.005	58Ha	
3.81	3.6(+1)	−3.7	25	C	μ=0.1M(NaCl)	60Ra	d
3.8	3.55(+1)	−3.4	25	T	μ=0.1	64Sd	
8.4	12.34(0)	−28.3	25	C	μ=0	66Cb	
9.7	12.35(0)	−23.9	25	C	μ=0	65Ib	

A48   ADENOSINE, 2'-deoxy $C_{10}H_{13}O_3N_5$ see A47

3.870	3.8(+1)	−4.4	25	C	μ=0.1	60Ra	d

A49   ADENOSINE, 2'-deoxy-5'-monophosphoric acid $C_{10}H_{14}O_6N_5P$ see A47

−1.04	6.65(−1)	−33.8	25	T	μ=0 98% pure dAMP	65Pe	
2.64	4.0(−1)	−9.5	25	C	μ=0.1	60Ra	d

A50   ADENOSINE, 5'-diphosphoric acid $C_{10}H_{15}O_{10}N_5P_2$ see A47

4.1	4.2(−1)	−5.4	25	C	μ=0	62Cc	
−1.3	7.0(−2)	−36	25	C	μ=0	62Cc	
−1.37	7.20(−2)	−37.5	25	T	μ=0	66Pf	
						65Pe,63Pc	

A51   ADENOSINE-5'-monophosphoric acid $C_{10}H_{14}O_7N_5P$ see A47

4.2	3.7(0)	−3.0	25	C	u=0	62Cc	
4.2	3.82(0)	−3.4	25	T	μ=0.1M(KNO$_3$)	74Ba	
−1.8	6.4(−1)	−36	25	C	μ=0	62Cc	
					pK value valid at μ=0.15		
−0.85	6.67(−1)	−33.4	25	T	μ=0	63Pc	
−1.1	6.21(−1)	−32	25	T	μ=0.1M(KNO$_3$)	74Ba	
					$\Delta H=10.228 - 1.278 \times 10^{-4} T^2$ (5−40°C)		
10.9	13.06(−2)	−23	25	C	μ=0	65Ib	

A52   ADENOSINE, 5'-triphosphoric acid $C_{10}H_{16}O_{13}N_5P_3$ see A47

3.7	4.0(−2)	−5.7	25	C	μ=0	62Cc	
4.1	4.05(−2)	−4.5	25	T	μ=0.1	66Ta	
−1.2	7.0(−3)	−36.0	25	C	μ=0	62Cc	
−1.68	7.68(−3)	−40.7	25	T	μ=0	66Pf	
						63Pc,65Pe	
0.5	6.52(−3)	−27.8	25	T	μ=0.1	66Ta	

A53   α-ALANINE $C_3H_7O_2N$ $CH_3CH(NH_2)CO_2H$

1.510	2.426(+1)	−5.6	1	T	μ=0	37Sa	
1.25	2.392(+1)	−6.5	10	C	μ=0	68Cd	
0.920	(+1)	-----	10	T	μ=0	34Hc	
1.210	2.383(+1)	−6.6	12.5	T	μ=0	37Sa	
0.913	2.350(+1)	−7.6	20	T	μ=0	33Na	
0.75	2.348(+1)	−8.2	25	C	μ=0	68Cd	
						67Ca	

ΔH, kcal /mole	pK	ΔS, cal /mole°K	T,°C	M	Conditions	Ref.	Remarks
A53, cont.							
0.615	2.35(+1)	-8.0	25	C	μ=0	42Sd	
0.805	2.350(+1)	-8.1	25	T	μ=0	39Hb	1:39Hb
					ΔCp=-35.3	37Sa	
0.720	2.340(+1)	-8.3	25	T	μ=0	33Na	
0.750	(+1)	-----	25	T	μ=0	34Hc	
0.538	2.332(+1)	-8.9	30	T	μ=0	33Na	
0.369	2.327(+1)	-9.4	35	T	μ=0	33Na	
0.320	2.330(+1)	-9.6	37.5	T	μ=0	37Sa	
0.23	2.328(+1)	-9.9	40	C	μ=0	68Cd	
0.215	2.324(+1)	-9.9	40	T	μ=0	33Na	
0.155	(+1)	-----	40	T	μ=0	34Hc	
0.7	2.37(20°,+1)	-8.6	0-40	T	μ=0	61Ib	c,e: CO_2H
0.070	2.322(+1)	-10.4	45	T	μ=0	33Na	
-0.250	2.332(+1)	-11.4	50	T	μ=0	37Sa	
0.65	2.35(+1)	-8.57	25	C	μ=0.2M(KCl)	73Ga	e:CO_2H
10.990	10.586(0)	-8.3	1.0	T	μ=0	37Sa	
11.68	10.29(0)	-5.8	10	C	μ=0	66Ad	
11.080	10.225(0)	-8.0	12.5	T	μ=0	37Sa	
10.900	10.006(0)	-8.6	20	T	μ=0	33Na	
10.838	9.83(0)	-8.0	25	C	μ=0	42Sd	
10.810	9.870(0)	-8.9	25	T	μ=0	33Na	
11.040	9.866(0)	-8.0	25	T	μ=0	37Sa	
10.740	9.740(0)	-9.1	30	T	μ=0	33Na	
10.680	9.615(0)	-9.3	35	T	μ=0	33Na	
10.890	9.548(0)	-8.6	37.5	T	μ=0	37Sa	
10.630	9.494(0)	-9.5	40	T	μ=0	33Na	
10.6	10.04(20°,0)	-9.8	0-40	T	μ=0	61Ib	c,e:NH_3^+
10.580	9.378(0)	-9.7	45	T	μ=0	33Na	
10.580	9.256(0)	-9.6	50	T	μ=0	37Sa	
10.06	(0)	-----	20	C	μ≈0.01	71Ma	e:NH_3^+
11.45	9.686(25°,0)	-5.92	0-25	T	μ=0.2	30Ba	b,c
						30Ma	
10.85	9.70(0)	-8.0	25	C	μ=0.1M(KNO_3)	72Ia	e:NH_3^+
10.40	9.916(25°,0)	-10.5	25-50	T	μ=0.1M(KCl)	70Ha	c,e:NH_3^+
10.62	9.60(0)	-8.3	25	C	μ=0.16(KNO_3)	70Md	
11.29	9.68(0)	-6.42	25	C	μ=0.2M(KCl)	73Ga	
10.70	9.870(25°,0)	-9.3	20-45	T	2N(HNO_3)	74Bd	c,e:NH_3^+
10.91	9.87(0)	-8.6	25	C	--	69Cc	
10.750	9.866(0)	-8.93	25	C	--	67Ac	
10.26	9.51(0)	-10.8	40	C	--	67Cc	

A54 β-ALANINE $C_3H_7O_2N$ $H_2NCH_2CH_2CO_2H$

ΔH, kcal /mole	pK	ΔS, cal /mole°K	T,°C	M	Conditions	Ref.	Remarks
0.62	3.606(+1)	-14.1	10	C	μ=0	68Cd	
1.08	3.551(+1)	-12.6	25	C	μ=0	67Ca	
0.77	3.55(+1)	-7.7	25	C	μ=0	68Cd	
1.18	3.55(+1)	-12	25	T	μ=0	63Ea	
0.38	3.517(+1)	-9.2	40	C	μ=0	68Cd	
1.4	3.551(25°,+1)	-11.6	0-40	T	2N(HNO_3)	74Bd	c,e: CO_2H
11.30	10.238(0)	-9.0	25	C	μ=0	69Cc	
11.06	10.764(0)	-9.46	5	T	μ=0.1(KCl)	73Ra	e:NH_3^+
11.73	10.162(0)	-7.12	25	T	μ=0.1(KCl)	73Ra	e:NH_3^+
12.25	9.610(0)	-5.45	45	T	μ=0.1(KCl)	73Ra	e:NH_3^+
11.27	10.333(25°,0)	-9.5	25-50	T	μ=0.1M(KCl)	70Ha	c,e: NH_3^+
11.28	10.235(25°,0)	-9.0	0-40	T	2N(HNO_3)	74Bd	c,e: NH_3^+
11.270	10.235(0)	-8.82	25	C	--	67Ac	

ΔH, kcal /mole	pK	ΔS, cal /mole°K	T, °C	M	Conditions	Ref.	Remarks
A55 β-ALANINE, methyl ester			$C_4H_9O_2N$		$CH_3CH(NH_2)CO_2CH_3$		
11.50	7.743(25°,0)	3.2	25-50	T	μ=0.1M(KCl)	70Ha	c,e:NH_3^+
A56 β-ALANINE, methyl ester			$C_4H_9O_2N$		$H_2NCH_2CH_2CO_2CH_3$		
12.62	9.170(25°,0)	0.4	25-50	T	μ=0.1M(KCl)	70Ha	c,e:NH_3^+
A57 β-ALANINE, N-acetyl-			$C_5H_9O_3N$		$CH_3CONHCH_2CH_2CO_2H$		
0.255	4.4452(+1)	-19.48	25	T	μ=0	56Kc	
A58 α-ALANINE(L), N-alanyl-(l)			$C_6H_{12}O_3N_2$		$CH_3CH(NH_2)CONHCH(CH_3)CO_2H$		
-0.134	3.30(+1)	-15.7	25	T	μ=0	56Ea	
1.0	3.21(+1)	-18.0	25	C	μ=0.1M(KNO_3)	72Bi	q:±0.2
-0.134	3.342(+1)	-15.7	25	T	μ=0.1	65Da	
2.291	3.447(+1)	-7.9	37	T	μ=0.1	65Da	
10.6	8.08(0)	-1.4	25	C	μ=0.1M(KNO_3)	72Bi	
8.371	8.17(0)	-9.4	25	T	μ=0.1	65Da	
5.930	8.28(0)	-17.5	37	T	μ=0.1	65Da	
A59 α-ALANINE(D), alanyl(L)-			$C_6H_{12}O_3N_2$		$H_2NCH(CH_3)CONHCH(CH_3)CO_2H$		
-1.023	3.119(+1)	-17.3	25	T	μ=0.10	56Ea	
-0.157	3.138(+1)	-14.7	37	T	μ=0.1	56Ea	
9.599	8.296(0)	-5.8	25	T	μ=0.10	56Ea	
9.829	8.021(0)	-5.1	37	T	μ=0.1	56Ea	
A60 α-ALANINE(D), alanyllysyl-			$C_{12}H_{24}O_4N_4$		$CH_3CH(NH_2)CONH(CH_2)_4CH(NH_2)$-$CONHCH(CH_3)CO_2H$		
10.4	8.23(25°,0)	-2.76	10-35	T	μ=2	59Gc	c,f
13.0	10.31(25°,-1)	-5.84	10-35	T	μ=2	59Gc	c,f
A61 α-ALANINE(D), carbamoyl-			$C_4H_8O_3N_2$		$NH_2CONHCH(CH_3)CO_2H$		
-0.232	3.8924(+1)	-18.59	25	T	μ=0	56Kb	
A62 β-ALANINE, carbamoyl-			$C_4H_8O_3N_2$		$NH_2CONHCH_2CH_2CO_2H$		
0.194	4.4873(+1)	-19.88	25	T	μ=0	56Kb	
A63 α-ALANINE, N-glycyl-			$C_5H_{10}O_3N_2$		$H_2NCH_2CONHCH(CH_3)CO_2H$		
-0.563	3.1532(+1)	-16.32	25	T	μ=0	57Ka	
0.0	3.06(+1)	-14.0	25	C	μ=0.1M(KNO_3)	72Bi	q:±0.2
10.660	8.331(0)	-2.3	25	T	μ=0	75Kb	e:NH_3^+
11.10	8.220(0)	-0.39	25	T	μ=0.02	30Ba	b
10.0	8.17(0)	-3.8	25	C	μ=0.1M(KNO_3)	72Bi	
-6.3	4.14(+1)	2.2	25	C	μ=0.1M(KNO_3) HA=[Cu(II)HNCH$_2$CONHCH-(CH$_3$)COO]$^{+1}$	72Bi	
A64 α-ALANINE(D), 3-methoxy-			$C_4H_9O_3N$		$CH_3OCH_2CH(NH_2)CO_2H$		
0.820	2.037(+1)	-6.56	25	T	μ=0	60Kb	
10.150	9.176(0)	-7.95	25	T	μ=0	60Kb	

ΔH, kcal /mole	pK	ΔS, cal /mole°K	T,°C	M	Conditions	Ref.	Remarks
A66 ALANINE, 3-phenyl-, methyl ester			$C_{10}H_{13}O_2N$		$C_6H_5CH_2CH(NH_2)CO_2CH_3$		
11.11	7.05(25°,+1)	5.0	25-50	T	μ=0.1M	67Hb	c,e:NH_3^+
A67 ALBUMIN, bovine serum							
11.5	10-10.3	-8	25	T	μ=0.15	52Ta	e:Phenolic group
A68 AMINE, butyl methyl			$C_5H_{13}N$		$CH_3CH_2CH_2CH_2NHCH_3$		
12.57	10.90(+1)	-7.7	25	C	μ=0	69Cc	
13.1	10.6(+1)	-5	25	C	μ=0.02	64Wa	
A69 AMINE, diallyl-			$C_6H_{11}N$		$HN(CH_2CH{:}CH_2)_2$		
11.67	9.29(+1)	-3.4	25	C	μ=0.02	69Cc	
A70 AMINE, dibenzyl-			$C_{14}H_{15}N$		$HN(CH_2C_6H_5)_2$		
11.33	8.52(+1)	-1.0	25	C	μ=0	69Cc	
A71 AMINE, di-n-butyl-			$C_8H_{19}N$		$HN(CH_2CH_2CH_2CH_3)_2$		
13.66	11.25(+1)	-5.7	25	C	μ=0	69Cc	
A72 AMINE, di-sec-butyl-			$C_8H_{19}N$		$HN(CH_2CH_2CH_2CH_3)_2$		
14.03	10.91(+1)	-2.8	25	C	μ=0	69Cc	
A73 AMINE, dicyclohexyl-			$C_{12}H_{23}N$		$HN(C_6H_{11})_2$		
14.21	11.25(+1)	-3.8	25	C	μ=0	69Cc	
A74 AMINE, dicyclopentyl-			$C_{10}H_{19}N$		$HN(C_5H_9)_2$		
14.17	10.93(+1)	-2.5	25	C	μ=0	69Cc	
A75 AMINE, diethyl-			$C_4H_{11}N$		$HN(CH_2CH_3)_2$		
12.73	10.93(+1)	-7.3	25	C	μ=0	69Cc	
10.3	10.76(+1)	-15.7	25	T	μ=0	55Fa	
12.77	10.489(40°,+1)	-7.2	20-40	T	μ=0	51Ea	c
12.83	10.489(40°,+1)	-7.0	20-40	T	μ=0.05	51Ea	c
12.85	10.503(40°,+1)	-7.0	20-40	T	μ=0.10	51Ea	c
12.98	10.507(40°,+1)	-6.6	20-40	T	μ=0.20	51Ea	c
15.34	12.04(+1)	-3.6	25	T	μ=0 S: Ethylene glycol ΔCp=-5	70Kc	
A76 AMINE, diethyl-2,2'-dihydroxy-			$C_4H_{11}O_2N$		$HN(C_2H_5OH)_2$		
10.01	8.88(+1)	-7.1	25	C	μ=0	69Cc	
13.42	10.45(+1)	-2.8	25	T	μ=0 S: Ethylene glycol ΔCp=-15	70Kc	

ΔH, kcal /mole	pK	ΔS, cal /mole°K	T, °C	M	Conditions	Ref.	Remarks
A77 AMINE, diethyl-2-hydroxy-			$C_4H_{11}ON$		$CH_3CH_2NHCH_2CH_2OH$		
11.401	9.961(25°,+1)	-7.3	0-60	T	μ=0	68Tb	c
A78 AMINE, N,N-diethyl-2-hydroxy-ethyl			$C_6H_{15}ON$		$(C_2H_5)_2NCH_2CH_2OH$		
9.719	9.804(25°,+1)	-12.3	0-60	T	μ=0	68Tb	c
A79 AMINE, diethyl methyl-,2,2'-dihydroxy-			$C_5H_{13}O_2N$		$(HOCH_2CH_2)_2NCH_3$		
9.2	8.52(+1)	-8	25	T	μ=0	59Sc	
A80 AMINE, diisobutyl-			$C_5H_{19}N$		$HN[CH_2CH(CH_3)_2]_2$		
13.38	10.82(+1)	-4.5	25	C	μ=0	69Cc	
A81 AMINE, diisopentyl-			$C_{10}H_{23}N$		$HN[CH_2CH_2CH(CH_3)_2]_2$		
13.39	10.92(+1)	-5.0	25	C	μ=0	69Cc	
A82 AMINE, diisopropyl-			$C_6H_{15}N$		$HN[CH(CH_3)_2]_2$		
13.55	11.20(+1)	-5.9	25	C	μ=0	69Cc	
A83 AMINE, dimethyl-			C_2H_7N		$HN(CH_3)_2$		
11.46	11.39(+1)	-10.9	5	T	μ=0	58Da	
11.65	11.08(+1)	-10.2	15	T	μ=0	58Da	
12.04	10.732(+1)	-8.7	25	C	μ=0	69Cc	
11.88	10.77(+1)	-9.5	25	T	μ=0	41Ea	
11.85	10.78(+1)	-9.6	25	T	μ=0	58Da	
12.05	10.49(+1)	-8.9	35	T	μ=0	58Da	
12.27	10.22(+1)	-8.2	45	T	μ=0	58Da	
11.892	8.792(+1)	-0.3	25	T	μ=0	63Tb	c
12.2	10.8(+1)	-8	25	C	μ=0.002	64Wa	
15.25	12.09(+1)	-4.1	25	T	μ=0 S: Ethylene glycol ΔCp=-11	70Kc	
12.030	9.828(+1)	-4.6	25	T	μ=0.10 S: 60 w % Methanol	52Ec	
A84 AMINE, diphenyl-, 2,4,6-trinitro-			$C_{12}H_8O_6N_4$		$H_5C_6NH[C_6H_2(NO_2)_3]$		
13.7	13.57(+1)	-16	25	T	S: Methanol	69Rd	
A85 AMINE, dipropyl-			$C_6H_{15}N$		$HN(CH_2CH_2CH_3)_2$		
12.69	(+1)	-----	0.50	C	μ=0	75Bb	
12.72	11.59(+1)	-7.7	5.00	C	μ=0	75Bb	
13.17	11.00(+1)	-6.2	25	C	μ=0	69Cc	
13.03	10.9(+1)	-6.2	25.00	C	μ=0	75Bb	
13.53	10.16(+1)	-4.6	50.00	C	μ=0 ΔCp=17.7 0<t °C<100 ΔH=12.632 + 0.01757t 0<t °C<100	75Bb	
13.85	9.49(+1)	-3.6	75.00	C	μ=0	75Bb	

ΔH, kcal /mole	pK	ΔS, cal /mole°K	T,°C	M	Conditions	Ref.	Remarks
A85, cont.							
14.46	8.89(+1)	-1.9	100.00	C	μ=0	75Bb	
A86 AMINE, di(2,4,6-trinitrophenyl)-			$C_{12}H_5O_{12}N_7$		$C_6H_5NHC_6H_2(NO_2)_3$		
13.99	13.570(+1)	-15.1	25	T	μ=0 S: Methanol ΔCp=0	70Bf	
A87 AMINE, ethyl isopropyl, 2,2'-dihydroxy-			$C_5H_{13}O_2N$		$COH(CH_3)_2NHCH_2CH_2OH$		
9.8	8.87(+1)	-7	25	T	μ=0	59Sc	
A88 AMINE, triallyl-			$C_9H_{15}N$		$N(CH_2CH:CH_2)_3$		
8.83	8.31(+1)	-8.5	25	C	μ=0	69Cc	
A89 AMINE, tri(2-aminoethyl)-			$C_6H_{18}N_4$		$N(CH_2CH_2NH_2)_3$		
10.3	7.71(30°,+1)	-2	10-40	T	μ=0	50Ba	c
12.15	8.41(+1)	2.3	25	C	--	63Pa	
12.3	9.00(30°,0)	0	10-40	T	μ=0	50Ba	c
12.85	9.42(0)	-0.2	25	C	--	63Pa	
11.0	9.91(30°,-1)	-9	10-40	T	μ=0	50Ba	c
11.70	10.14(-1)	-7.0	25	C	--	63Pa	
A90 AMINE, triethyl-			$C_6H_{15}N$		$N(CH_2CH_3)_3$		
9.33	(+1)	-----	0.50	C	μ=0	75Bb	
9.44	11.235(+1)	-17.7	5.00	C	μ=0	75Bb	
9.75	11.036(+1)	-16.0	12.50	C	μ=0	75Bb	
10.31	10.715(+1)	-14.4	25.00	C	μ=0	75Bb	
10.32	10.715(+1)	-14.4	25	C	μ=0	69Cc	
10.6	10.715(+1)	-13.6	25	T	μ=0	61Bb	
9.7	10.67(+1)	-16.3	25	T	μ=0	55Fa	
11.10	10.110(+1)	-11.9	50.00	C	μ=0 ΔCp=42.50 - 0.256t $0<t$ °C<100 ΔH=9.275 + 0.0425t - 0.000128t^2 $0<t$ °C<100	75Bb	
11.69	9.556(+1)	-10.1	75.00	C	μ=0	75Bb	
12.23	9.052(+1)	-8.6	100.00	C	μ=0	75Bb	
10.38	10.75(+1)	-14.4	25	C	0.1 M KCl	65Pb	
10.6	10.715(+1)	-13.6	25	-	--	61Bb	
13.41	11.45(+1)	-7.4	25	T	μ=0 S: Ethylene Glycol ΔCp=-9	70Kc	
A91 AMINE, tri(2-hydroxyethyl)-			$C_6H_{15}O_3N$		$N(CH_2CH_2OH)_3$		
7.696	8.290(+1)	-9.75	0	T	μ=0	60Ba	
7.753	8.173(+1)	-9.54	5	T	μ=0	60Ba	
7.813	8.067(+1)	-9.32	10	T	μ=0	60Ba	
7.873	7.963(+1)	-9.13	15	T	μ=0	60Ba	
7.932	7.861(+1)	-8.91	20	T	μ=0	60Ba	
8.16	7.762(+1)	-8.2	25	C	μ=0	69Cc	
7.994	7.762(+1)	-8.70	25	T	μ=0	60Ba	
8.057	7.666(+1)	-8.48	30	T	μ=0	60Ba	
8.121	7.570(+1)	-8.29	35	T	μ=0	60Ba	
8.186	7.477(+1)	-8.08	40	T	μ=0	60Ba	
8.253	7.387(+1)	-7.86	45	T	μ=0	60Ba	

ΔH, kcal /mole	pK	ΔS, cal /mole°K	T, °C	M	Conditions	Ref.	Remarks

A91, cont.

ΔH, kcal /mole	pK	ΔS, cal /mole°K	T, °C	M	Conditions	Ref.	Remarks
8.296	7.299(+1)	-7.65	50	T	μ=0	60Ba	
11.54	9.57(+1)	-5.1	25	T	μ=0 S: Ethylene Glycol ΔCp=-14	70Kc	

A92 AMINE, trimethyl- C_3H_9N $N(CH_3)_3$

ΔH, kcal /mole	pK	ΔS, cal /mole°K	T, °C	M	Conditions	Ref.	Remarks
7.97	10.24(+1)	-18.2	5	T	μ=0	58Da	
8.37	10.02(+1)	-16.8	15	T	μ=0	58Da	
8.80	9.752(+1)	-15.1	25	C	μ=0	69Cc	
8.815	9.79(+1)	-15.2	25	T	μ=0	41Fa	
8.82	9.80(+1)	-15.3	25	T	μ=0	65La	
8.79	9.80(+1)	-15.3	25	T	μ=0	58Da	
9.22	9.59(+1)	-13.9	35	T	μ=0	58Da	
9.67	9.38(+1	-12.5	45	T	μ=0	58Da	
8.86	9.79(+1)	-15.1	25	C	0.1 M KCl	65Pb	
13.17	10.92(+1)	-5.8	25	T	μ=0 S: Ethylene Glycol ΔCp=-13	70Kc	
9.970	8.744(+1)	-6.6	25	T	μ=0.10 S: 60 w % Methanol	52Ec	

A93 AMINE, tripropyl- $C_9H_{21}N$ $N(CH_2CH_2CH_3)_3$

ΔH, kcal /mole	pK	ΔS, cal /mole°K	T, °C	M	Conditions	Ref.	Remarks
10.5	10.66(+1)	-14.7	25	C	μ=0	69Cc	

A94 AMINE, tripropyl-2,2',2"-tri-hydroxy- $C_9H_{21}O_3N$ $N(CH_2CH(OH)CH_3]_3$

ΔH, kcal /mole	pK	ΔS, cal /mole°K	T, °C	M	Conditions	Ref.	Remarks
8.9	7.86(+1)	-6	25	T	μ=0	59Sc	

A95 AMMONIA H_3N NH_3

ΔH, kcal /mole	pK	ΔS, cal /mole°K	T, °C	M	Conditions	Ref.	Remarks
12.33	10.070(+1)	-0.93	0	C	μ=0	75Oa	
12.56	10.0813(+1)	-0.17	0	T	μ=0	49Ba	
12.54	9.90(+1)	-0.21	5	T	μ=0	58Da	
12.50	9.57(+1)	-0.34	15	T	μ=0	58Da	
12.51	9.242(+1)	-0.3	25	C	μ=0	69Cc	
12.417	(+1)	-----	25	C	μ=0	72Va	a: 13.337
12.40	9.241(+1)	-0.69	25	C	μ=0	75Oa	
12.54	9.24(+1)	-0.22	25	C	μ=0	61Ta	
12.38	9.24(+1)	-0.75	25	T	μ=0	65La	
12.48	9.25(+1)	-0.45	25	T	μ=0	58Da	
12.48	9.2449(+1)	-0.45	25	T	μ=0	49Ba	
12.44	8.95(+1)	-0.56	35	T	μ=0	58Da	
12.41	8.67(+1)	-0.67	45	T	μ=0	58Da	
12.47	8.536(+1)	-0.49	50	C	μ=0	75Oa	
12.39	8.5387(+1)	-0.73	50	T	μ=0	49Ba	
12.54	7.928(+1)	-0.24	75	C	μ=0	75Oa	
12.60	7.400(+1)	-0.07	100	C	μ=0	75Oa	
12.67	6.935(+1)	0.10	125	C	μ=0 ΔH=11.59 + 0.00271T 273<T °K<573 0-573°K	75Oa	
12.74	6.523(+1)	0.23	150	C	μ=0 ΔCp=2.72 (0-300°C)		
12.88	5.824(+1)	0.58	200	-	μ=0	75Oa	
13.02	5.253(+1)	0.86	250	-	μ=0	75Oa	
13.14	4.776(+1)	1.05	300	-	μ=0	75Oa	

ΔH, kcal /mole	pK	ΔS, cal /mole°K	T,°C	M	Conditions	Ref.	Remarks
A95, cont.							
11.98	(+1)	-----	20	C	μ≈0.01(NH_4Cl)	71Ma	
12.43	9.29(+1)	-0.8	25	C	0.1M(KCl)	65Pb	
16.29	11.34(+1)	2.7	25	T	μ=0 S: Ethylene Glycol ΔCp=6	70Kc	
12.560	8.626(+1)	2.65	25	T	μ=0 S: 60 w % Methanol	52Ec	
26.1	32.5(+1)	-40.1	-58	T	S: Ammonia	59Ca	
1.240	4.66	-17.6	18	C	μ=0 R: $NH_3 + H_2O = NH_4OH$	68Va	
0.875	4.75	-18.8	25	C	μ=0 R: $NH_3 + H_2O = NH_4OH$	68Va	
0.600	4.82	-19.7	30	C	μ=0 R: $NH_3 + H_2O = NH_4OH$	68Va	
0.330	4.90	-20.6	35	C	μ=0 R: $NH_3 + H_2O = NH_4OH$	68Va	
0.080	4.97	-21.4	40	C	μ=0 R: $NH_3 + H_2O = NH_4OH$	68Va	
1.192	4.33	-16.1	18	C	μ=0.5 $NaNO_3$ R: $NH_3 + H_2O = NH_4OH$	68Va	
0.850	4.41	-17.3	25	C	μ=0.5 $NaNO_3$ R: $NH_3 + H_2O = NH_4OH$	68Va	
0.600	4.47	-18.1	30	C	μ=0.5 $NaNO_3$ R: $NH_3 + H_2O = NH_4OH$	68Va	
0.350	4.55	-18.9	35	C	μ=0.5 $NaNO_3$ R: $NH_3 + H_2O = NH_4OH$	68Va	
0.081	4.62	-19.8	40	C	μ=0.5 $NaNO_3$ R: $NH_3 + H_2O = NH_4OH$	68Va	
1.020	4.27	-16.5	18	C	μ=1.0 $NaNO_3$ R: $NH_3 + H_2O = NH_4OH$	68Va	
0.700	4.36	-17.6	25	C	μ=1.0 $NaNO_3$ R: $NH_3 + H_2O = NH_4OH$	68Va	
0.462	4.42	-18.4	30	C	μ=1.0 $NaNO_3$ R: $NH_3 + H_2O = NH_4OH$	68Va	
0.222	4.49	-19.2	35	C	μ=1.0 $NaNO_3$ R: $NH_3 + H_2O = NH_4OH$	68Va	
-0.025	4.57	-20.0	40	C	μ=1.0 $NaNO_3$ R: $NH_3 + H_2O = NH_4OH$	68Va	
0.361	4.19	-18.4	18	C	μ=3.0 $NaNO_3$ R: $NH_3 + H_2O = NH_4OH$	68Va	
0.109	4.28	-19.2	25	C	μ=3.0 $NaNO_3$ R: $NH_3 + H_2O = NH_4OH$	68Va	
-0.075	4.36	-19.8	30	C	μ=3.0 $NaNO_3$ R: $NH_3 + H_2O = NH_4OH$	68Va	
-0.266	4.43	-20.5	35	C	μ=3.0 $NaNO_3$ R: $NH_3 + H_2O = NH_4OH$	68Va	
8.32	5.00(+3)	5.5	25	C	μ=0.1M HA=$[Cr(III)(NH_3)_5OH_2]^{3+}$	70Cf	
6	6.24(35°,+3)	-9	9.4&35.0	T	μ=0.2M HA=$[Rh(III)(NH_3)_5OH_2]^{3+}$	67Pa	b,c,q: +1
9.44	6.14(+3)	4.1	25	C	μ=0.1M HA=$[Rh(III)(NH_3)_5OH_2]^{3+}$	70Cf	
9.03	6.07(+3)	2.9	25	C	μ=0.1M HA=$[Co(III)(NH_3)_5OH_2]^{3+}$	70Cf	
9.98	5.576(+3)	7.9	25	C	μ=0 HA=$[Co(III)(NH_3)_5OH_2]^{3+}$	74La	

A97 ANGIOTENSIN II

6.3	10.37	-27	25	T	μ=0	63Vb	e: Tyrosyl group

A98 ANILINE C_6H_7N $C_6H_5NH_2$

ΔH, kcal /mole	pK	ΔS, cal /mole°K	T,°C	M	Conditions	Ref.	Remarks
7.08	4.78(+1)	2.68	15	T	μ=0	71Ga	
5.18	4.44(+1)	-1.7	15	T	μ=0	66Aa	
7.16	4.70(+1)	2.95	20	T	μ=0	71Ga	
6.74	3.36(+1)	1.66	25	C	μ=0	59Za	n
7.24	4.60(+1)	3.3	25	C	μ=0	68Oa	
7.345	4.580(+1)	3.68	25	C	μ=0	70Ka	
6.74	4.58(+1)	1.66	25	C	μ=0	59Za	
7.43	4.60(+1)	3.89	25	C	μ=0	73Lb	
7.1	4.59(+1)	2.8	25	T	μ=0	65La	
5.78	4.61(+1)	-1.7	25	T	μ=0	66Aa	
6.52	4.60(+1)	0.82	25	T	μ=0	61Bc	
7.23	4.615(+1)	3.20	25	T	μ=0 ΔCp=13.7	71Ga	
7.110	4.580(+1)	2.89	25	T	μ=0 ΔCp=0.03	70Ka	
7.257	4.60(+1)	3.3	25	C	μ≃0	71Va 70Va	
7.30	4.51(+1)	3.41	30	T	μ=0	71Ga	
5.78	4.78(+1)	-1.7	35	T	μ=0	66Aa	
7.35	4.425(+1)	3.59	35	T	μ=0	71Ga	
7.40	4.345(+1)	3.75	40	T	μ=0	71Ga	
7.44	4.27(+1)	3.88	45	T	μ=0	71Ga	
7.377	4.596(25°,+1)	3.72	10-50	T	μ=0	67Ba	c
5.9	4.19(20°,+1)	1.0	0-20	T	C=0.01 S: 50 w % Ethanol	59Eb	c
7.490	(+1)	-----	10	C	C=0.06M	49La	
7.370	(+1)	-----	20	C	C=0.06M	49La	
7.276	4.60(+1)	3.4	25	C	C=0.06M	49La	
7.180	(+1)	-----	30	C	C=0.06M	49La	
7.24	4.60(+1)	3.3	25	C	0.095M(HCl)	67Oa	
7.64	4.87(+1)	3.35	25	C	μ=0.5M(KNO_3)	73Be	
7.11	4.596(+1)	2.8	25	-	--	59Ba	
7.105	4.592(+1)	2.8	25	T	--	39Ea	
6.52	4.606(25°,+1)	0.80	20-40	T	--	61Bc	c
7.565	4.49(+1)	4.8	25	C	μ≃0 S: 0.034 mole fraction Ethanol in H_2O	70Va	
7.975	4.39(+1)	6.6	25	C	μ≃0 S: 0.074 mole fraction Ethanol in H_2O	70Va	
8.105	4.36(+1)	7.2	25	C	μ≃0 S: 0.097 mole fraction Ethanol in H_2O	70Va	
8.089	4.36(+1)	7.2	25	C	μ≃0 S: 0.10 mole fraction Ethanol in H_2O	71Va	
8.050	4.32(+1)	7.2	25	C	μ≃0 S: 0.131 mole fraction Ethanol in H_2O	70Va	
7.680	4.30(+1)	6.1	25	C	μ≃0 S: 0.171 mole fraction Ethanol in H_2O	70Va	

A99 ANILINE, 2-bromo- C_6H_6NBr $C_6H_4NH_2(Br)$

ΔH, kcal /mole	pK	ΔS, cal /mole°K	T,°C	M	Conditions	Ref.	Remarks
4.99	2.53(+1)	5.2	25	C	μ=0	69Cc	
5.807	2.527(+1)	7.91	25	T	μ=0 ΔCp=-18	69Be	
5.129	2.53(+1)	5.6	25	C	μ≃0	71Va	
3.9	2.53(+1)	1.5	25	-	--	69La	f
6.143	2.31(+1)	10.0	25	C	μ≃0 S: 0.10 mole fraction Ethanol in H_2O	71Va	

A100 ANILINE, 3-bromo- C_6H_6NBr $C_6H_4NH_2(Br)$

ΔH, kcal /mole	pK	ΔS, cal /mole°K	T,°C	M	Conditions	Ref.	Remarks
6.59	3.53(+1)	5.97	25	C	μ=0	73Lb	
6.23	3.53(+1)	4.7	25	C	μ=0	69Cc	
6.248	3.527(+1)	4.82	25	T	μ=0	68Bb	

ΔH, kcal /mole	pK	ΔS, cal /mole°K	T, °C	M	Conditions	Ref.	Remarks
A100, cont.							
5.8	3.5(+1)	3.4	25	-	--	69La	f
A101 ANILINE, 4-bromo-			C_6H_6NBr		$C_6H_4NH_2(Br)$		
6.72	3.88(+1)	4.8	25	C	μ=0	69Cc	
6.50	3.88(+1)	4.02	25	C	μ=0	73Lb	
6.704	3.888(+1)	4.69	25	T	μ=0 ΔCp=-7	69Bd	
6.0	3.87(+1)	2.4	25	-	--	69La	f
A102 ANILINE, 2-bromo-4,6-dinitro-			$C_6H_4O_4N_3Br$		$C_6H_2NH_2(Br)(NO_2)_2$		
-6.320	-6.944(+1)	9.4	25	T	μ=0	70Be	g:69Ja
A103 ANILINE, 3-bromo-2,4,6-trinitro-			$C_6H_3O_6N_4Br$		$C_6HNH_2(Br)(NO_2)_3$		
-8.113	-9.518(+1)	15.8	25	T	μ=0	70Be	g:69Ja
A104 ANILINE, 2-chloro-			C_6H_6NCl		$C_6H_4NH_2(Cl)$		
4.41	2.70(+1)	2.71	20	T	μ=0	71Ga	f,g: 61Bc
6.0	2.63(+1)	8.1	25	T	μ=0	65La	
6.002	2.632(+1)	8.1	25	T	μ=0	33Ha	
5.902	2.661(+1)	7.60	25	T	μ=0 ΔCp=-9	69Be	
4.58	2.64(+1)	3.62	25	T	μ=0 ΔCp=54.6	71Ga	
4.96	2.54(+1)	4.55	30	T	μ=0	71Ga	
5.25	2.52(+1)	5.49	35	T	μ=0	71Ga	
5.55	2.46(+1)	6.46	40	T	μ=0	71Ga	
5.423	2.70(+1)	5.8	25	C	μ≃0	71Va	
6.00	2.634(+1)	8.0	25	-	--	59Ba	
6.405	2.49(+1)	10.1	25	C	μ≃0 S: 0.10 mole fraction Ethanol in H_2O	71Va	
A105 ANILINE, 3-chloro-			C_6H_6NCl		$C_6H_4NH_2(Cl)$		
5.94	3.60(+1)	3.79	20	T	μ=0	71Ga	g:61Bc
6.45	3.52(+1)	5.53	25	C	μ=0	73Lb	
5.79	3.53(+1)	3.30	25	T	μ=0 ΔCp=-30.9	71Ga	
6.268	3.521(+1)	4.91	25	T	μ=0	68Bb	
6.305	3.517(+1)	5.0	25	C	μ≃0	70Va	
5.63	3.45(+1)	2.76	30	T	μ=0	71Ga	
5.44	3.39(+1)	2.15	35	T	μ=0	71Ga	
5.24	3.33(+1)	1.49	40	T	μ=0	71Ga	
5.8	3.53(+1)	3.3	25	-	--	69La	
7.070	3.317(+1)	8.5	25	C	μ≃0 S: 0.10 mole fraction Ethanol in H_2O	70Va	
A106 ANILINE, 4-chloro-			C_6H_6NCl		$C_6H_4NH_2(Cl)$		
6.28	4.06(+1)	2.81	20	T	μ=0	71Ga	g:61Bc
6.67	3.99(+1)	4.13	25	C	μ=0	73Lb	
6.38	3.99(+1)	3.15	25	T	μ=0 ΔCp=19.5	71Ga	
6.633	3.982(+1)	4.01	25	T	μ=0 ΔCp=9	69Bd	
6.420	3.977(+1)	3.3	25	C	μ≃0	70Va	
6.47	3.91(+1)	3.47	30	T	μ=0	71Ga	
6.56	3.83(+1)	3.77	35	T	μ=0	71Ga	

ΔH, kcal /mole	pK	ΔS, cal /mole°K	T, °C	M	Conditions	Ref.	Remarks
A106, cont.							
6.65	3.75(+1)	4.04	40	T	μ=0	71Ga	
6.4	3.99(+1)	3.2	25	-	--	69La	
7.240	3.766(+1)	7.0	25	C	μ≃0 S: 0.10 mole fraction Ethanol in H_2O	53Ka	
A107 ANILINE, 2-chloro-6-nitro-			$C_6H_5O_2N_2Cl$		$C_6H_3NH_2(Cl)(NO_2)$		
-0.063	-2.414(+1)	10.7	25	T	μ=0	70Be	g:69Ja
A108 ANILINE, 4-chloro-2-nitro-			$C_6H_5O_2N_2Cl$		$C_6H_3NH_2(Cl)(NO_2)$		
0.836	-1.104(+1)	7.7	25	T	μ=0	70Be	g:69Ja
A109 ANILINE, 3-cyano-			$C_7H_6N_2$		$C_6H_4NH_2(CN)$		
5.74	2.75(+1)	6.67	25	C	μ=0	73Lb	
A110 ANILINE, 4-cyano-			$C_7H_6N_2$		$C_6H_4NH_2(CN)$		
4.58	1.74(+1)	7.41	25	C	μ=0	73Lb	
A111 ANILINE, 3,5-dibromo-			$C_6H_5NBr_2$		$C_6H_3NH_2(Br)_2$		
5.112	2.355(+1)	6.37	25	T	μ=0 ΔCp=-8	70Bd	
A112 ANILINE, 3,5-dichloro-			$C_6H_5NCl_2$		$C_6H_3NH_2(Cl)_2$		
5.215	2.371(+1)	6.63	25	T	μ=0 ΔCp=-10	70Bd	
A113 ANILINE, 2,4-dichloro-6-nitro			$C_6H_4O_2N_2Cl_2$		$C_6H_2NH_2(Cl)_2(NO_2)$		
-0.14	-3.00(25°,+1)	13.24	15-45	T	μ=0	74Ac	c
0.27	(+1)	-----	15-45	T	4.75 m $HClO_4$	74Ac	
0.42	(+1)	-----	15-45	T	5.60 m $HClO_4$	74Ac	
0.56	(+1)	-----	15-45	T	6.00 m $HClO_4$	74Ac	
0.69	(+1)	-----	15-45	T	6.45 m $HClO_4$	74Ac	
1.11	(+1)	-----	15-45	T	6.90 m $HClO_4$	74Ac	
1.46	(+1)	-----	15-45	T	7.25 m $HClO_4$	74Ac	
1.31	(+1)	-----	15-45	T	7.65 m $HClO_4$	74Ac	
1.53	(+1)	-----	15-45	T	8.10 m $HClO_4$	74Ac	
2.08	(+1)	-----	15-45	T	8.50 m $HClO_4$	74Ac	
2.51	(+1)	-----	15-45	T	9.00 m $HClO_4$	74Ac	
2.92	(+1)	-----	15-45	T	9.45 m $HClO_4$	74Ac	
3.06	(+1)	-----	15-45	T	9.95 m $HClO_4$	74Ac	
2.77	(+1)	-----	15-45	T	10.4 m $HClO_4$	74Ac	
3.49	(+1)	-----	15-45	T	10.9 m $HClO_4$	74Ac	
A114 ANILINE, 2,5-dichloro-4-nitro-			$C_6H_4O_2N_2Cl_2$		$C_6H_2NH_2(Cl)_2(NO_2)$		
0.194	-1.743(+1)	8.7	25	T	μ=0	70Be	g:69Ja
1.53	-1.78(25°,+1)	13.28	15-45	T	μ=0	74Ac	c
1.53	(+1)	-----	15-45	T	1.70 m $HClO_4$	74Ac	
1.80	(+1)	-----	15-45	T	1.98 m $HClO_4$	74Ac	
1.80	(+1)	-----	15-45	T	2.30 m $HClO_4$	74Ac	
2.08	(+1)	-----	15-45	T	2.60 m $HClO_4$	74Ac	
2.08	(+1)	-----	15-45	T	3.50 m $HClO_4$	74Ac	

ΔH, kcal /mole	pK	ΔS, cal /mole°K	T, °C	M	Conditions	Ref.	Remarks
A114, cont.							
2.22	(+1)	-----	15-45	T	4.10 m $HClO_4$	74Ac	
1.80	(+1)	-----	15-45	T	4.75 m $HClO_4$	74Ac	
2.08	(+1)	-----	15-45	T	5.60 m $HClO_4$	74Ac	
1.94	(+1)	-----	15-45	T	6.00 m $HClO_4$	74Ac	
2.18	(+1)	-----	15-45	T	6.45 m $HClO_4$	74Ac	
2.36	(+1)	-----	15-45	T	6.90 m $HClO_4$	74Ac	
2.36	(+1)	-----	15-45	T	7.25 m $HClO_4$	74Ac	
A115 ANILINE, 2,6-dichloro-4-nitro-			$C_6H_4O_2N_2Cl_2$		$C_6H_5NH_2(Cl)_2(NO_2)$		
-1.024	-3.311(+1)	11.6	25	T	μ=0	70Be	g:69Ja
A116 ANILINE, 3,5-diodo-			$C_6H_5NI_2$		$C_6H_3NH_2(I)_2$		
5.084	2.368(+1)	6.22	25	T	μ=0 ΔCp=0	70Bd	
A117 ANILINE, 3,5-dimethoxy-			$C_8H_{11}O_2N$		$C_6H_3NH_2(OCH_3)_2$		
6.695	3.860(+1)	4.79	25	T	μ=0 ΔCp=-23	70Bd	
A118 ANILINE, N,N-dimethyl-			$C_8H_{11}N$		$C_6H_5N(CH_3)_2$		
4.920	5.150(+1)	-7.06	25	C	μ=0	70Ka	
A119 ANILINE, 2,4-dinitro-			$C_6H_5O_4N_3$		$C_6H_3NH_2(NO_2)_2$		
-3.155	-4.486(+1)	8.9	25	T	μ=0	70Be	g:69Ja
-2.08	-4.25(25°,+1)	12.47	15-45	T	μ=0	74Ac	c
0.83	(+1)	-----	15-45	T	9.45 m $HClO_4$	74Ac	
1.04	(+1)	-----	15-45	T	9.95 m $HClO_4$	74Ac	
1.28	(+1)	-----	15-45	T	10.40 m $HClO_4$	74Ac	
1.53	(+1)	-----	15-45	T	10.90 m $HClO_4$	74Ac	
2.77	(+1)	-----	15-45	T	11.50 m $HClO_4$	74Ac	
2.92	(+1)	-----	15-45	T	12.07 m $HClO_4$	74Ac	
2.77	(+1)	-----	15-45	T	12.60 m $HClO_4$	74Ac	
3.06	(+1)	-----	15-45	T	13.25 m $HClO_4$	74Ac	
3.87	(+1)	-----	15-45	T	13.90 m $HClO_4$	74Ac	
4.30	(+1)	-----	15-45	T	14.55 m $HClO_4$	74Ac	
A120 ANILINE, 2,6-dinitro-			$C_6H_5O_4N_3$		$C_6H_5N(NO_2)_2$		
-4.860	-5.562(+1)	8.3	25	T	μ=0	70Be	g:69Ja
-4.16	-5.23(25°,+1)	9.96	15-45	T	μ=0	74Ac	c
1.04	(+1)	-----	15-45	T	11.50 m $HClO_4$	74Ac	
1.39	(+1)	-----	15-45	T	12.07 m $HClO_4$	74Ac	
1.39	(+1)	-----	15-45	T	12.60 m $HClO_4$	74Ac	
1.39	(+1)	-----	15-45	T	13.25 m $HClO_4$	74Ac	
1.53	(+1)	-----	15-45	T	13.90 m $HClO_4$	74Ac	
1.53	(+1)	-----	15-45	T	14.55 m $HClO_4$	74Ac	
2.22	(+1)	-----	15-45	T	15.25 m $HClO_4$	74Ac	
1.74	(+1)	-----	15-45	T	15.90 m $HClO_4$	74Ac	
3.06	(+1)	-----	15-45	T	16.65 m $HClO_4$	74Ac	
A121 ANILINE, 3,5-dinitro-			$C_6H_5O_4N_3$		$C_6H_3NH_2(NO_2)_2$		
2.687	0.229(+1)	7.96	25	T	μ=0 ΔCp=0	70Bd	

ΔH, kcal /mole	pK	ΔS, cal /mole°K	T, °C	M	Conditions	Ref.	Remarks
A122 ANILINE, 2-ethoxy-			$C_8H_{11}ON$		$C_6H_4NH_2(OCH_2CH_2)$		
7.405	4.47(+1)	4.4	25	C	μ≃0	71Va	
9.001	4.24(+1)	10.8	25	C	μ≃0 S: 0.10 mole fraction Ethanol	71Va	
A123 ANILINE, 3-ethoxy-			$C_8H_{11}ON$		$C_6H_4NH_2(OCH_2CH_3)$		
7.05	4.16(+1)	4.60	25	C	μ=0	73Lb	
6.660	4.167(+1)	3.3	25	C	μ≃0	70Va	
7.313	3.977(+1)	6.3	25	C	μ≃0 S: 0.10 mole fraction Ethanol	70Va	
A124 ANILINE, 4-ethoxy-			$C_8H_{11}ON$		$C_6H_4NH_2(OCH_2CH_3)$		
8.53	5.25(+1)	4.60	25	C	μ=0	73Lb	
8.153	5.245(+1)	3.3	25	C	μ≃0	70Va	
8.543	5.026(+1)	5.6	25	C	μ≃0 S: 0.10 mole fraction Ethanol	70Va	
A125 ANILINE, 2-fluoro-			C_6H_6NF		$C_6H_4NH_2(F)$		
5.879	3.20(+1)	5.1	25	C	μ≃0	71Va	
5.4	3.22(+1)	3.4	25	-	--	69La	f
7.170	2.96(+1)	10.5	25	C	μ≃0 S: 0.10 mole fraction Ethanol	71Va	
A126 ANILINE, 3-fluoro-			C_6H_6NF		$C_6H_4NH_2(F)$		
6.56	3.58(+1)	5.60	25	C	μ=0	73Lb	
6.226	3.587(+1)	4.5	25	C	μ≃0	70Va	
7.2	3.57(+1)	7.8	25	-	--	69La	f
7.189	3.377(+1)	8.6	25	C	μ≃0 S: 0.10 mole fraction Ethanol	70Va	
A127 ANILINE, 4-fluoro-			C_6H_6NF		$C_6H_4NH_2(F)$		
7.76	4.65(+1)	4.76	25	C	μ=0	73Lb	
7.450	4.646(+1)	3.7	25	C	μ≃0	70Va	
6.8	4.66(+1)	1.5	25	-	--	69La	f
8.490	4.376(+1)	8.4	25	C	μ≃0 S: 0.10 mole fraction Ethanol	70Va	
A128 ANILINE, 2-iodo-			C_6H_6NI		$C_6H_4NH_2(I)$		
5.816	2.537(+1)	7.88	25	T	μ=0 ΔCp=-10	69Be	
4.9	2.56(+1)	4.7	25	-	--	69La	f
A129 ANILINE, 3-iodo			C_6H_6NI		$C_6H_4NH_2(I)$		
6.78	3.58(+1)	6.34	25	C	μ=0	73Lb	
6.326	3.583(+1)	4.81	25	T	μ=0	68Bb	
5.7	3.59(+1)	2.7	25	-	--	69La	f
A130 ANILINE, 4-iodo-			C_6H_6NI		$C_6H_4NH_2(I)$		
6.50	3.82(+1)	4.33	25	C	μ=0	73Lb	
6.552	3.812(+1)	4.51	25	T	μ=0 ΔCp=14	69Bd	

ΔH, kcal /mole	pK	ΔS, cal /mole°K	T,°C	M	Conditions	Ref.	Remarks
A130, cont.							
6.3	3.77(+1)	3.9	25	-	--	69La	f
A131 ANILINE, 2-methoxy-			C_7H_9ON		$C_6H_4NH_2(OCH_3)$		
7.622	4.52(+1)	4.9	25	C	μ≈0	71Va	
7.411	4.527(+1)	4.12	25	T	μ=0 ΔCp=0	69Be	
6.9	4.53(+1)	2.4	25	-	--	69La	f
8.979	4.30(+1)	10.5	25	C	μ≈0 S: 0.10 mole fraction Ethanol	71Va	
A132 ANILINE, 3-methoxy-			C_7H_9ON		$C_6H_4NH_2(OCH_3)$		
7.11	4.20(+1)	4.63	25	C	μ=0	73Lb	
7.013	4.204(+1)	4.29	25	T	μ=0	68Bb	
6.887	4.226(+1)	3.8	25	C	μ≈0	70Va	
5.9	4.22(+1)	0.5	25	-	--	69La	f
7.220	4.11(+1)	5.4	25	C	μ≈0 S: 0.034 mole fraction Ethanol	70Va	
7.405	4.03(+1)	6.4	25	C	μ≈0 S: 0.074 mole fraction Ethanol	70Va	
7.440	3.99(+1)	6.7	25	C	μ≈0 S: 0.097 mole fraction Ethanol	70Va	
7.440	3.956(+1)	6.8	25	C	μ≈0 S: 0.10 mole fraction Ethanol	70Va	
7.385	3.95(+1)	6.7	25	C	μ≈0 S: 0.131 mole fraction Ethanol	70Va	
7.140	3.93(+1)	6.0	25	C	μ≈0 S: 0.171 mole fraction Ethanol	70Va	
A133 ANILINE, 4-methoxy-			C_7H_9ON		$C_6H_4NH_2(OCH_3)$		
8.51	5.36(+1)	4.02	25	C	μ=0	73Lb	
8.341	5.357(+1)	3.44	25	T	μ=0 ΔCp=0	69Bd	
8.210	5.335(+1)	3.1	25	C	μ≈0	70Va	
8.777	5.126(+1)	6.0	25	C	μ≈0 S: 0.10 mole fraction Ethanol	70Va	
A134 ANILINE, 2-nitro-			$C_6H_6O_2N_2$		$C_6H_6N(NO_2)$		
1.957	-0.246(+1)	7.69	25	T	μ=0 ΔCp=25	69Be	
1.675	-0.287(+1)	7.1	25	T	μ=0	70Be	g:69Ja
2.63	-0.28(25°,+1)	10.09	15-45	T	μ=0	74Ac	c
2.77	(+1)	-----	15-45	T	0.38 m $HClO_4$	74Ac	c
3.18	(+1)	-----	15-45	T	0.52 m $HClO_4$	74Ac	c
2.94	(+1)	-----	15-45	T	0.80 m $HClO_4$	74Ac	c
3.06	(+1)	-----	15-45	T	1.10 m $HClO_4$	74Ac	c
2.92	(+1)	-----	15-45	T	1.40 m $HClO_4$	74Ac	c
2.51	(+1)	-----	15-45	T	1.70 m $HClO_4$	74Ac	c
2.51	(+1)	-----	15-45	T	1.98 m $HClO_4$	74Ac	c
2.92	(+1)	-----	15-45	T	2.30 m $HClO_4$	74Ac	c
2.70	(+1)	-----	15-45	T	2.60 m $HClO_4$	74Ac	c
1.6	-0.29(+1)	6.7	25	-	--	69La	f
A135 ANILINE, 3-nitro-			$C_6H_6O_2N_2$		$C_6H_4NH_2(NO_2)$		
5.37	2.46(+1)	6.78	25	C	μ=0	73Lb	
4.5	(+1)	-----	25	C	μ=0	67Oa	
4.980	2.460(+1)	5.45	25	T	μ=0	68Bb	
4.5	2.45(+1)	3.9	25	-	--	69La	f

ΔH, kcal /mole	pK	ΔS, cal /mole°K	T, °C	M	Conditions	Ref.	Remarks
A136 ANILINE, 4-nitro-			$C_6H_6O_2N_2$		$C_6H_6N(NO_2)$		
4.12	1.02(+1)	9.16	25	C	μ=0	73Lb	
3.417	1.016(+1)	6.79	25	T	μ=0 ΔCp=22	69Bd	
3.085	0.997(+1)	5.8	25	T	μ=0	70Be	g:69Ja
3.49	0.997(25°,+1)	7.14	15-45	T	μ=0	74Ac	c
3.54	(+1)	-----	15-45	T	0.065 m $HClO_4$	74Ac	c
3.49	(+1)	-----	15-45	T	0.08 m $HClO_4$	74Ac	c
3.44	(+1)	-----	15-45	T	0.14 m $HClO_4$	74Ac	c
3.61	(+1)	-----	15-45	T	0.25 m $HClO_4$	74Ac	c
3.35	(+1)	-----	15-45	T	0.38 m $HClO_4$	74Ac	c
3.58	(+1)	-----	15-45	T	0.52 m $HClO_4$	74Ac	c
3.92	(+1)	-----	15-45	T	0.65 m $HClO_4$	74Ac	c
3.78	(+1)	-----	15-45	T	0.80 m $HClO_4$	74Ac	c
2.9	$\bar{1}$.01(+1)	5.1	25	-	--	69La	f
A137 ANILINE, 2,4,6-trinitro-			$C_6H_4O_6N_4$		$C_6H_2NH_2(NO_2)_3$		
-9.704	-10.228(+1)	13.5	25	T	μ=0	70Be	g:69Ja
A138 ANSERINE			$C_{10}H_{16}O_3N_4$		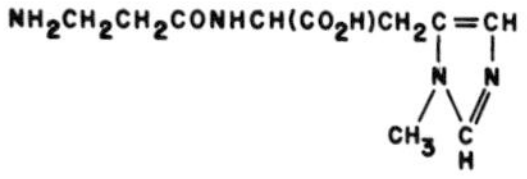		
7.5	6.8(37°,0)	-6.9	22-37	T	μ=0.3	56Bb	c
A139 ANTHRACENE, 1-amino-			$C_{14}H_{11}N$		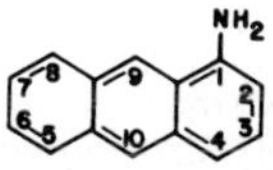		
5.2	3.22(20°,+1)	2.9	0-20	T	C=0.005 S: 50 w % Ethanol	59Eb	c
A140 ANTIPYRINE, 4-amino-			$C_{11}H_{13}ON_3$		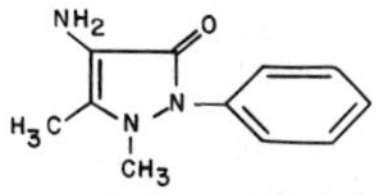		
3.847	4.941(+1)	-9.72	25	C	μ=0	70Ka	
3.738	4.941(+1)	-10.01	25	T	μ=0 ΔCp=6.37	70Ka	
A141 ANTIPYRINE, 4-(dimethylamino)-			$C_{13}H_{17}ON_3$		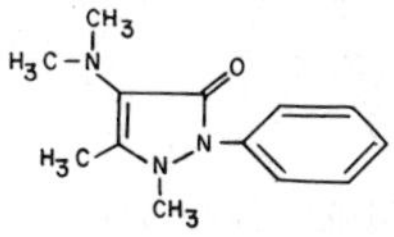		
6.329	4.183(+1)	2.09	25	C	μ=0	70Ka	
6.098	4.183(+1)	1.31	25	T	μ=0 ΔCp=9.31	70Ka	

ΔH, kcal /mole	pK	ΔS, cal /mole°K	T,°C	M	Conditions	Ref.	Remarks
A142	**AQUOCOBALAMINE COMPLEX**						
10.5	10.25	−12	25	T	μ=0	66Ha	e: Imino group
A143	**ARABINOSE**		$C_5H_{10}O_5$		$CH_2CH(OH)CH(OH)CH(OH)CHOH$ $\llcorner$——O——$\lrcorner$		
9.96	12.79(0)	−23.4	10	C	μ=0	70Cd	
9.55	12.34($\overline{0}$)	−24.4	25	C	μ=0	72Cc	
8.3	12.54($\overline{0}$)	−29.6	25	C	μ=0	64Wb	
A144	**ARGININE**		$C_6H_{14}O_2N_4$		$H_2NC(:NH)NH(CH_2)_3CH(NH_2)CO_2H$		
1.63	1.914(+$\underline{1}$)	−2.8	0	T	μ=0	61Da	
1.50	1.8847(+$\overline{1}$)	−3.3	5	T	μ=0	61Da	
1.37	1.8697(+$\underline{1}$)	−3.7	10	T	μ=0	61Da	
1.25	1.8488(+$\underline{1}$)	−4.2	15	T	μ=0	61Da	
1.12	1.8374(+$\underline{1}$)	−4.6	20	T	μ=0	61Da	
0.98	1.8217(+$\underline{1}$)	−5.0	25	T	μ=0	61Da	
0.85	1.8138(+$\underline{1}$)	−5.5	30	T	μ=0	61Da	
0.71	1.8013(+$\underline{1}$)	−6.0	35	T	μ=0	61Da	
0.57	1.7995(+$\underline{1}$)	−6.4	40	T	μ=0	61Da	
0.40	1.975(25°,+$\underline{1}$)	−7.7	10–40	T	μ=0	60Pb	c
0.43	1.7852(+$\underline{1}$)	−6.8	45	T	μ=0	61Da	
0.28	1.787(+$\underline{1}$)	−7.3	50	T	μ=0	61Da	
10.95	9.7178($\overline{0}$)	−4.4	0	T	μ=0	61Da	
10.91	9.5626($\overline{0}$)	−4.5	5	T	μ=0	61Da	
10.86	9.4073($\overline{0}$)	−4.7	10	T	μ=0	61Da	
10.82	9.2669($\overline{0}$)	−4.9	15	T	μ=0	61Da	
10.77	9.1226($\overline{0}$)	−5.0	20	T	μ=0	61Da	
10.73	8.9936($\overline{0}$)	−5.2	25	T	μ=0	61Da	
10.73	8.99($\overline{0}$)	−5.2	25	T	μ=0	54Da	
10.68	8.8589(0)	−5.3	30	T	μ=0	61Da	
10.64	8.7390(0)	−5.5	35	T	μ=0	61Da	
10.59	8.614(0)	−5.6	40	T	μ=0	61Da	
11.0	9.05(25°,0)	−4.5	10–40	T	μ=0	60Pb	c
10.54	8.5040(0)	−5.8	45	T	μ=0	61Da	
10.49	8.3852($\overline{0}$)	−5.9	50	T	μ=0	61Da	
10.44	8.2824($\overline{0}$)	−6.1	55	T	μ=0	61Da	
10.90	(0)	-----	20	C	μ≈0.01	71Ma	e:α-NH_3^+
10.25	9.00(0)	−6.7	25	T	μ=0.0237	54Da	
12.400	12.48(−$\underline{1}$)	−15.5	0–25	T	μ=0.01	33Ga	c
12.400	12.46(−$\underline{1}$)	−15.5	25	T	μ=0.02	30Ma	
A145	**ARGININE, phenylalanyl-**		$C_{15}H_{23}O_3N_5$		$NH_2C(:NH)NH(CH_2)_3CH(CO_2H)-$ $NHCOCH(NH_2)CH_2C_6H_5$		
0.450	2.66(25°,+$\underline{1}$)	−10.7	0&25	T	μ=0.01	33Ga	b,c,e: carboxyl
10.150	7.57(25°,$\underline{0}$)	−0.6	0&25	T	μ=0.01	33Ga	b,c,e: amino
11.950	12.40(25°,−$\underline{1}$)	−5.0	0&25	T	μ=0.01	33Ga	b,c,e: guan- idine
A146	**ARGININE, tyrosyl-**		$C_{15}H_{23}O_4N_5$		$NH_2C(:NH)NH(CH_2)_3CH(CO_2H)-$ $NHCOCH(NH_2)CH_2C_6H_4OH$		
0.0	2.65(+$\underline{1}$)	−12.1	0&25	T	μ=0.01	33Ga	b,c,e: carboxyl

ΔH, kcal /mole	pK	ΔS, cal /mole°K	T, °C	M	Conditions	Ref.	Remarks
A146, cont.							
10.5	7.39(0)	1.41	0&25	T	μ=0.01	33Ga	b,c,e: amino
6.0	9.36(-1)	-22.7	0&25	T	μ=0.01	33Ga	b,c,e: oxy-phenol
13.0	11.62(-2)	-9.6	0&25	T	μ=0.01	33Ga	b,c,e: guan-idine
A147 ARSENIC ACID, ortho-			O_4H_3As		H_3AsO_4		
-1.69	2.25(0)	-16.0	25	C	μ=0.1-0.3	64Sb	
0.77	6.77(-1)	-28.4	25	C	μ=0.1-0.3	64Sb	
4.35	11.53(-2)	-38.5	25	C	μ=0.1-0.3	64Sb	
A148 ARSENOUS ACID			O_3H_3As		H_3AsO_3		
6.58	9.23(0)	-20.2	25	C	μ=0.1-0.3	64Sb	
16.4	-0.114(11.1°, +4)	57.3	2.4-11.1	T	μ=1.00 HA=$[CeH_3AsO_3]^{4+}$	71Ea	
A149 ASPARAGINE			$C_4H_8O_3N_2$		$H_2NCOCH_2CH(NH_2)CO_2H$		
1.13	2.14(+1)	-6.0	25	C	μ=0.2(KCl)	75Ga	
1.22	2.586(+1)	-7.74	25	C	μ=3.0M($NaClO_4$)	72Ga 74Bc	e:CO_2H
10.97	10.241(0)	-7.40	5	T	μ=0.1(KCl)	73Ra	e:NH_3^+
9.75	(0)	-----	25	C	μ=0.10M	71Ba	e:NH_3^+
10.49	9.674(0)	-9.06	25	T	μ=0.1(KCl)	73Ra	e:NH_3^+
9.54	9.210(0)	-12.14	45	T	μ=0.1(KCl)	73Ra	e:NH_3^+
9.89	8.74(0)	-6.7	25	C	μ=0.2(KCl)	75Ga	
12.07	9.303(0)	-2.1	25	C	μ=3.0M($NaClO_4$)	72Ga	e:NH_3^+
12.1	9.30(0)	-2.13	25	C	μ=3.00M($NaClO_4$)	74Bc	e:NH_3^+
A150 ASPARAGINE, glycyl-			$C_6H_{11}O_4N_3$		$NH_2CH_2CONHCOCH_2CH(NH_2)CO_2H$		
0.133	2.9420(+1)	-13.01	25	T	μ=0	57Ka	
A151 ASPARTIC ACID			$C_4H_7O_4N$		$HO_2CCH_2CH(NH_2)CO_2H$		
1.85	2.05(+1)	-3.2	25	C	μ=0	67Ca	
2.324	2.122(+1)	-1.3	1	T	μ=0	42Sb	
2.094	2.054(+1)	-2.1	12.5	T	μ=0	42Sb	
1.783	1.990(+1)	-3.1	25	T	μ=0	42Sb	
1.373	1.945(+1)	-4.5	37.5	T	μ=0	42Sb	
0.889	1.907(+1)	-6.0	50	T	μ=0	42Sb	
1.4	1.95(+1)	-4	25	C	μ=0.2M(KCl)	74Na	e: α-CO_2H
1.7	1.95(25°,+1)	-3	20-35	T	μ=0.2M(KCl)	74Na	e: α-CO_2H
1.74	2.345(+1)	-4.9	25	C	μ=3.0M($NaClO_4$)	72Ga	e:CO_2H
0.96	3.87(0)	-14.5	25	C	μ=0	67Ca	
1.761	4.006(0)	-11.9	1	T	μ=0	42Sb	
1.448	3.944(0)	-13.0	12.5	T	μ=0	42Sb	
1.110	3.900(0)	-14.2	25	T	μ=0	42Sb	
0.619	3.878(0)	-15.8	37.5	T	μ=0	42Sb	
0.057	3.870(0)	-17.8	50	T	μ=0	42Sb	
1.530	(0)	-----	25	C	μ=0.01	59Ka	
1.760	3.65(0)	-10.8	25	T	μ=0.02	30Ma	

ΔH, kcal /mole	pK	ΔS, cal /mole°K	T, °C	M	Conditions	Ref.	Re-marks
A151, cont.							
1.1	3.68(0)	-13	25	C	μ=0.2M(KCl)	74Na	e: β-CO_2H
5.71	4.067(0)	0.55	25	C	μ=3.0M($NaClO_4$)	72Ga	e: β-CO_2H
9.280	10.604(-1)	-14.6	1	T	μ=0	42Sb	
9.220	10.304(-1)	-14.8	12.5	T	μ=0	42Sb	
9.025	10.002(-1)	-15.5	25	T	μ=0	42Sb	
8.695	9.742(-1)	-16.6	37.5	T	μ=0	42Sb	
8.220	9.512(-1)	-18.1	50	T	μ=0	42Sb	
9.22	(-1)	-----	20	C	μ≈0.01	71Ma	e:NH_3^+
10.50	9.60(-1)	-8.7	25	T	μ=0.02	30Ma	
9.4	9.63(-1)	-12	25	C	μ=0.2M(KCl)	74Na	e:NH_3^+
9.8	9.63(25°,-1)	-11	20-35	T	μ=0.2M(KCl)	74Na	c,e:NH_3^+
5.64	10.007(-1)	-26.8	25	C	μ=3.0M($NaClO_4$)	72Ga	e:NH_3^+
2.4	3.68(+1)	-9	25	C	μ=0.2M(KCl) HA=[Cu(II)HO_2CCH_2CH-(NH_2)CO_2]$^+$	74Na	
2.5	3.68(25°,+1)	-8	20-35	T	μ=0.2M(KCl) HA=[Cu(II)HO_2CCH_2CH-(NH_2)CO_2]$^+$	74Na	c
3.5	4.43(25°,+1)	-9	1.6-40	T	μ=0.15M(KCl) HA=Aspartic acid residue 52; Asp-52 residue of lysozyme	72Pe	c,q: ±1.4
4.5	5.22(25°,+1)	-9	1.6-40	T	μ=0.15M(KCl) HA=Aspartic acid residue 52; Asp-52 residue of lysozyme	72Pe	c,q: ±3.2

A152 1-AZACYCLOHEXANE, N-methyl $C_6H_{13}N$

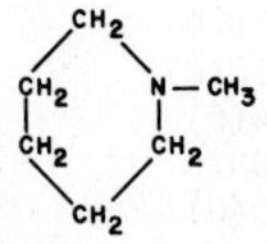

11.5	8.99(25°,+1)	-2.6	15-35	T	S: 60 w % Methanol	38Wa	c

A153 1-AZACYCLONONANE $C_8H_{17}N$ $HNCH_2(CH_2)_6CH_2$

12.9	9.66(25°,+1)	-0.93	25	T	S: 60 w % Methanol	38Wa	c

A154 1-AZACYCLOOCTADECANE $C_{17}H_{35}N$ $HNCH_2(CH_2)_{15}CH_2$

12.0	9.41(25°,+1)	-2.81	25	T	S: 60 w % Methanol	38Wa	

A155 AZETIDINE C_3H_7N $CH_2CH_2CH_2NH$

12.52	11.29(+1)	-9.7	25	C	μ=0	71Ca 68Ca	

A156 AZIRIDINE C_2H_5N CH_2CH_2NH

9.06	8.04(+1)	-6.4	25	C	μ=0	69Cc	
9.02	8.04(+1)	-6.5	25	C	μ=0	71Ca 68Ca	
10.5	(+1)	-----	25	T	μ=0	56Sa	

ΔH, kcal /mole	pK	ΔS, cal /mole°K	T,°C	M	Conditions	Ref.	Re-marks
			B				
B1 BARBITURIC ACID			$C_4H_4O_3N_2$		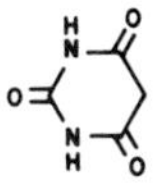		
-0.834	3.969(+1)	-21.1	15	T	μ=0	69Bf	
-0.815	3.980(+1)	-21.0	20	T	μ=0	69Bf	
0.06	4.04(+1)	-18.8	25	C	μ=0	65Ma	
-0.795	3.990(+1)	-20.9	25	T	μ=0	69Bf	
0.007	4.024(+1)	-18.4	25	T	μ=0	74Kd	
-0.775	3.999(+1)	-20.9	30	T	μ=0	69Bf	
-0.756	4.008(+1)	-20.8	35	T	μ=0	69Bf	
-0.737	4.017(+1)	-20.7	40	T	μ=0	69Bf	
-0.717	4.025(+1)	-20.7	45	T	μ=0	69Bf	
-0.697	4.032(+1)	-20.6	50	T	μ=0	69Bf	
4.124	8.493(0)	-24.6	15	T	μ=0	69Bf	
4.775	8.435(0)	-22.3	20	T	μ=0	69Bf	
5.425	8.372(0)	-20.1	25	T	μ=0	69Bf	
6.076	8.302(0)	-18.0	30	T	μ=0	69Bf	
6.727	8.227(0)	-15.8	35	T	μ=0	69Bf	
7.378	8.147(0)	-13.7	40	T	μ=0	69Bf	
8.028	8.063(0)	-11.7	45	T	μ=0	69Bf	
8.679	7.974(0)	-9.61	50	T	μ=0	69Bf	
B2 BARBITURIC ACID, 5-allyl-			$C_7H_8O_3N_2$		$C_4H_3O_3N_2(CH_2CH{:}CH_2)$		
-1.89	4.78(+1)	-28.2	25	C	μ=0	65Ma	
B3 BARBITURIC ACID, 5-allyl-5(1-methylbutyl)			$C_{12}H_{18}O_3N_2$		$C_4H_2O_3N_2(CH_2CH{:}CH_2)[CH(CH_3)$-$CH_2CH_2CH_3]$		
5.00	8.08(0)	-20.2	25	C	μ=0	65Ma	
B4 BARBITURIC ACID, 5-amino-N,N-diacetic acid			$C_8H_9O_7N_3$		$C_4H_3O_3N_2[N(CH_2CO_2H)_2]$		
6.9	9.63(20°,-2)	-19	20-39	T	μ=0.1M[$(CH_3)_4NNO_3$]	63Ia	c
B5 BARBITURIC ACID, 5-amyl-5-ethyl-			$C_{11}H_{18}O_3N_2$		$C_4H_2O_3N_2(C_5H_{11})(C_2H_5)$		
4.674	8.089(0)	-20.8	15	T	μ=0	69Bf	
5.074	8.025(0)	-19.4	20	T	μ=0	69Bf	
5.474	7.960(0)	-18.1	25	T	μ=0	69Bf	
5.875	7.891(0)	-16.7	30	T	μ=0	69Bf	
6.275	7.820(0)	-15.4	35	T	μ=0	69Bf	
6.675	7.747(0)	-14.1	40	T	μ=0	69Bf	
7.075	7.671(0)	-12.9	45	T	μ=0	69Bf	
7.476	7.594(0)	-11.6	50	T	μ=0	69Bf	
B6 BARBITURIC ACID, 5,5-diallyl-			$C_{10}H_{12}O_3N_2$		$C_4H_2O_3N_2(CH_2CH{:}CH_2)_2$		
4.92	7.78(0)	-19.1	25	C	μ=0	65Ma	
B7 BARBITURIC ACID, 5,5-diethyl-			$C_8H_{12}O_3N_2$		$C_4H_2O_3N_2(C_2H_5)_2$		
5.788	8.169(0)	-17.3	15	T	μ=0	69Bf	

ΔH, kcal /mole	pK	ΔS, cal /mole°K	T,°C	M		Conditions	Ref.	Remarks
B7, cont.								
5.861	8.094(0)	-17.0	20	T	μ=0		69Bf	
5.81	7.98(0)	-17.0	25	C	μ=0		65Ma	
5.934	8.020(0)	-16.8	25	T	μ=0		69Bf	
6.006	7.948(0)	-16.6	30	T	μ=0		69Bf	
6.079	7.877(0)	-16.3	35	T	μ=0		69Bf	
6.152	7.808(0)	-16.1	40	T	μ=0		69Bf	
6.224	7.740(0)	-15.9	45	T	μ=0		69Bf	
6.297	7.673(0)	-15.6	50	T	μ=0		69Bf	

B8 BARBITURIC ACID, 1,3-dimethyl- $C_6H_8O_3N$ $C_4H_2O_3N(CH_3)_2$

-0.05	4.68(+1)	-21.6	25	C	μ=0		65Ma	

B9 BARBITURIC ACID, 5-ethyl- $C_6H_8O_3N_2$ $C_4H_3O_3N_2(C_2H_5)$

-2.06	3.691(+1)	-23.8	25	T	μ=0		74Kd	

B10 BARBITURIC ACID, 5-ethyl-5-(1-methylbutyl) $C_{11}H_{18}O_3N_2$ $C_4H_2O_3N_2(CH_2CH_3)[CH(CH_3)CH_2CH_2CH_3]$

5.81	8.11(0)	-17.6	25	C	μ=0		65Ma	

B11 BARBITURIC ACID, 5-ethyl-5-pentyl- $C_{11}H_{18}O_3N_2$ $C_4H_2O_3N_2(CH_2CH_3)(CH_2CH_2CH_2CH_2CH_3)$

5.22	7.95(0)	-18.6	25	C	μ=0		65Ma	

B12 BARBITURIC ACID, 5-ethyl-5-phenyl- $C_{12}H_{12}O_3N_2$ $C_4H_2O_3N_2(C_2H_5)(C_6H_5)$

5.804	7.592(0)	-14.6	15	T	μ=0		69Bf	
5.751	7.517(0)	-14.8	20	T	μ=0		69Bf	
4.60	7.44(0)	-18.6	25	C	μ=0		65Ma	
5.697	7.445(0)	-15.0	25	T	μ=0		69Bf	
5.643	7.377(0)	-15.1	30	T	μ=0		69Bf	
5.590	7.311(0)	-15.3	35	T	μ=0		69Bf	
5.536	7.248(0)	-15.5	40	T	μ=0		69Bf	
5.483	7.188(0)	-15.7	45	T	μ=0		69Bf	
5.429	7.130(0)	-15.8	50	T	μ=0		69Bf	

ΔH, kcal /mole	pK	ΔS, cal /mole°K	T, °C	M	Conditions	Ref.	Remarks
B13 BARBUTURIC ACID, 5-isopropyl-			$C_7H_{10}O_3N_2$		$C_4H_3O_3N_2[CH(CH_3)_2]$		
0.21	4.907(+1)	-21.7	25	T	μ=0	74Kd	
B14 BARBITURIC ACID, 5-methyl-			$C_5H_6O_3N_2$		$C_4H_3O_3N_2(CH_3)$		
-1.28	4.40(+1)	-24.4	25	C	μ=0	65Ma	
-1.60	3.386(+1)	-20.9	25	T	μ=0	74Kd	
B15 BARBITURIC ACID, 5-methyl-5-			$C_{11}H_{10}O_3N_2$		$C_4H_2O_3N_2(CH_3)(C_6H_5)$		
3.620	8.104(0)	-24.5	15	T	μ=0	69Bf	
3.653	8.057(0)	-24.4	20	T	μ=0	69Bf	
3.687	8.011(0)	-24.3	25	T	μ=0	69Bf	
3.720	7.966(0)	-24.2	30	T	μ=0	69Bf	
3.754	7.922(0)	-24.1	35	T	μ=0	69Bf	
3.787	7.879(0)	-24.0	40	T	μ=0	69Bf	
3.821	7.838(0)	-23.9	45	T	μ=0	69Bf	
3.854	7.797(0)	-23.8	50	T	μ=0	69Bf	
B16 BARBITURIC ACID, 5-phenyl-			$C_{10}H_8O_3N_2$		$C_4H_3O_3N_2(C_6H_5)$		
-2.29	2.544(+1)	-19.3	25	T	μ=0	74Kd	
B17 BENZALDEHYDE, 4-dimethylamino-			$C_9H_{11}ON$		$C_7H_5O[N(CH_3)_2]$		
1.800	1.647(+1)	-1.5	25	T	μ=0	59Rb	
B18 BENZALDEHYDE, 2-hydroxy-			$C_7H_6O_2$		$C_7H_5O(OH)$		
5.15	8.37(0)	-21.0	25	C	μ=0	64Mc	
6.3	8.80(0)	-20	25	T	μ=3	64Sc	
B19 BENZALDEHYDE, 3-hydroxy-			$C_7H_6O_2$		$C_7H_5O(OH)$		
5.17	9.02(0)	-23.9	25	C	μ=0	64Mc	
4.993	8.983(0)	-24.36	25	T	μ=0 ΔCp=-33.3	67Bb	

ΔH, kcal /mole	pK	ΔS, cal /mole°K	T,°C	M	Conditions	Ref.	Remarks
B20 BENZALDEHYDE, 4-hydroxy-			$C_7H_6O_2$		$C_7H_5O(OH)$		
4.29	7.620(0)	-20.5	25	C	μ=0	75Pa	
4.26	7.62(0)	-20.6	25	C	μ=0	64Mc	
4.357	7.616(25°,0)	-20.2	14.9-55.9		0.0017M	67Lb	c
4.992	8.982(25°,0)	-24.36	5-60	T	μ=0.04	67Bb	c
4.42	7.664(0)	-20.2	25	C	μ=0 S: 0.059 mole fraction Methanol	75Pa	
4.29	7.832(0)	-21.4	25	C	μ=0 S: 0.123 mole fraction Methanol	75Pa	
3.90	7.989(0)	-23.5	25	C	μ=0 S: 0.194 mole fraction Methanol	75Pa	
3.76	8.177(0)	-24.8	25	C	μ=0 S: 0.273 mole fraction Methanol	75Pa	
3.52	8.368(0)	-26.5	25	C	μ=0 S: 0.360 mole fraction Methanol	75Pa	
3.35	8.503(0)	-27.6	25	C	μ=0 S: 0.458 mole fraction Methanol	75Pa	
3.11	8.743(0)	-29.6	25	C	μ=0 S: 0.567 mole fraction Methanol	75Pa	
2.88	9.076(0)	-31.8	25	C	μ=0 S: 0.692 mole fraction Methanol	75Pa	
2.97	9.692(0)	-34.4	25	C	μ=0 S: 0.835 mole fraction Methanol	75Pa	
7.68	11.907(0)	-28.7	25	C	μ=0 S: Methanol	75Pa	
8.00	11.906(0)	-27.7	25	T	μ=0 S: Methanol ΔCp=-24	70Bf	
7.7	12.01(0)	-29	25	T	S: Methanol	69Rd	
B21 BENZALDEHYDE, 2-hydroxy-4-amino-			$C_7H_7O_2N$		$C_7H_4O(OH)(NH_2)$		
11.0	13.74(25°,0)	-27	15-35	T	3.0M($NaClO_4$)	56Aa	c
B22 BENZALDEHYDE, 2-hydroxy-3-methoxy-			$C_8H_8O_3$		$C_7H_4O(OH)(OCH_3)$		
4.13	7.91(0)	-22.3	25	C	μ=0	64Mc	
4.0	7.4(0)	-2	25	T	μ=0.10	69Da	
B23 BENZALDEHYDE, 3-hydroxy-4-methoxy-			$C_8H_8O_3$		$C_7H_4O(OH)(OCH_3)$		
4.62	8.89(0)	-25.2	25	C	μ=0	64Mc	
B24 BENZALDEHYDE, 4-hydroxy-3-methoxy-			$C_8H_8O_3$		$C_7H_4O(OCH_3)(OH)$		
3.75	7.40(0)	-21.3	25	C	μ=0	64Mc	
B25 BENZENE, 1-allyl-2-hydroxy-			$C_9H_{10}O$		$C_6H_4(CH_2CH:CH_2)(OH)$		
6.01	10.28(0)	-26.9	25	C	μ=0	63Oa	
B26 BENZENE, 1-amino-2,3-dimethyl			$C_8H_{11}N$		$C_6H_3(NH_2)(CH_3)_2$		
8.79	4.57(+1)	8.59	25	C	μ=0	59Za	
6.84	4.72(+1)	1.40	25	T	μ=0	73Pa	

ΔH, kcal /mole	pK	ΔS, cal /mole°K	T, °C	M	Conditions	Ref.	Remarks
B27 BENZENE, 1-amino-2,4-dimethyl- $C_8H_{11}N$					$C_6H_3(NH_2)(CH_3)_2$		
7.90	5.0(+1)	3.63	25	C	μ=0	59Za	
5.79	4.86(+1)	-2.80	25	T	μ=0	73Pa	
B28 BENZENE, 1-amino-2,5-dimethyl- $C_8H_{11}N$					$C_6H_3(NH_2)(CH_3)_2$		
6.53	4.6(+1)	0.87	25	C	μ=0	59Za	
6.07	4.53(+1)	-0.34	25	T	μ=0	73Pa	
B29 BENZENE, 1-amino-2,6-dimethyl- $C_8H_{11}N$					$C_6H_3(NH_2)(CH_3)_2$		
5.24	4.20(+1)	-1.63	25	C	μ=0	59Za	
5.75	3.93(+1)	1.35	25	T	μ=0	73Pa	
B30 BENZENE, 1-amino-3,4-dimethyl- $C_8H_{11}N$					$C_6H_3(NH_2)(CH_3)_2$		
8.04	5.15(+1)	3.42	25	C	μ=0	59Za	
B31 BENZENE, 1-amino-3,5-dimethyl- $C_8H_{11}N$					$C_6H_3(NH_2)(CH_3)_2$		
9.04	4.75(+1)	8.75	25	C	μ=0	59Za	
7.393	4.765(+1)	2.99	25	T	μ=0 ΔCp=-15	70Bd	
B32 BENZENE, 1-amino-2-ethyl- $C_8H_{11}N$					$C_6H_4(NH_2)(CH_2CH_3)$		
6.024	4.42(+1)	2.7	25	C	μ≃0	71Va	
8.039	4.19(+1)	7.8	25	C	μ≃0 S: 0.10 mole fraction Ethanol	71Va	
B33 BENZENE, 1-amino-3-ethyl- $C_8H_{11}N$					$C_6H_4(NH_2)(C_2H_5)$		
7.31	4.70(+1)	3.02	25	C	μ=0	73Lb	
B34 BENZENE, 1-amino-4-ethyl- $C_8H_{11}N$					$C_6H_4(NH_2)(C_2H_5)$		
7.74	5.00(+1)	3.09	25	C	μ=0	73Lb	
B35 BENZENE, 1(t-butyl)-2-hydroxy- $C_{10}H_{14}O$					$C_6H_4[C(CH_3)_3](OH)$		
5.242	10.623(0)	-31.02	25	T	μ=0 ΔCp=-41	72Be	
11.6	16.46(0)	-36	25	T	S: Methanol	69Rd	
11.87	16.357(0)	-35	25	T	μ=0 S: Methanol ΔCp=-44	70Bf	
B36 BENZENE, 1-(t-butyl)-3-hydroxy- $C_{10}H_{14}O$					$C_6H_4[C(CH_3)_3](OH)$		
5.501	10.119(0)	-27.85	25	T	μ=0 ΔCp=-59	72Be	
B37 BENZENE, 1(t-butyl)-4-hydroxy- $C_{10}H_{14}O$					$C_6H_4[C(CH_3)_3](OH)$		
4.41	10.230(0)	-32.0	25	C	μ=0	75Pa	
5.219	10.390(0)	-30.03	25	T	μ=0 ΔCp=-43	72Be	
4.85	10.381(0)	-31.2	25	C	μ=0 S: 0.059 mole fraction Methanol	75Pa	

ΔH, kcal /mole	pK	ΔS, cal /mole°K	T,°C	M	Conditions	Ref.	Remarks
B37, cont.							
5.18	10.537(0)	-30.8	25	C	μ=0 S: 0.123 mole fraction Methanol	75Pa	
5.51	10.763(0)	-30.8	25	C	μ=0 S: 0.194 mole fraction Methanol	75Pa	
5.43	10.973(0)	-32.0	25	C	μ=0 S: 0.273 mole fraction Methanol	75Pa	
5.26	11.175(0)	-33.5	25	C	μ=0 S: 0.360 mole fraction Methanol	75Pa	
5.12	11.308(0)	-34.6	25	C	μ=0 S: 0.458 mole fraction Methanol	75Pa	
4.64	11.569	-37.4	25	C	μ=0 S: 0.567 mole fraction Methanol	75Pa	
4.26	11.844(0)	-39.9	25	C	μ=0 S: 0.692 mole fraction Methanol	75Pa	
4.20	12.352(0)	-42.4	25	C	μ=0 S: 0.835 mole fraction Methanol	75Pa	
8.42	14.544(0)	-38.3	25	C	μ=0 S: Methanol	75Pa	
8.85	14.543(0)	-36.8	25	T	μ=0 S: Methanol ΔCp=-67	70Bf	
8.8	14.65(0)	-38	25	T	S: Methanol	69Rd	
B38 BENZENE, 4-chloro-1,2-dihydroxy-		$C_6H_5O_2Cl$			$C_6H_3(Cl)(OH)_2$		
6.85	8.522(0)	-16.00	25	C	μ=0.100M(KNO_3)	72Jb	
B39 BENZENE, 1-chloro-2,6-dimethyl-4-hydroxy-		C_8H_9OCl			$C_6H_2(Cl)(CH_3)_2(OH)$		
5.560	9.702(0)	-25.73	25	T	μ=0 ΔCp=-33	70Bc	
B40 BENZENE, 1-chloro-4-hydroxy-6-methyl-		C_7H_7OCl			$C_6H_3(Cl)(OH)(CH_3)$		
5.462	9.549(0)	-25.38	25	T	μ=0 ΔCp=-32	70Bc	
B41 BENZENE, 1,2-diamino-		$C_6H_8N_2$			$C_6H_4(NH_2)_2$		
6.3	4.74(25°,+1)	-0.6	15-40	T	μ=0.3M($NaClO_4$)	71Mb	c
B42 BENZENE, 1,2-diamino-4-methyl-		$C_7H_{10}N_2$			$C_6H_3(NH_2)_2(CH_3)$		
6.3	4.97(25°,+1)	-1.7	15-40	T	μ=0.3M($NaClO_4$)	71Mb	c
B43 BENZENE, 1,3-di(t-butyl)-5-formyl-2-hydroxy-		$C_{15}H_{22}O_2$			$C_6H_2[C(CH_3)_3]_2(CHO)(OH)$		
7.92	12.267(0)	-29.6	25	T	μ=0 S: Methanol ΔCp=-52	70Bf	
7.7	12.37(0)	-31	25	T	S: Methanol	69Rd	
B44 BENZENE, 1,3-di(t-butyl)-2-hydroxy-		$C_{14}H_{22}O$			$C_6H_3[C(CH_3)_3]_2(OH)$		
8.35	17.201(0)	-50.7	25	T	μ=0 S: Methanol ΔCp=0	70Bf	
8.4	17.30(0)	-51	25	T	S: Methanol	69Rd	

ΔH, kcal /mole	pK	ΔS, cal /mole°K	T,°C	M	Conditions	Ref.	Remarks
B45 BENZENE, 1,3-di(t-butyl)-4-hydroxy			$C_{14}H_{22}O$		$C_6H_3[C(CH_3)_3]_2(OH)$		
11.85	16.643(0)	-36.3	25	T	μ=0 S: Methanol ΔCp=-35	70Bf	
11.7	16.75(0)	-37	25	T	S: Methanol	69Rd	
B46 BENZENE, 1,3-di(t-butyl)-5-hydroxy-			$C_{14}H_{22}O$		$C_6H_3[C(CH_3)_3]_2(OH)$		
5.246	10.293(0)	-29.51	25	T	μ=0 ΔCp=-42	72Be	
9.02	14.763(0)	-37.3	25	T	μ=0 S: Methanol ΔCp=-63	70Bf	
8.9	14.87(0)	-38	25	T	S: Methanol	69Rd	
B47 BENZENE, 1,3-di(t-butyl)-2-hydroxy-5-nitro-			$C_{14}H_{21}O_3N$		$C_6H_2[C(CH_3)_3]_2(OH)(NO_2)$		
8.66	10.888(0)	-20.7	25	T	μ=0 S: Methanol ΔCp=0	70Bf	
8.4	10.99(0)	-22	25	T	S: Methanol	69Rd	
B48 BENZENE, 1,2-dichloro-			$C_6H_4Cl_2$		$C_6H_4(Cl)_2$		
14.5	(0)	-----	25	C	$C_L \sim 10^{-3}M$ S: Fluorosulfuric acid	74Ab	
B49 BENZENE, 1,3-dichloro 2,5-dihydroxy-			$C_6H_4O_2Cl_2$		$C_6H_2(Cl)_2(OH)_2$		
2.3	7.30(26.1°,0)	-26	13-25	T	μ=0.65	53Bb	c
5.1	9.99(25.1°, -1)	-29	13-25	T	μ=0.65	53Bb	c
B50 BENZENE, 1,2-dihydroxy-			$C_6H_6O_2$		$C_6H_4(OH)_2$		
8.20	9.356(0)	-15.70	25	T	μ=0 ΔCp=-141	72Ja	
6.0	9.19(0)	-22	25	C	μ=0.1(KNO_3)	70Kb	
8.24	9.195(0)	-14.40	25	C	μ=0.100M(KNO_3)	72Ja	
8.13	9.195(0)	-14.80	25	T	μ=0.100M(KNO_3) ΔCp=-87.7	72Ja	
5.02	12.98(-1)	-42.1	25	T	μ=0.100M(KNO_3)	72Ja	
B51 BENZENE, 1,4-dihydroxy-			$C_6H_6O_2$		$C_6H_4(OH)_2$		
3.8	9.85(25°,0)	-32	13-25	T	μ=0.65	53Bb	c
3.0	11.39(25.9°, -1)	-42	13-25		μ=0.65	53Bb	c
B52 BENZENE, 1,4-dihydroxy-2-methyl-			$C_7H_8O_2$		$C_6H_3(OH)_2(CH_3)$		
5.0	10.05(25.1°,0)	-29	13-25	T	μ=0.65	53Bb	c
3.3	11.64(25.9°, -1)	-42	13-25	T	μ=0.65	53Bb	c

ΔH, kcal /mole	pK	ΔS, cal /mole°K	T,°C	M	Conditions	Ref.	Remarks
B53 BENZENE, 1,2-dihydroxy-4-nitro-			$C_6H_5O_4N$		$C_6H_3(OH)(NO_2)$		
5.73	6.701(0)	-11.50	25	C	μ=0.100M(KNO_3)	72Jb	
B54 BENZENE, 2,4-dihydroxy-1-phenylazo-			$C_{12}H_{10}O_2N_2$		$C_6H_3(OH)_2(N{:}N\text{-}C_6H_5)$		
8.69	11.98(25°,-1)	-25.5	5-35	T	μ=0.1	72Rb	c
B55 BENZENE, 2,4-dihydroxy-1(m-nitrophenylazo)-			$C_{12}H_9O_4N_3$		$C_6H_3(OH)_2[N{:}N\text{-}C_6H_4(NO_2)]$		
7.15	11.51(25°,-1)	-28.7	5-35	T	μ=0.1	72Rb	c
B56 BENZENE, 2,4-dihydroxy-1(p-nitrophenylazo)-			$C_{12}H_9O_4N_3$		$C_6H_3(OH)_2[N{:}N\text{-}C_6H_4(NO_2)]$		
6.86	11.67(25°,-1)	-30.4	5-35	T	μ=0.1	72Rb	c
B57 BENZENE, 2,4-dihydroxy-1(p-sulphophenylazo)-			$C_{12}H_{10}O_5N_2S$		$C_6H_5(OH)_2[N{:}N\text{-}C_6H_4(SO_3H)]$		
6.23	11.95(25°,-2)	-33.7	5-35	T	μ=0.1	72Rb	c
B58 BENZENE, 1,4-dihydroxy-2,3,5,6-tetramethyl-			$C_{10}H_{14}O_2$		$C_6(OH)_2(CH_3)_4$		
7.8	11.25(25°,0)	-25	12-30	T	μ=0.65	53Bb	c
12.6	12.70(29.8°,-1)	-17	12-30	T	μ=0.65	53Bb	c
B59 BENZENE, 5(dimethylamino)-1-hydroxy-2-(2-thiazolylazo)-			$C_{11}H_{12}ON_4S$		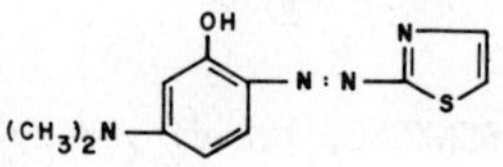		
1.68	8.90(0)	-35.18	15-35	T	μ=0.1(KNO_3)	75Ka	c
B60 BENZENE, 1,2-dimethyl-3-hydroxy-			$C_8H_{10}O$		$C_6H_3(CH_3)_2(OH)$		
5.70	10.54(0)	-29.1	25	T	μ=0 ΔCp=-29.4	62Cb	
5.640	10.53(0)	-29.3	25	T	μ=0	60Rb	
6.61	10.36(0)	-25.3	25	C	μ=0	59Pa	h
B61 BENZENE, 1,2-dimethyl-4-hydroxy-			$C_8H_{10}O$		$C_6H_3(CH_3)_2(OH)$		
5.37	10.36(0)	-29.4	25	T	μ=0 ΔCp=-29.7	62Cb	
5.360	10.35(0)	-29.4	25	T	μ=0	60Rb	
5.377	10.345(0)	-29.3	25	T	μ=0	70Bc	g:60Rb
5.358	10.356(0)	-29.4	25	T	μ=0 ΔCp=-32	70Bc	g:62Cb
8.24	10.16(0)	-18.9	25	C	μ=0	59Pa	h

ΔH, kcal /mole	pK	ΔS, cal /mole°K	T, °C	M	Conditions		Ref.	Re-marks
B62 BENZENE, 1,3-dimethyl-2-hydroxy-			$C_8H_{10}O$		$C_6H_3(CH_3)_2(OH)$			
5.46	10.62(0)	-30.3	25	T	μ=0	ΔCp=-30.7	62Cb	
5.530	10.59(0)	-29.9	25	T	μ=0		60Rb	
4.95	10.59(0)	-31.9	25	C	μ=0		59Pa	h
B63 BENZENE, 1,3-dimethyl-4-hydroxy-			$C_8H_{10}O$		$C_6H_3(CH_3)_2(OH)$			
5.76	10.60(0)	-29.2	25	T	μ=0	ΔCp=-29.3	62Cb	
5.760	10.55(0)	-28.9	25	T	μ=0		60Rb	
7.68	10.48(0)	-22.1	25	C	μ=0		59Pa	h
B64 BENZENE, 1,3-dimethyl-5-hydroxy-			$C_8H_{10}O$		$C_6H_3(CH_3)_2(OH)$			
5.34	10.20(0)	-28.8	25	T	μ=0	ΔCp=-28.8	62Cb	
5.490	10.17(0)	-28.1	25	T	μ=0		60Rb	
7.51	10.02(0)	-20.7	25	C	μ=0		59Pa	h
B65 BENZENE, 1,4-dimethyl-2-hydroxy			$C_8H_{10}O$		$C_6H_3(CH_3)_2(OH)$			
5.58	10.40(0)	-28.9	25	T	μ=0	ΔCp=-28.8	62Cb	
5.640	10.38(0)	-28.6	25	T	μ=0		60Rb	
6.22	10.19(0)	-25.7	25	C	μ=0		59Pa	h
B66 BENZENE, 1-ethyl-3-hydroxy-			$C_8H_{10}O$		$C_6H_4(CH_2CH_3)(OH)$			
5.294	10.069(0)	-28.31	25	T	μ=0	ΔCp=-47.5	67Bb	
B67 BENZENE, 1-hydroxy-2,4-dihydroxymethyl-			$C_8H_{10}O_3$		$C_6H_3(OH)(CH_2OH)_2$			
4.9	9.79(0)	-28.4	25	T	μ=0		67Za	
B68 BENZENE, 1-hydroxy-2-hydroxymethyl-			$C_7H_8O_2$		$C_6H_4(OH)(CH_2OH)$			
5.4	9.95(0)	-27.4	25	T	μ=0		67Za	
B69 BENZENE, 1-hydroxy-4-hydroxymethyl-			$C_7H_8O_2$		$C_6H_4(OH)(CH_2OH)$			
4.5	9.84(0)	-30.0	25	T	μ=0		67Za	
B70 BENZENE, 1-hydroxy-4-methyl-2(p-sulphophenylazo)-			$C_{12}H_{12}O_4N_2S$		$C_6H_5(OH)(CH_3)[N:N-C_6H_4(SO_3H)]$			
4.96	9.20(25°,-1)	-25.4	5-35	T	μ=0.1		72Rb	c
B71 BENZENE, 1-hydroxy-2-propyl-			$C_9H_{12}O$		$C_6H_4(OH)(CH_2CH_2CH_3)$			
6.00	10.50(0)	-28.0	25	C	μ=0		63Oa	

ΔH, kcal /mole	pK	ΔS, cal /mole°K	T,°C	M	Conditions	Ref.	Remarks
B72	**BENZENE, 1-hydroxy-2,4,6-tri-(t-butyl)-**		$C_{18}H_{30}O$		$C_6H_2(OH)[C(CH_3)_3]_3$		
10.07	17.512(0)	-46.4	25	T	μ=0 S: Methanol ΔCp=0	70Bf	g:69Rd
10.1	17.62(0)	-47	25	T	S: Methanol	69Rd	
B73	**BENZENE, 1-hydroxy-2,3,4-trimethyl-**		$C_9H_{12}O$		$C_6H_2(OH)(CH_3)_3$		
5.80	10.59(0)	-29.0	25	-	--	69La	
B74	**BENZENE, 1-hydroxy-2,3,5-trimethyl-**		$C_9H_{12}O$		$C_6H_2(OH)(CH_3)_3$		
5.750	10.61(0)	-29.3	25	T	μ=0	60Rb	
6.00	10.67(0)	-28.7	25	C	μ=0	64Ka	
B75	**BENZENE, 1-hydroxy-2,4,5-trimethyl-**		$C_9H_{12}O$		$C_6H_2(OH)(CH_3)_3$		
6.40	10.57(0)	-26.9	25	C	μ=0	64Ka	
B76	**BENZENE, 1-hydroxy-2,4,6-trimethyl-**		$C_9H_{12}O$		$C_6H_2(OH)(CH_3)_3$		
5.670	10.91(0)	-30.9	25	T	μ=0	60Rb	
5.44	10.89(0)	-31.6	25	C	μ=0	60Ka	
B77	**BENZENE, 1-hydroxy-3,4,5-trimethyl-**		$C_9H_{12}O$		$C_6H_2(OH)(CH_3)_3$		
5.68	10.25(0)	-27.9	25	C	μ=0	64Ka	
5.510	10.25(0)	-29.3	25	C	μ=0	70Bc	g:64Ka
B78	**BENZENE, 1-hydroxy-2,4,6-trihydroxymethyl**		$C_9H_{12}O_4$		$C_6H_2(OH)(CH_2OH)_3$		
4.5	9.56(0)	-28.6	25	T	μ=0	67Za	
B79	**BENZENE, mercapto-**		C_6H_6S		$C_6H_5(SH)$		
4.02	6.49(0)	-16.2	25	C	μ=0	74Lc	
B80	**BENZENE, 1,3,5-triamino-**		$C_6H_9N_3$		$C_6H_3(NH_2)_3$		
14.6	4.57(25°,+1)	28.5	6-24	T	Moderately acidic solution	70Ya	c,e:NH_3^+
27.2	5.44(24°,+1)	-15.7	6-24	T	Moderately acidic solution	70Ya	c,e: Ring Carbon, i
B81	**BENZENEBORONIC ACID**		$C_6H_7O_2B$		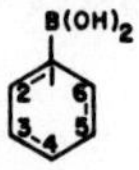		

ΔH, kcal /mole	pK	ΔS, cal /mole°K	T, °C	M	Conditions	Ref.	Remarks
B81, cont.							
1.9	8.84(25°,0)	-34	20-38.5	T	μ=0	61Ea	c
B82 1,3-BENZENEDICARBOXYLIC ACID	$C_8H_6O_4$		$C_6H_4(CO_2H)_2$				
0	3.50(0)	-15.9	25	T	μ=0	72Pf	
-0.44	4.50(-1)	-22.0	25	T	μ=0	72Pf	
B83 1,3-BENZENEDICARBOXYLIC ACID, mononitrile	$C_8H_5O_2N$		$C_6H_4(CO_2H)(CN)$				
0.727	3.608(0)	-13.9	15	T	μ=0	67Wa	
0.454	3.600(0)	-14.9	20	T	μ=0	67Wa	
0.182	3.596(0)	-15.84	25	T	μ=0	67Wa	q: ±0.026
-0.04	3.598(0)	-16.6	25	T	μ=0	59Ba	
-0.090	3.596(0)	-16.7	30	T	μ=0	67Wa	
-0.362	3.598(0)	-16.4	35	T	μ=0	67Wa	
-0.634	3.604(0)	-18.5	40	T	μ=0	67Wa	
B84 1,4-BENZENEDICARBOXYLIC ACID, mononitrile	$C_8H_5O_2N$		$C_6H_4(CO_2H)(CN)$				
0.440	3.557(0)	-14.7	15	T	μ=0	67Wa	
0.238	3.552(0)	-15.4	20	T	μ=0	67Wa	
0.036	3.550(0)	-16.12	25	T	μ=0	67Wa	q: ±0.030
0.03	3.551(0)	-16.1	25	T	μ=0	59Ba	
-0.165	3.551(0)	-16.7	30	T	μ=0	67Wa	
-0.367	3.554(0)	-17.4	35	T	μ=0	67Wa	
-0.569	3.560(0)	-18.1	40	T	μ=0	67Wa	
B85 1,3-BENZENEDISULFONIC ACID, 4,5-dihydroxy-	$C_6H_6O_8S_2$		$C_6H_2(HSO_3)_2(OH)_2$				
5.0	7.61	-18	25	C	μ=0.1(KNO_3)	70Kb	
B86 BENZENEHEXACARBOXYLIC ACID	$C_{12}H_6O_{12}$		$C_6(CO_2H)_6$				
-7	0.68(0)	-27	25	T	μ=0	72Pf	
-3.56	2.21(-1)	-21.9	25	T	μ=0	72Pf	
-2.75	3.52(-2)	-25.3	25	T	μ=0 ΔCp=-64	72Pf	
-1.13	5.09(-3)	-27.1	25	T	μ=0 ΔCp=-58	72Pf	
0.05	6.32(-4)	-28.7	25	T	μ=0 ΔCp=-48	72Pf	
-0.27	7.49(-5)	-35.5	25	T	μ=0 ΔCp=-61	72Pf	
B87 BENZENESULFONIC ACID, 2-amino	$C_6H_7O_3NS$		$C_6H_4(HSO_3)(NH_2)$				
2.374	2.4580(0)	-3.28	25	T	μ=0	65Cb 57Da	
2.43	2.458(0)	-3.1	25	C	μ=0.02	69Cc	
B88 BENZENESULFONIC ACID, 3-amino-	$C_6H_7O_3NS$		$C_6H_4(HSO_3)(NH_2)$				
4.64	3.808(0)	-1.9	25	C	μ=0	68Ce	
5.02	3.738(0)	-0.3	25	C	μ=0.012	69Cc	

ΔH, kcal /mole	pK	ΔS, cal /mole°K	T, °C	M	Conditions	Ref.	Remarks
B89 BENZENESULFONIC ACID, 4-amino-		$C_6H_7O_3NS$			$C_6H_4(HSO_3)(NH_2)$		
4.2	3.23(0)	-0.7	25	C	μ=0	65Hc	
4.252	3.2316(0)	-0.52	25	T	μ=0	65Cb	
4.29	3.227(0)	-0.37	25	T	μ=0	51Ma	
4.44	3.232(0)	0.1	25	C	μ=0.02	69Cc	
B90 BENZENESULFONIC ACID, 3-amino-4-methyl-		$C_7H_9O_3NS$			$C_6H_3(HSO_3)(NH_2)(CH_3)$		
5.053	3.6327(0)	0.33	25	T	μ=0	65Cb 69La	
B91 BENZENESULFONIC ACID, 4-amino-3-methyl-		$C_7H_9O_3NS$			$C_6H_3(HSO_3)(NH_2)(CH_3)$		
4.494	3.1239(0)	0.78	25	T	μ=0	65Cb	
4.489	3.126(0)	0.75	25	T	μ=0	63Ma	
B92 BENZENESULFONIC ACID, 4-hydroxy-		$C_6H_6O_4S$			$C_6H_4(HSO_3)(OH)$		
4.036	9.055(-1)	-27.9	25	T	μ=0	43Bb	
B93 1,2,4,5-BENZENETETRACARBOXYLIC ACID		$C_{10}H_6O_8$			$C_6H_2(CO_2H)_4$		
-3.11	1.70(0)	-18.2	25	T	μ=0	72Pf	
-1.57	3.12(-1)	-19.6	25	T	μ=0	72Pf	
-0.79	4.92(-2)	-25.2	25	T	μ=0 ΔCp=-86	72Pf	
-1.60	6.23(-3)	-33.9	25	T	μ=0 ΔCp=-73	72Pf	
B94 1,2,3-BENZENETRICARBOXYLIC ACID		$C_9H_6O_6$			$C_6H_3(CO_2H)_3$		
-1.06	2.88(0)	-16.7	25	T	μ=0 ΔCp=-22	72Pf	
0.21	4.75(-1)	-21.1	25	T	μ=0 ΔCp=-65	72Pf	
0.37	7.13(-2)	-31.4	25	T	μ=0 ΔCp=-57	72Pf	
B95 1,2,4-BENZENETRICARBOXYLIC ACID		$C_9H_6O_6$			$C_6H_3(CO_2H)_3$		
-1.24	2.48(0)	-15.5	25	T	μ=0 ΔCp=-32	72Pf	
-0.09	4.04(-1)	-18.5	25	T	μ=0 ΔCp=-46	72Pf	
-0.95	5.54(-2)	-28.6	25	T	μ=0 ΔCp=-91	72Pf	
B96 1,3,5-BENZENETRICARBOXYLIC ACID		$C_9H_6O_6$			$C_6H_3(CO_2H)_3$		
0.87	3.12(0)	-11.4	25	T	μ=0 ΔCp=-53	72Pf	
-0.49	4.10(-1)	-20.4	25	T	μ=0 ΔCp=-49	72Pf	
-1.17	5.18(-2)	-27.7	25	T	μ=0 ΔCp=-66	72Pf	
B97 BENZIMIDAZOLE, 2-ethyl-		$C_9H_{10}N_2$			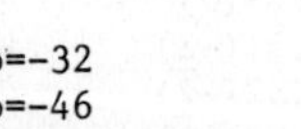		

ΔH, kcal /mole	pK	ΔS, cal /mole°K	T,°C	M	Conditions	Ref.	Remarks
B97, cont.							
9.3	6.27(+1)	2.5	4-35	T	μ=0.16	60La	c
B98 BENZIMIDAZOLE, 2-methyl-			$C_8H_8N_2$		see B97		
9.8	6.29(+1)	4.1	4-35	T	μ=0.16	60La	c
B99 BENZIMIDAZOLE, 2-(2-pyridyl)-			$C_{12}H_9N_3$		see B97		
8.7	5.58(+1)	-6	25	T	μ=0.16	60La	
3.8	3.44(25°,+1)	-3	0-40	T	μ=0.005 S: 50 w % Dioxane	55Ha	c
B100 BENZOHYDROXAMIC ACID, N-m-tolyl-			$C_{14}H_{13}O_2N$		$C_6H_5CON(OH)C_6H_4CH_3$		
4.21	13.41(25°,0)	-47.40	25&35	T	S: 70 v % Dioxane	75Sa	b,c
B101 BENZOIC ACID			$C_7H_6O_2$		$C_6H_5CO_2H$		
0.959	4.231(0)	-15.9	5	C	μ=0	74Mb	
0.736	(0)	-----	5	C	μ=0	70Ld	
0.622	(0)	-----	10	C	μ=0	48Ca	
0.785	(0)	-----	10	C	μ=0	59Fb	
0.726	4.220(0)	-16.7	10	C	μ=0 ΔCp=-47.4	74Mb	
-----	(0)	-----	10	C	μ=0 ΔCp=-27.6	70Ld	
0.601	(0)	-----	12.5	C	μ=0	74Mb	
0.499	4.212(0)	-17.5	15	C	μ=0 ΔCp=-43.8	74Mb	
0.460	(0)	-----	15	C	μ=0	70Ld	
0.974	4.217(0)	-15.9	15	T	μ=0	70Wc 67Wa	
0.295	(0)	-----	20	C	μ=0	48Ca	
0.288	4.206(0)	-18.3	20	C	μ=0 ΔCp=-40.6	74Mb	
0.338	4.205(0)	-18.1	20	T	μ=0	66Ta	
0.562	4.207(0)	-17.3	20	T	μ=0	70Wc 67Wa	
-----	(0)	-----	20	C	μ=0 ΔCp=-34.2	70Ld	
0.15	4.201(0)	-18.7	25	C	μ=0	67Ca	
0.105	4.20(0)	-18.9	25	C	μ=0	59Fb	
0.10	4.19(0)	-18.8	25	C	μ=0	69Ab	
0.110	4.204(0)	-18.87	25	C	μ=0 ΔCp=-37.9	74Mb 74Mc	
0.110	4.214(0)	-18.9	25	C	μ=0	48Ca	
0.09	4.21(0)	-18.9	25	C	μ=0	58Ca 59Zb	a:13.5
0.10	4.18(0)	-18.8	25	C	μ=0	71Ac	
0.118	(0)	-----	25	C	μ=0	70Ld	
0.10	(0)	-----	25	C	μ=0	71Ad 70Ab	
0.104	4.201(0)	-18.9	25	T	μ=0	66Ta	
-0.86	4.213(0)	-17.9	25	T	μ=0	59Ba	
0.0	4.2(0)	-19.1	25	T	μ=0	63Hd	
-0.067	4.202(0)	-19.44	25	T	μ=0 ΔCp=-42	72Bd	q:±0.019

ΔH, kcal /mole	pK	ΔS, cal /mole°K	T,°C	M	Conditions	Ref.	Remarks
B101, cont.							
0.060	4.204(0)	-19.0	25	T	μ=0 ΔCp=-33 $\Delta H=R(2AT^3 + BT^2)$ $\Delta Cp=2R(3AT^2 + BT)$ T=°C $A=-7.6728 \times 10^{-5}$ $B=4.612 \times 10^{-2}$	75Ta	
0.150	4.203(0)	-18.73	25	T	μ=0	70Wc 67Wa	
-0.068	(0)	-----	30	C	μ=0	48Ca	
-0.092	(0)	-----	30	C	μ=0 ΔCp=-35.6	74Mb	
-0.262	4.204(0)	-20.1	30	T	μ=0	70Wc 67Wa	
-0.143	4.203(0)	-19.7	30	T	μ=0	66Ta	
-----	(0)	-----	30	C	μ=0 ΔCp=-39.0	70Ld	
-0.265	4.206(0)	-20.1	35.0	C	μ=0 ΔCp=-33.6	74Mb	
-0.272	(0)	-----	35	C	μ=0	70Ld	
-0.674	4.209(0)	-21.4	35	T	μ=0	70Wc 67Wa	
-0.390	4.205(0)	-20.5	35	T	μ=0	66Ta	
-0.465	(0)	-----	40	C	μ=0	59Fb	
-0.455	(0)	-----	40	C	μ=0 ΔCp=-31.9	74Mb	
-1.090	4.219(0)	-22.8	40	T	μ=0	70Wc 67Wa	
-----	(0)	-----	40	C	μ=0 ΔCp=-38.5	70Ld	
-0.657	(0)	-----	45	C	μ=0	70Ld	
-0.585	(0)	-----	45	C	μ=0	74Mb	
-0.787	4.223(0)	-21.7	50	C	μ=0 ΔCp=-29.6	74Mb	
-0.777	(0)	-----	50	T	μ=0	75Ta	
-----	(0)	-----	50	C	μ=0 ΔCp=-37.2	70Ld	
-1.029	(0)	-----	55	C	μ=0	70Ld	
-0.881	(0)	-----	55	C	μ=0	74Mb	
-1.024	4.240(0)	-22.5	60	C	μ=0 ΔCp=-28.3	74Mb	
-1.145	(0)	-----	65	C	μ=0	74Mb	
-1.422	4.275(0)	-23.6	75	C	μ=0 ΔCp=-28.3	74Mb	
-1.713	(0)	-----	85	C	μ=0	74Mb	
-1.882	4.318(0)	-24.9	90	C	μ=0 ΔCp=-30.1	74Mb	
-2.170	4.351(0)	-25.7	100	C	μ=0 ΔCp=-32.2	74Mb	
-3.091	4.447(0)	-28.1	125	C	μ=0 ΔCp=-40.2	74Mb	
-4.226	4.565(0)	-30.9	150	C	μ=0 ΔCp=-51.1	74Mb	
-5.666	4.706(0)	-34.2	175	C	μ=0 ΔCp=-64.5	74Mb	
-7.465	4.874(0)	-38.1	200	C	μ=0 ΔCp=-79.7	74Mb	
-9.665	5.071(0)	-42.6	225	C	μ=0 ΔCp=-96.5	74Mb	
0.10	(0)	-----	25	C	$C_L=3\text{-}5\times10^{-3}M$	74Ra	
0.539	4.165(35°,0)	-17	0&35	T	--	39Wa	b,c
0.040	4.164(0)	-18.9	25	T	--	39Ea	
0.05	4.54(0)	-20.6	25	C	S: 10 w % Acetone	71Ac	
0.02	4.71(0)	-21.5	25	C	S: 15 w % Acetone	71Ac	
-0.04	4.88(0)	-22.5	25	C	S: 20 w % Acetone	71Ac	
0.21	5.32(0)	-23.7	25	C	S: 30 w % Acetone	71Ac	
0.40	5.74(0)	-24.9	25	C	S: 40 w % Acetone	71Ac	
0.39	5.97(0)	-26.0	25	C	S: 45 w % Acetone	71Ac	
0.37	6.21(0)	-27.2	25	C	S: 50 w % Acetone	71Ac	
0.38	6.77(0)	-29.7	25	C	S: 60 w % Acetone	71Ac	
0.35	7.39(0)	-32.7	25	C	S: 70 w % Acetone	71Ac	
-0.29	(0)	-----	25	C	μ=0 S: 10 w % tert-Butanol	71Ad 70Ab	
-0.84	(0)	-----	25	C	μ=0 S: 20 w % tert-Butanol	71Ad 70Ab	
0.1	(0)	-----	25	C	μ=0 S: 22 w % tert-Butanol	71Ad 70Ab	
0.74	(0)	-----	25	C	μ=0 S: 25 w % tert-Butanol	71Ad 70Ab	
1.5	(0)	-----	25	C	μ=0 S: 30 w % tert-Butanol	71Ad 70Ab	

ΔH, kcal /mole	pK	ΔS, cal /mole°K	T,°C	M	Conditions	Ref.	Re-marks
B101, cont.							
2.1	(0)	-----	25	C	μ=0 S: 38 w % tert-Butanol	71Ad 70Ab	
2.1	(0)	-----	25	C	μ=0 S: 40 w % tert-Butanol	71Ad 70Ab	
1.8	(0)	-----	25	C	μ=0 S: 48 w % tert-Butanol	71Ad 70Ab	
1.7	(0)	-----	25	C	μ=0 S: 50 w % tert-Butanol	71Ad 70Ab	
1.2	(0)	-----	25	C	μ=0 S: 60 w % tert-Butanol	71Ad 70Ab	
8.30	12.31(0)	-28.0	20	T	S: Dimethylformamide	68Pc	
8.58	12.20(0)	-27.1	25	T	S: Dimethylformamide	68Pc	
8.87	12.10(0)	-26.1	30	T	S: Dimethylformamide	68Pc	
9.17	12.00(0)	-25.1	35	T	S: Dimethylformamide	68Pc	
9.47	11.89(0)	-24.2	40	T	S: Dimethylformamide	68Pc	
0.48	(0)	-----	25	C	C_L=3-5x10^{-3}M 0.1 Molar ratio of Dimethylsulfoxide (DMSO) to Water	74Ra	
1.21	(0)	-----	25	C	C_L=3-5x10^{-3}M 0.2 Molar ratio of DMSO to Water	74Ra	
1.88	(0)	-----	25	C	C_L=3-5x10^{-3}M 0.3 Molar ratio of DMSO to Water	74Ra	
2.43	(0)	-----	25	C	C_L=3-5x10^{-3}M 0.4 Molar ratio of DMSO to Water	74Ra	
3.38	(0)	-----	25	C	C_L=3-5x10^{-3}M 0.5 Molar ratio of DMSO to Water	74Ra	
3.30	(0)	-----	25	C	C_L=3-5x10^{-3}M 0.6 Molar ratio of DMSO to Water	74Ra	
4.29	(0)	-----	25	C	C_L=3-5x10^{-3}M 0.7 Molar ratio of DMSO to Water	74Ra	
6.07	(0)	-----	25	C	C_L=3-5x10^{-3}M 0.8 Molar ratio of DMSO to Water	74Ra	
-0.06	4.50(0)	-20.8	25	C	μ=0 S: 10 w % Ethanol	67Ad	
-0.15	4.68(0)	-21.9	25	C	μ=0 S: 17 w % Ethanol	67Ad	
-0.15	4.76(0)	-22.3	25	C	μ=0 S: 20 w % Ethanol	67Ad	
0.05	4.94(0)	-22.4	25	C	μ=0 S: 26 w % Ethanol	67Ad	
0.47	5.04(0)	-21.5	25	C	μ=0 S: 30 w % Ethanol	67Ad	
0.77	5.20(0)	-21.2	25	C	μ=0 S: 35 w % Ethanol	67Ad	
1.05	5.35(0)	-21.0	25	C	μ=0 S: 40 w % Ethanol	67Ad	
1.02	5.66(0)	-22.5	25	C	μ=0 S: 50 w % Ethanol	67Ad	
0.86	5.98(0)	-24.5	25	C	μ=0 S: 60 w % Ethanol	67Ad	
0.32	6.32(0)	-27.8	25	C	μ=0 S: 70 w % Ethanol	67Ad	
-0.08	6.74(0)	-31.1	25	C	μ=0 S: 80 w % Ethanol	67Ad	
0.0	(0)	-----	25	C	μ=0 S: 10 w % Isopropanol	71Ad 70Ab	

ΔH, kcal /mole	pK	ΔS, cal /mole°K	T, °C	M	Conditions	Ref.	Remarks
B101, cont.							
-0.29	(0)	-----	25	C	μ=0 S: 15 w % Isopropanol	71Ad 70Ab	
-0.50	(0)	-----	25	C	μ=0 S: 20 w % Isopropanol	71Ad 70Ab	
0.41	(0)	-----	25	C	μ=0 S: 25 w % Isopropanol	71Ad 70Ab	
1.2	(0)	-----	25	C	μ=0 S: 30 w % Isopropanol	71Ad 70Ab	
1.8	(0)	-----	25	C	μ=0 S: 35 w % Isopropanol	71Ad 70Ab	
1.7	(0)	-----	25	C	μ=0 S: 40 w % Isopropanol	71Ad 70Ab	
1.4	(0)	-----	25	C	μ=0 S: 50 w % Isopropanol	71Ad 70Ab	
0.96	(0)	-----	25	C	μ=0 S: 60 w % Isopropanol	71Ad 70Ab	
0.15	4.29(0)	-19.1	25	C	S: 5 w % Methanol	71Ac	
0.25	4.36(0)	-19.2	25	C	S: 10 w % Methanol	71Ac	
0.20	4.58(0)	-20.2	25	C	S: 20 w % Methanol	71Ac	
0.45	4.83(0)	-20.6	25	C	S: 30 w % Methanol	71Ac	
0.65	5.10(0)	-21.2	25	C	S: 40 w % Methanol	71Ac	
0.70	5.39(0)	-22.3	25	C	S: 50 w % Methanol	71Ac	
0.55	5.70(0)	-24.2	25	C	S: 60 w % Methanol	71Ac	
0.40	6.03(0)	-26.3	25	C	S: 70 w % Methanol	71Ac	
0.30	6.46(0)	-28.6	25	C	S: 80 w % Methanol	71Ac	
14.369	11.93(-90°,0)	24.5	-95	C	μ=0 S: Methanol ΔCp=-4.9	70Le	
14.127	11.54(-80°,0)	20.4	-85	C	μ=0 S: Methanol ΔCp=-5.3	70Le	
13.882	(0)	-----	-75	C	μ=0 S: Methanol	70Le	
13.626	11.04(0)	14.9	-65	C	μ=0 S: Methanol ΔCp=-6.8	70Le	
13.349	(0)	-----	-55	C	μ=0 S: Methanol	70Le	
13.059	10.48(0)	9.3	-45	C	μ=0 S: Methanol ΔCp=-10.8	70Le	
12.743	(0)	-----	-35	C	μ=0 S: Methanol	70Le	
12.394	10.03(0)	4.0	-25	C	μ=0 S: Methanol ΔCp=-17.1	70Le	
12.039	9.75(-10°,0)	1.1	-15	C	μ=0 S: Methanol ΔCp=-23.3	70Le	
11.638	(0)	-----	-5	C	μ=0 S: Methanol ΔCp=-28.1	70Le	
11.208	9.60(0°,0)	-2.9	5	C	μ=0 S: Methanol	70Le	
10.748	9.46(10°,0)	-5.3	15	C	μ=0 S: Methanol ΔCp=-33.5	70Le	
10.227	9.28(0)	-8.2	25	C	μ=0 S: Methanol ΔCp=-39.4	70Le	
9.932	9.25(27.5°,0)	-9.4	30	C	μ=0 S: Methanol ΔCp=-44.3	70Le	
9.611	9.21(32.5°,0)	-10.7	35	C	μ=0 S: Methanol ΔCp=-47.6	70Le	

B102 BENZOIC ACID, amide, N,N-dimethyl- $C_9H_{11}ON$ $C_6H_5CON(CH_3)_2$

ΔH, kcal /mole	pK	ΔS, cal /mole°K	T, °C	M	Conditions	Ref.	Remarks
29.1	(0)	-----	25	C	$C_L \sim 10^{-3}M$ S: Fluorosulfuric acid	74Ab	

B103 BENZOIC ACID, 2-amino- $C_7H_7O_2N$ $C_6H_4CO_2H(NH_2)$

ΔH, kcal /mole	pK	ΔS, cal /mole$^{\circ}K$	T,$^{\circ}$C	M	Conditions	Ref.	Remarks

B103, cont.

3.79	2.09($\underline{+1}$)	3.0	25	C	μ=0	67Cb	
6.091	5.125($\underline{0}$)	-2.64	20	T	μ=0	65Sa	
2.79	4.79($\underline{0}$)	-13.4	25	C	μ=0	67Cb	
2.242	5.025($\underline{0}$)	-15.63	30	T	μ=0	65Sa	
-1.793	5.030($\underline{0}$)	-28.74	40	T	μ=0	65Sa	
-6.243	5.100($\underline{0}$)	-41.85	50	T	μ=0	65Sa	
-10.23	5.200($\underline{0}$)	-54.87	60	T	μ=0	65Sa	

B104 BENZOIC ACID, 3-amino- $C_7H_7O_2N$ $C_6H_4CO_2H(NH_2)$

2.56	3.07($\underline{+1}$)	-5.4	25	C	μ=0	67Cb	
4.17	4.79($\underline{0}$)	-7.9	25	C	μ=0	67Cb	
9.665	4.90($\underline{0}$)	10.17	20	T	--	65Sa	
6.229	4.75($\underline{0}$)	-1.283	30	T	--	65Sa	
2.655	4.675($\underline{0}$)	-12.92	40	T	--	65Sa	
-1.013	4.600($\underline{0}$)	-24.39	50	T	--	65Sa	
-4.817	4.70($\underline{0}$)	-39.02	60	T	--	65Sa	

Reactions below refer to micro species

0.0	3.30($\underline{+1}$)	-15.1	25	C	μ=0 R: $^+H_3NC_6H_4COOH$ = $^+H_3NC_6H_4COO^-$ + H$^+$	67Cb	
6.5	3.45($\underline{+1}$)	6.0	25	C	μ=0 R: $^+H_3NC_6H_4COOH$ = $H_2NC_6H_4COOH$ + H$^+$	67Cb	
6.5	4.56($\underline{+1}$)	0.9	25	C	μ=0 R: $^+H_3NC_6H_4COO^-$ = $H_2NC_6H_4COO^-$ + H$^+$	67Cb	
0.1	4.41($\underline{+1}$)	-19.8	25	C	μ=0 R: $H_2NC_6H_4COOH$ = $H_2NC_6H_4COO^-$ + H$^+$	67Cb	
-6.5	-0.15($\underline{0}$)	-21.1	25	C	μ=0 R: $H_2NC_6H_4COOH$ = $^+H_3NC_6H_4COO^-$	67Cb	

B105 BENZOIC ACID, 4-amino- $C_7H_7O_2N$ $C_6H_4CO_2H(NH_2)$

4.97	2.41($\underline{+1}$)	5.6	25	C	μ=0	67Cb	
0.70	4.85($\underline{0}$)	-19.8	25	C	μ=0	67Cb	
3.938	4.95($\underline{0}$)	-9.17	20	T	--	65Sa	
-0.142	4.90($\underline{0}$)	-22.82	30	T	--	65Sa	
-3.93	4.95($\underline{0}$)	-36.67	40	T	--	65Sa	
-8.758	5.10($\underline{0}$)	-50.42	50	T	--	65Sa	
-13.25	5.25($\underline{0}$)	-64.16	60	T	--	65Sa	

Reactions below refer to microspecies

0.1	3.49($\underline{+1}$)	-15.6	25	C	μ=0 R: $^+H_3NC_6H_4COOH$ = $^+H_3NC_6H_4COO^-$ + H$^+$	67Cb	
5.1	2.45($\underline{+1}$)	5.9	25	C	μ=0 R: $^+H_3NC_6H_4COOH$ = $H_2NC_6H_4COOH$ + H$^+$	67Cb	
5.1	3.79($\underline{+1}$)	-0.2	25	C	μ=0 R: $^+H_3NC_6H_4COO^-$ = $H_2NC_6H_4COO^-$ + H$^+$	67Cb	
0.4	4.83($\underline{0}$)	-20.8	25	C	μ=0 R: $H_2NC_6H_4COOH$ = $H_2NC_6H_4COO^-$ + H$^+$	67Cb	
-5.0	1.04($\underline{0}$)	-21.5	25	C	μ=0 R: $H_2NC_6H_4COOH$ = $^+H_3NC_6H_4COO^-$	67Cb	

B106 BENZOIC ACID, 2-amino-N,N-dimethyl- $C_9H_{11}O_2N$ $C_6H_4CO_2H[N(CH_3)_2]$

0.94	1.63($\underline{+1}$)	-4.4	25	C	μ=0.023	69Cc	
5.18	8.42($\underline{0}$)	-21.1	25	C	μ=0	69Cc	

ΔH, kcal /mole	pK	ΔS, cal /mole°K	T, °C	M	Conditions	Ref.	Re-marks
B107 BENZOIC ACID, 4-amino-2-hydroxy-, amide			$C_7H_8O_2N_2$		$C_6H_3CONH_2(NH_2)(OH)$		
5.1	9.16(25°,0)	-25	15-35	T	μ=3($NaClO_4$)	56Aa	c
B108 BENZOIC ACID, 5-amino-2-hydroxy-, amide			$C_7H_8O_2N_2$		$C_6H_3CONH_2(NH_2)(OH)$		
11.0	13.74(0)	-27	25	T	μ=3	64Sc	
B109 BENZOIC ACID, 2-amino-, methyl ester			$C_8H_9O_2N$		$C_6H_4CO_2CH_3(NH_2)$		
4.57	2.36(+1)	4.5	25	C	μ=0	67Cb	
B110 BENZOIC ACID, 3-amino-, methyl ester			$C_8H_9O_2N$		$C_6H_4CO_2CH_3(NH_2)$		
6.50	3.58(+1)	5.4	25	C	μ=0	67Cb	
B111 BENZOIC ACID, 4-amino-, methyl ester			$C_8H_9O_2N$		$C_6H_4CO_2CH_3(NH_2)$		
5.09	2.45(+1)	5.9	25	C	μ=0	67Cb	
B112 BENZOIC ACID, 2-amino-5-sulpho-			$C_7H_7O_5NS$		$C_6H_3CO_2H(NH_2)(SO_3H)$		
1.68	4.68(20°,-1)	-15.8	20-40	T	μ=0.1M	69Bc	c
B113 BENZOIC ACID, 3-bromo-			$C_7H_5O_2Br$		$C_6H_4CO_2H(Br)$		
0.521	3.818(0)	-15.7	15	T	μ=0	67Wa	
0.345	3.813(0)	-16.3	20	T	μ=0	67Wa	
0.197	3.810(0)	-16.77	25	T	μ=0 ΔCp=-45	72Bd	q: ±0.017
0.168	3.809(0)	-16.86	25	T	μ=0	67Wa	q: ±0.016
-0.06	3.809(0)	-17.6	25	T	μ=0	59Ba	
0.180	3.810(0)	-16.83	25	-	μ=0	74Mc	1
-0.0076	3.808(0)	-17.4	30	T	μ=0	67Wa	
-0.184	3.810(0)	-18.0	35	T	μ=0	67Wa	
-0.360	3.813(0)	-18.6	40	T	μ=0	67Wa	
0.731	3.810(35°,0)	-15	0&35	T	--	39Wa	b,c
B114 BENZOIC ACID, 4-bromo-			$C_7H_5O_2Br$		$C_6H_4CO_2H(Br)$		
0.502	4.011(0)	-16.6	15	T	μ=0	67Wa	
0.346	4.005(0)	-17.1	20	T	μ=0	67Wa	
0.191	4.002(0)	-17.67	25	T	μ=0	67Wa	
0.11	4.002(0)	-17.9	25	T	μ=0	59Ba	
0.107	3.961(0)	-17.76	25	T	μ=0 ΔCp=-35	72Bd	q: ±0.028
0.150	3.980(0)	-17.71	25	-	μ=0	74Mc	1
0.036	4.001(0)	-18.2	30	T	μ=0	67Wa	
-0.119	4.001(0)	-18.7	35	T	μ=0	67Wa	
-0.274	4.003(0)	-19.2	40	T	μ=0	67Wa	

ΔH, kcal /mole	pK	ΔS, cal /mole°K	T, °C	M		Conditions	Ref.	Remarks
B115 BENZOIC ACID, 3-tert-butyl-			$C_{11}H_{14}O_2$			$C_6H_4CO_2H[C(CH_3)_3]$		
2.750	4.266(0)	-10.0	15	T	μ=0		70Wc	
2.610	4.231(0)	-10.5	20	T	μ=0		70Wc	
2.470	4.199(0)	-10.9	25	T	μ=0		70Wc	
2.340	4.170(0)	-11.4	30	T	μ=0		70Wc	
2.200	4.143(0)	-11.8	35	T	μ=0		70Wc	
2.060	4.119(0)	-12.3	40	T	μ=0		70Wc	
B116 BENZOIC ACID, 4-tert-butyl-			$C_{11}H_{14}O_2$			$C_6H_4CO_2H[C(CH_3)_3]$		
2.880	4.463(0)	-10.4	15	T	μ=0		70Wc	
2.890	4.425(0)	-10.4	20	T	μ=0		70Wc	
2.900	4.389(0)	-10.4	25	T	μ=0		70Wc	
2.910	4.354(0)	-10.3	30	T	μ=0		70Wc	
2.920	4.320(0)	-10.3	35	T	μ=0		70Wc	
2.930	4.287(0)	-10.3	40	T	μ=0		70Wc	
B117 BENZOIC ACID, 2-chloro-			$C_7H_5O_2Cl$			$C_6H_4CO_2H(Cl)$		
-0.673	2.924(35°,0)	-11	0&35	T	--		39Wa	b,c
-2.470	2.877(0)	-21.5	25	T	--		39Ea	f
B118 BENZOIC ACID, 3-chloro-			$C_7H_5O_2Cl$			$C_6H_4CO_2H(Cl)$		
0.634	3.838(0)	-15.4	15	T	μ=0		67Wa	
0.433	3.831(0)	-16.0	20	T	μ=0		67Wa	
0.178	3.837(0)	-16.95	25	T	μ=0	ΔCp=-38	72Bd	q: ±0.017
0.233	3.826(0)	-16.72	25	T	μ=0		67Wa	q:±0.46
-0.18	3.827(0)	-18.1	25	T	μ=0		59Ba	
0.200	3.834(0)	-16.87	25	-	μ=0		74Mc	1
0.032	3.825(0)	-17.4	30	T	μ=0		67Wa	
-0.168	3.826(0)	-18.0	35	T	μ=0		67Wa	
-0.368	3.829(0)	-18.7	40	T	μ=0		67Wa	
0.019	3.83(0)	-17.4	25	T	--		61Db	
0.421	3.812(35°,0)	-16	0&35	T	--		39Wa	b,c
B119 BENZOIC ACID, 4-chloro-			$C_7H_5O_2Cl$			$C_6H_4CO_2H(Cl)$		
0.773	4.000(0)	-15.6	15	T	μ=0		67Wa	
0.581	3.991(0)	-16.3	20	T	μ=0		67Wa	
0.100	3.986(0)	-17.90	25	T	μ=0	ΔCp=-44	72Bd	q: ±0.026
0.388	3.985(0)	-16.93	25	T	μ=0		67Wa	
0.23	3.986(0)	-17.4	25	T	μ=0		59Ba	
0.240	3.986(0)	-17.43	25	-	μ=0		74Mc	1
0.195	3.981(0)	-17.5	30	T	μ=0		67Wa	
0.002	3.980(0)	-18.2	35	T	μ=0		67Wa	
-0.191	3.981(0)	-18.8	40	T	μ=0		67Wa	
0.226	3.98(0)	-17.5	25	T	--		61Db	
1.014	4.168(35°,0)	-16	0&35	T	--		39Wa	b,c
B120 BENZOIC ACID, 2,3-dihydroxy-			$C_7H_6O_4$			$C_6H_3CO_2H(OH)_2$		
1.926	2.98(0)	-7.03	30	T	μ=0.01	ΔCp=24.9	72Pc	
1.360	3.280(35°,0)	-20	0&35	T	--		39Wa	b,c
-2.030	10.14(-1)	-52.97	30	T	μ=0.01	ΔCp=-11.1	72Pc	

ΔH, kcal /mole	pK	ΔS, cal /mole°K	T, °C	M		Conditions	Ref.	Remarks
B121 BENZOIC ACID, 2,5-dihydroxy-			$C_7H_6O_4$			$C_6H_3CO_2H(OH)_2$		
5.067	3.00(0)	2.72	30	T	μ=0.01	ΔCp=-201.8	72Pb	
1.029	2.889(35°,0)	-10	0&35	T	--		39Wa	b,c
-2.065	10.25(-1)	-53.67	30	T	μ=0.01	ΔCp=38.8	72Pb	
B122 BENZOIC ACID, 2,4-dihydroxy-, ethyl ester			$C_9H_{10}O_4$			$C_6H_3CO_2C_2H_5(OH)_2$		
1.5	8.43(0)	-33.5	25	T	μ=0		63Hd	
B123 BENZOIC ACID, 2,6-dihydroxy-, ethyl ester			$C_9H_{10}O_4$			$C_6H_3CO_2C_2H_5(OH_2)$		
7.1	10.0(0)	-22.1	25	T	μ=0		63Hd	
B124 BENZOIC ACID, 2,3-dimethyl-			$C_9H_{10}O_2$			$C_6H_3CO_2H(CH_3)_2$		
-1.772	3.663(0)	-22.9	15	T	μ=0		67Wa	
-1.880	3.687(0)	-23.3	20	T	μ=0		67Wa	
-2.008	3.771(0)	-23.71	25	T	μ=0		67Wa	q: ±0.192
-2.126	3.736(0)	-24.1	30	T	μ=0		67Wa	
-2.244	3.762(0)	-24.5	35	T	μ=0		67Wa	
-2.362	3.788(0)	-24.9	40	T	μ=0		67Wa	
B125 BENZOIC ACID, 2,4-dimethyl-			$C_9H_{10}O_2$			$C_6H_3CO_2H(CH_3)_2$		
-2.657	4.154(0)	-28.2	15	T	μ=0		67Wa	
-2.488	4.187(0)	-27.6	20	T	μ=0		67Wa	
-2.320	4.217(0)	-27.07	25	T	μ=0		67Wa	
-2.151	4.244(0)	-26.5	30	T	μ=0		67Wa	
-1.982	4.268(0)	-26	35	T	μ=0		67Wa	
-1.813	4.290(0)	-25.4	40	T	μ=0		67Wa	
B126 BENZOIC ACID, 2,5-dimethyl-			$C_9H_{10}O_2$			$C_6H_3CO_2H(CH_3)_2$		
-3.498	3.911(0)	-30.0	15	T	μ=0		67Wa	
-3.106	3.954(0)	-28.7	20	T	μ=0		67Wa	
-2.714	3.990(0)	-27.36	25	T	μ=0		67Wa	
-2.322	4.020(0)	-26.1	30	T	μ=0		67Wa	
-1.930	4.045(0)	-24.8	35	T	μ=0		67Wa	
-1.538	4.065(0)	-23.5	40	T	μ=0		67Wa	
B127 BENZOIC ACID, 2,6-dimethyl-			$C_9H_{10}O_2$			$C_6H_3CO_2H(CH_3)_2$		
-5.817	3.234(0)	-35.0	15	T	μ=0		67Wa	
-5.043	3.304(0)	-32.3	20	T	μ=0		67Wa	
-4.269	3.362(0)	-29.70	25	T	μ=0		67Wa	
-3.495	3.409(0)	-27.1	30	T	μ=0		67Wa	
-2.721	3.445(0)	-24.6	35	T	μ=0		67Wa	
-1.947	3.472(0)	-22.1	40	T	μ=0		67Wa	
B128 BENZOIC ACID, 3,5-dimethyl-			$C_9H_{10}O_2$			$C_6H_3CO_2H(CH_3)_2$		
-0.367	4.294(0)	-20.9	15	T	μ=0		67Wa	
-0.298	4.299(0)	-20.7	20	T	μ=0		67Wa	
-0.229	4.302(0)	-20.45	25	T	μ=0		67Wa	q: ±0.096

ΔH, kcal /mole	pK	ΔS, cal /mole°K	T, °C	M	Conditions	Ref.	Remarks
B128, cont.							
-0.160	4.304(0)	-20.2	30	T	μ=0	67Wa	
-0.091	4.306(0)	-20.0	35	T	μ=0	67Wa	
-0.022	4.306(0)	-19.8	40	T	μ=0	67Wa	

B129 BENZOIC ACID, 3,5-dinitro- $C_7H_4O_6N_2$ $C_6H_3CO_2H(NO_2)_2$

ΔH, kcal /mole	pK	ΔS, cal /mole°K	T, °C	M	Conditions	Ref.	Remarks
-4.240	2.60(0)	-26.9	10	T	0	65Gb	
-4.546	2.73(0)	-28.0	20	T	0	65Gb	
-4.861	2.85(0)	-29.1	30	T	0	65Gb	
-5.187	2.96(0)	-30.1	40	T	0	65Gb	
-5.524	3.07(0)	-31.1	50	T	0	65Gb	
0.965	2.785(35°,0)	-10	0&35	T	--	39Wa	b,c

B130 BENZOIC ACID, 3-ethoxy- $C_9H_{10}O_3$ $C_6H_4CO_2H(OCH_2CH_3)$

ΔH, kcal /mole	pK	ΔS, cal /mole°K	T, °C	M	Conditions	Ref.	Remarks
0.495	3.879(35°,0)	-16	0&35	T	--	39Wa	b,c

B131 BENZOIC ACID, 3-fluoro- $C_7H_5O_2F$ $C_6H_4CO_2HF$

ΔH, kcal /mole	pK	ΔS, cal /mole°K	T, °C	M	Conditions	Ref.	Remarks
0.220	3.865(0)	-16.94	25	C	μ=0	74Mc	

B132 BENZOIC ACID, 2-hydroxy- $C_7H_6O_3$ $C_6H_4CO_2H(OH)$

ΔH, kcal /mole	pK	ΔS, cal /mole°K	T, °C	M	Conditions	Ref.	Remarks
0.73	2.973(0)	-11.2	25	C	μ=0	64Ec	
0.800	2.98(0)	-10.9	25	C	μ=0	67Vc	
1.5	3.0(0)	-8.7	1-45	T	μ=0	63Hd	c
0.888	2.792(0)	-9.8	25	C	μ=0.05($NaNO_3$)	67Vc 67Vb	
0.456	2.771(0)	-11.2	35	C	μ=0.05($NaNO_3$)	67Vb	
0.209	2.768(0)	-12.0	40	C	μ=0.05($NaNO_3$)	67Vb	q: ±0.006
-0.028	2.778(0)	-12.8	45	C	μ=0.05($NaNO_3$)	67Vb	
-0.257	2.777(0)	-13.5	50	C	μ=0.05($NaNO_3$)	67Vb	
-0.580	2.892(0)	-15.0	55	C	μ=0.05($NaNO_3$)	67Vb	
0.91	2.89(0)	-10	25	C	μ=0.1(KNO_3)	70Kb	
0.925	2.689(0)	-9.2	25	C	μ=0.1($NaNO_3$)	67Vb 67Vc	
0.453	2.660(0)	-10.7	35	C	μ=0.1($NaNO_3$)	67Vb	
0.267	2.656(0)	-11.3	40	C	μ=0.1($NaNO_3$)	67Vb	
0.046	2.654(0)	-12.0	45	C	μ=0.1($NaNO_3$)	67Vb	q: ±0.008
-0.190	2.6473(0)	-12.7	50	C	μ=0.1($NaNO_3$)	67Vb	
-0.376	2.656(0)	-13.3	55	C	μ=0.1($NaNO_3$)	67Vb	
1.159	2.336(0)	-6.8	25	C	μ=0.5($NaNO_3$)	67Vb	
0.461	2.332(0)	-9.2	40	C	μ=0.5($NaNO_3$)	67Vb	
0.273	2.351(0)	-9.9	45	C	μ=0.5($NaNO_3$)	67Vb	
0.046	2.326(0)	-10.5	50	C	μ=0.5($NaNO_3$)	67Vb	q: ±0.04
-0.182	2.327(0)	-11.2	55	C	μ=0.5($NaNO_3$)	67Vb	
1.375	2.429(0)	-6.5	25	C	μ=1.0($NaNO_3$)	67Vb 67Vc	
0.620	2.400(0)	-9.0	40	C	μ=1.0($NaNO_3$)	67Vb	
0.193	2.382(0)	-10.3	50	C	μ=1.0($NaNO_3$)	67Vb	
-0.055	2.389(0)	-11.1	55	C	μ=1.0($NaNO_3$)	67Vb	
-0.513	2.422(0)	-12.6	65	C	μ=1.0($NaNO_3$)	67Vb	
2.622	(0)	-----	10	C	μ=3.0($NaNO_3$)	67Vb	
2.391	(0)	-----	18	C	μ=3.0($NaNO_3$)	67Vb	
2.177	1.88(0)	-1.3	25	C	μ=3.0($NaNO_3$)	67Vb 67Vc	

ΔH, kcal /mole	pK	ΔS, cal /mole°K	T,°C	M	Conditions	Ref.	Remarks
B132, cont.							
0.432	(0)	-----	55	C	μ=3.0($NaNO_3$)	67Vb	
1.169	2.975(35°,0)	-10	0&35	T	--	39Wa	b,c
1.000	2.972(0)	-10.2	25	T	--	39Ea	
0.500	3.000(25°,0)	-12.0	30	T	--	25Ba	
3.21	7.89(0)	-25.2	20	T	S: Dimethylformamide	68Pc	
3.32	7.85(0)	-24.8	25	T	S: Dimethylformamide	68Pc	
3.43	7.82(0)	-24.4	30	T	S: Dimethylformamide	68Pc	
3.54	7.77(0)	-24.1	35	T	S: Dimethylformamide	68Pc	
3.66	7.72(0)	-23.7	40	T	S: Dimethylformamide	68Pc	
4.11	4.95(25°,0)	-8.9	25-45	T	μ=0.1M S: 70 w % Dioxane	69Pb	
6.28	6.31(25°,0)	-7.8	25-45	T	μ=0.1M S: 82 w % Dioxane	69Pb	
-0.200	3.180(25°,0)	-15.2	30	T	S: 16.2 w % Ethanol	25Ba	
-0.300	3.585(25°,0)	-17.4	30	T	S: 33.2 w % Ethanol	25Ba	
-0.400	4.056(25°,0)	-19.8	30	T	S: 52.0 w % Ethanol	25Ba	
-1.200	4.745(25°,0)	-25.0	30	T	S: 73.5 w % Ethanol	25Ba	
-2.200	5.346(25°,0)	-31.8	30	T	S: 85.7 w % Ethanol	25Ba	
-1.700	6.398(25°,0)	-35.0	30	T	S: 95.0 w % Ethanol	25Ba	
0.800	(0)	-----	30	T	S: Ethanol	25Ba	
-5.180	(0)	-----	17	C	S: Methanol	27Wa	
8.51	13.596(-1)	-33.7	25	C	μ=0	64Ec	
8.600	12.70(-1)	-29.2	25	C	μ=0	67Vc	
10.5	13.1(-1)	-24.8	25	T	μ=0	63Hd	
7.1	10.04(-1)	-22	25	T	0.01M Buffer C=1.1 x 10^{-4}M	63Hd	
13.2	13.6(-1)	-17.9	25	C	μ=0.1(KNO_3)	70Kb	
9.465	13.31(-1)	-27.4	10	C	μ=0.5($NaNO_3$)	67Vb 66Vc	
9.229	13.15(-1)	-28.4	18	C	μ=0.5($NaNO_3$)	67Vb 66Vc	
8.965	12.94(-1)	-29.1	25	C	μ=0.5($NaNO_3$)	67Vb 66Vc	
8.390	12.64(-1)	-31.0	40	C	μ=0.5($NaNO_3$)	67Vb 66Vc	
8.215	12.45(-1)	-31.6	50	C	μ=0.5($NaNO_3$)	67Vb 66Vc	
9.498	13.189(-1)	-26.8	10	C	μ=1.0($NaNO_3$)	67Vb	
9.271	12.966(-1)	-27.7	18	C	μ=1.0($NaNO_3$)	67Vb	
9.068	12.811(-1)	-28.2	25	C	μ=1.0($NaNO_3$)	67Vb 67Vc	
8.702	12.499(-1)	-29.4	40	C	μ=1.0($NaNO_3$)	67Vb	
8.612	(-1)	-30.9	55	C	μ=1.0($NaNO_3$)	67Vb	
9.871	13.171(-1)	-25.4	10	C	μ=3.0($NaNO_3$)	67Vb	
9.760	12.938(-1)	-25.9	18	C	μ=3.0($NaNO_3$)	67Vb	
9.660	12.764(-1)	-26.0	25	C	μ=3.0($NaNO_3$)	67Vb	
9.412	12.338(-1)	-26.4	40	C	μ=3.0($NaNO_3$)	67Vb	
8.6	13.12(25°,-1)	-32	15-35	T	μ=3.0M($NaClO_4$)	56Aa	c

B133 BENZOIC ACID, 2-hydroxy-, acetate $C_8H_8O_2$ $C_6H_4(OH)(COCH_3)$

-0.835	3.506(0)	-18.75	35	C	μ=0.005	42Sc	
-0.625	3.565(35°,0)	-18	0&35	T	--	39Wa	b,c

B134 BENZOIC ACID, 2-hydroxy-, amide $C_7H_7O_2N$ $C_6H_4(OH)(CONH_2)$

6.7	8.89(25°,0)	-18	15-35	T	μ=3.0($NaClO_4$)	56Aa	c

ΔH, kcal /mole	pK	ΔS, cal /mole°K	T, °C	M	Conditions	Ref.	Remarks
B135 BENZOIC ACID, 2-hydroxy-, methyl ester			$C_8H_8O_3$		$C_6H_4(OH)(CO_2CH_3)$		
7.5	10.19(25°,0)	-22	15-35	T	μ=3.0($NaClO_4$)	56Aa	c
B136 BENZOIC ACID, 3-hydroxy-			$C_7H_6O_3$		$C_6H_4(CO_2H)(OH)$		
0.348	(0)	-----	20	C	μ=0	48Ca	
0.159	4.075(0)	-18.1	25	C	μ=0	48Ca	
0.160	4.076(0)	-18.12	25	-	μ=0	74Mc	1
-0.029	(0)	-----	30	C	μ=0	48Ca	
0.170	4.076(0)	-18.1	25	T	--	39Ea	
0.404	4.103(35°,0)	-17	0&35	T	--	39Wa	b,c
B137 BENZOIC ACID, 4-hydroxy-			$C_7H_6O_3$		$C_6H_4(CO_2H)(OH)$		
0.945	(0)	-----	10	C	μ=0	48Ca	
0.852	4.596(0)	-18.0	15	T	μ=0	67Wa	
0.561	(0)	-----	20	C	μ=0	48Ca	
0.619	4.586(0)	-18.8	20	T	μ=0	67Wa	
0.363	4.595(0)	-9.8	25	C	μ=0	48Ca	
0.54	4.582(0)	-19.1	25	T	μ=0	59Ba	
0.030	4.200(0)	-19.1	25	T	μ=0	60Pa	
0.386	4.580(0)	-19.66	25	T	μ=0	67Wa	q: ±0.034
0.370	4.582(0)	-19.72	25	-	μ=0	74Mc	1
0.167	(0)	------	30	C	μ=0	48Ca	
0.153	4.577(0)	-20.4	30	T	μ=0	67Wa	
-0.080	4.576(0)	-21.2	35	T	μ=0	67Wa	
-0.313	4.578(0)	-21.9	40	T	μ=0	67Wa	
0.025	4.746(0)	-18.9	40	T	μ=0	60Pa	
0.640	4.542(35°,0)	-19	0&35	T	--	39Wa	b,c
3.4	9.2(0)	-30.8	25	T	μ=0	63Hd	
B138 BENZOIC ACID, 4-hydroxy-, ethyl ester			$C_9H_{10}O_3$		$C_6H_4CO_2C_2H_5(OH)$		
1.5	8.43(0)	-34	25	T	0.01M Buffer	63Hd	
B139 BENZOIC ACID, 4-hydroxy-3-methoxy-			$C_8H_8O_4$		$C_6H_3(CO_2H)(OH)(OCH_3)$		
0.145	4.355(0)	-19.50	25	T	μ=0	64Ca	
0.804	4.450(35°,0)	-18	0&35	T	--	39Wa	b,c
B219 BENZOIC ACID, 2-hydroxy-, nitrile			C_7H_5ON		$C_6H_4(OH)(CN)$		
4.0	6.8(0)	-18	25	T	μ=0.10	69Da	
B220 BENZOIC ACID, 3-hydroxy-, nitrile			C_7H_5ON		$C_6H_4(OH)(CN)$		
5.20	8.57(0)	-21.8	25	C	μ=0	64Ka	
5.3	8.3(0)	-20	25	T	μ=0.10	69Da	
B221 BENZOIC ACID, 4-hydroxy-, nitrile			C_7H_5ON		$C_6H_4(OH)(CN)$		

ΔH, kcal /mole	pK	ΔS, cal /mole°K	T, °C	M	Conditions	Ref.	Remarks
B221, cont.							
4.92	7.97(0)	-20.0	25	C	μ=0	64Ka	
4.8	7.6(0)	-19	25	T	μ=0.10	69Da	
B140 BENZOIC ACID, 2-hydroxy-5-(m-nitrophenylazo)-			$C_{13}H_9O_5N_3$		$C_6H_3(CO_2H)(OH)[N:N-C_6H_4(NO_2)]$		
6.71	11.00(25°,-1)	-27.8	5-35	T	μ=0	72Rb	c
B141 BENZOIC ACID, 2-hydroxy-5-(p-nitrophenylazo)-			$C_{13}H_9O_5N_3$		$C_6H_3(CO_2H)(OH)[N:N-C_6H_4(NO_2)]$		
5.85	10.76(25°,-1)	-29.6	5-35	T	μ=0.1	72Rb	c
B142 BENZOIC ACID, 2-hydroxy-5-sulfo-			$C_7H_6O_6S$		$C_6H_3(CO_2H)(OH)(SO_3H)$		
6.1	2.62(0)	8.4	25	C	μ=0.1(KNO_3)	70Kb	
3.68	4.77(25°,0)	-9.4	25-45	T	μ=0.1M S: 70 w % Dioxane	69Pb	c
6.79	5.58(25°,0)	-4.7	25-45	T	μ=0.1M S: 82 w % Dioxane	69Pb	c
7.1	11.74(25°,0)	-30	15-35	T	μ=3.0M($NaClO_4$)	56Aa	c
B143 BENZOIC ACID, 2-iodo-			$C_7H_5O_2I$		$C_6H_4(CO_2H)(I)$		
-3.250	2.860(0)	-24.0	25	T	--	39Ea	f
B144 BENZOIC ACID, 3-iodo-			$C_7H_5O_2I$		$C_6H_4(CO_2H)(I)$		
0.190	3.856(0)	-17.00	25	-	μ=0	74Mc	1
0.190	3.86(0)	-17.0	25	T	--	61Db	
0.190	3.855(0)	-17.0	25	T	--	39Ea	
B145 BENZOIC ACID, 4-iodo-			$C_7H_5O_2I$		$C_6H_4(CO_2H)(I)$		
0.078	3.995(0)	-18.01	25	T	μ=0 ΔCp=-32	72Bd 74Mc	1,q: ±0.015
B146 BENZOIC ACID, 2-mercapto			$C_7H_6O_2S$		$C_6H_4(CO_2H)(SH)$		
3.08	4.05(20°,0)	-8.0	20-40	T	μ=1 x 10^{-3}M	72La	c
5.32	4.42(30°,0)	-2.6	20-40	T	μ=1 x 10^{-3}M S: 25 w % Dioxan	72La	c
5.03	4.54(30°,0)	-4.0	20-40	T	μ=1 x 10^{-3}M S: 30 w % Dioxan	72La	c
4.69	4.84(30°,0)	-6.6	20-40	T	μ=1 x 10^{-3}M S: 40 w % Dioxan	72La	c
5.72	5.19(30°,0)	-4.8	20-40	T	μ=1 x 10^{-3}M S: 50 w % Dioxan	72La	c
4.39	5.48(30°,0)	-10.4	20-40	T	μ=1 x 10^{-3}M S: 60 w % Dioxan	72La	c
4.39	4.55(30°,0)	-6.3	20-40	T	μ=1 x 10^{-3}M S: 42.6 w % Methanol	72La	c
5.54	4.76(30°,0)	-3.3	20-40	T	μ=1 x 10^{-3}M S: 52.4 w % Methanol	72La	c
2.93	4.25(30°,0)	-9.8	20-40	T	μ=1 x 10^{-3}M S: 21 w % Ethanol	72La	c

B146, cont.

ΔH, kcal /mole	pK	ΔS, cal /mole°K	T, °C	M	Conditions	Ref.	Remarks
3.84	4.35(30°,0)	-7.3	20-40	T	μ=1 x 10^{-3}M S: 25.3 w % Ethanol	72La	c
3.77	4.50(30°,0)	-8.0	20-40	T	μ=1 x 10^{-3}M S: 34.3 w % Ethanol	72La	c
4.43	4.70(30°,0)	-6.8	20-40	T	μ=1 x 10^{-3}M S: 43.7 w % Ethanol	72La	c
4.58	4.77(30°,0)	-6.7	20-40	T	μ=1 x 10^{-3}M S: 53.5 w % Ethanol	72La	c
4.47	4.44(30°,0)	-5.4	20-40	T	μ=1 x 10^{-3}M S: 21 w % Isopropanol	72La	c
3.96	4.64(30°,0)	-8.0	20-40	T	μ=1 x 10^{-3}M S: 25 w % Isopropanol	72La	c
4.58	4.89(30°,0)	-7.0	20-40	T	μ=1 x 10^{-3}M S: 34 w % Isopropanol	72La	c
4.98	4.98(30°,0)	-6.0	20-40	T	μ=1 x 10^{-3}M S: 43.5 w % Isopropanol	72La	c
4.58	5.03(30°,0)	-7.8	20-40	T	μ=1 x 10^{-3}M S: 53 w % Isopropanol	72La	c
4.39	4.08(30°,0)	-4.2	20-40	T	μ=1 x 10^{-3}M S: 20 w % Methanol	72La	c
4.48	4.18(30°,0)	-4.3	20-40	T	μ=1 x 10^{-3}M S: 24.34 w % Methanol	72La	c
4.16	4.35(30°,0)	-6.0	20-40	T	μ=1 x 10^{-3}M S: 33.24 w % Methanol	72La	c
5.72	8.88(-1)	-21.4	25	C	μ=0.10	64Ia	

B147 BENZOIC ACID, 2-methoxy- $C_8H_8O_3$ $C_6H_4(CO_2H)(OCH_3)$

ΔH, kcal /mole	pK	ΔS, cal /mole°K	T, °C	M	Conditions	Ref.	Remarks
-1.60	4.09(0)	-24.1	25	C	μ=0	59Zb	

B148 BENZOIC ACID, 3-methoxy- $C_8H_8O_3$ $C_6H_4(CO_2H)(OCH_3)$

ΔH, kcal /mole	pK	ΔS, cal /mole°K	T, °C	M	Conditions	Ref.	Remarks
0.049	4.093(0)	-18.56	25	C	μ=0	74Mc	
0.06	4.09(0)	-18.5	25	C	μ=0	59Zb	
0.022	4.093(0)	-18.65	25	T	μ=0 ΔCp=-23	72Bd	q: ±0.012

B149 BENZOIC ACID, 4-methoxy- $C_8H_8O_3$ $C_6H_4(CO_2H)(OCH_3)$

ΔH, kcal /mole	pK	ΔS, cal /mole°K	T, °C	M	Conditions	Ref.	Remarks
0.600	4.486(0)	-18.51	25	-	μ=0	74Mc	1
0.57	4.47(0)	-18.5	25	C	μ=0	59Zb	
0.258	4.494(0)	-19.69	25	T	μ=0 ΔCp=-45	72Bd	q: ±0.026
0.800	4.478(0)	-17.8	25	T	--	39Ea	
0.601	4.481(35°,0)	-18	0&35	T	--	39Wa	b,c

B150 BENZOIC ACID, 2-methyl- $C_8H_8O_2$ $C_6H_4(CO_2H)(CH_3)$

ΔH, kcal /mole	pK	ΔS, cal /mole°K	T, °C	M	Conditions	Ref.	Remarks
-1.034	3.866(0)	-21.3	15	T	μ=0	67Wa	
-1.028	3.880(0)	-21.3	20	T	μ=0	67Wa	
-1.50	3.91(0)	-22.9	25	C	μ=0	59Zb	
-1.022	3.893(0)	-21.24	25	T	μ=0	67Wa	q: ±0.117
-1.016	3.905(0)	-21.2	30	T	μ=0	67Wa	
-1.011	3.917(0)	-21.2	35	T	μ=0	67Wa	
-1.005	3.928(0)	-21.2	40	T	μ=0	67Wa	
-1.400	3.873(0)	-22.4	25	T	--	39Ea	
-1.15	3.903(35°,0)	-22	0&35	T	--	39Wa	b,c

ΔH, kcal /mole	pK	ΔS, cal /mole°K	T,°C	M	Conditions	Ref.	Remarks

B151 BENZOIC ACID, 3-methyl- $C_8H_8O_2$ $C_6H_4(CO_2H)(CH_3)$

ΔH, kcal /mole	pK	ΔS, cal /mole°K	T,°C	M	Conditions	Ref.	Remarks
1.490	4.304(0)	-14.5	15	T	μ=0	70Wc 67Wa	
1.350	4.285(0)	-15.0	20	T	μ=0	70Wc 67Wa	
0.070	4.252(0)	-19.22	25	C	μ=0	74Mc	1
0.07	4.24(0)	-19.2	25	C	μ=0	59Zb	
-0.091	4.252(0)	-19.75	25	T	μ=0 ΔCp=-34	72Bd	q: ±0.013
1.200	4.269(0)	-15.30	25	T	μ=0	70Wc 67Wa	
1.060	4.256(0)	-16.0	30	T	μ=0	70Wc 67Wa	
0.914	4.244(0)	-16.4	35	T	μ=0	70Wc 67Wa	
0.770	4.235(0)	-16.9	40	T	μ=0	70Wc 67Wa	
0.349	4.256(35°,0)	-18	0&35	T	--	39Wa	b,c
0.070	4.240(0)	-19.2	25	T	--	39Ea	

B152 BENZOIC, 4-methyl- $C_8H_8O_2$ $C_6H_4(CO_2H)(CH_3)$

ΔH, kcal /mole	pK	ΔS, cal /mole°K	T,°C	M	Conditions	Ref.	Remarks
1.040	4.390(0)	-16.5	15	T	μ=0	70Wc 67Wa	
1.070	4.376(0)	-16.4	20	T	μ=0	70Wc 67Wa	
0.24	4.34(0)	-19.0	25	C	μ=0	59Zb	
1.100	4.362(0)	-16.27	25	T	μ=0	70Wc 67Wa	
-0.134	4.370(0)	-20.44	25	T	μ=0 ΔCp=-39	72Bd	
0.250	4.362(0)	-19.12	25	-	μ=0	74Mc	1
1.130	4.349(0)	-16.2	30	T	μ=0	70Wc 67Wa	
1.160	4.336(0)	-16.1	35	T	μ=0	70Wc 67Wa	
1.190	4.322(0)	-16.0	40	T	μ=0	70Wc 67Wa	
0.300	4.341(0)	-18.9	25	T	--	39Ea	
0.630	4.359(35°,0)	-18	0&35	T	--	39Wa	b,c

B153 BENZOIC ACID, 2-nitro- $C_7H_5O_4N$ $C_6H_4(CO_2H)(NO_2)$

ΔH, kcal /mole	pK	ΔS, cal /mole°K	T,°C	M	Conditions	Ref.	Remarks
-3.355	2.184(0)	-21.3	25	T	--	39Ea	f

B154 BENZOIC ACID, 3-nitro- $C_6H_5O_4N$ $C_6H_4(CO_2H)(NO_2)$

ΔH, kcal /mole	pK	ΔS, cal /mole°K	T,°C	M	Conditions	Ref.	Remarks
0.380	3.460(0)	-14.56	25	C	μ=0	74Mc	
0.421	3.460(0)	-14.41	25	T	μ=0 ΔCp=-41	72Bd	
0.320	3.447(0)	-14.7	25	T	--	39Ea	
0.142	3.467(35°,0)	-15	0&35	T	--	39Wa	b,c

B155 BENZOIC ACID, 4-nitro- $C_7H_5O_4N$ $C_6H_4(CO_2H)(NO_2)$

ΔH, kcal /mole	pK	ΔS, cal /mole°K	T,°C	M	Conditions	Ref.	Remarks
0.432	3.448(0)	-14.2	15	T	μ=0	67Wa	
0.277	3.444(0)	-14.8	20	T	μ=0	67Wa	
0.432	3.424(0)	-14.21	25	T	μ=0 ΔCp=-25	72Bd	
0.122	3.441(0)	-15.33	25	T	μ=0	67Wa	q: ±0.023
0.432	3.424(0)	-14.22	25	-	μ=0	74Mc	1
-0.032	3.441(0)	-15.8	30	T	μ=0	67Wa	
-0.188	3.442(0)	-16.4	35	T	μ=0	67Wa	

ΔH, kcal /mole	pK	ΔS, cal /mole°K	T, °C	M	Conditions	Ref.	Remarks
B155, cont.							
-0.342	3.445(0)	-16.8	40	T	μ=0	67Wa	
0.641	3.383(35°,0)	-13	0&35	T	--	39Wa	b,c
B156 BENZOIC ACID, 2,3,5,6-tetramethyl-			$C_{11}H_{14}O_2$		$C_6H(CO_2H)(CH_3)_4$		
-4.744	3.310(0)	-31.6	15	T	μ=0	67Wa	
-4.116	3.367(0)	-29.4	20	T	μ=0	67Wa	
-3.489	3.415(0)	-27.32	25	T	μ=0	67Wa	
-2.861	3.453(0)	-25.2	30	T	μ=0	67Wa	
-2.233	3.483(0)	-23.2	35	T	μ=0	67Wa	
-1.606	3.505(0)	-21.1	40	T	μ=0	67Wa	
B157 BENZOIC ACID, 3,4,5-trihydroxy-			$C_7H_6O_5$		$C_6H_2(CO_2H)(OH)_3$		
0.733	4.404(35°,0)	-18	0&35	T	--	39Wa	b,c
-1.83	4.19(0)	-25.33	20	T	--	62Ca	f
-2.98	4.30(0)	-29.19	30	T	--	62Ca	f
-4.17	4.38(0)	-33.06	40	T	--	62Ca	f
-5.39	4.53(0)	-36.93	50	T	--	62Ca	f
-6.66	4.65(0)	-40.79	60	T	--	62Ca	f
B158 BENZOIC ACID, 2,4,6-trimethyl-			$C_{10}H_{12}O_2$		$C_6H_2(CO_2H)(CH_3)_3$		
-5.300	3.325(0)	-33.6	15	T	μ=0	67Wa	
-4.832	3.391(0)	-32.0	20	T	μ=0	67Wa	
-4.364	3.448(0)	-30.41	25	T	μ=0	67Wa	
-3.896	3.498(0)	-28.9	30	T	μ=0	67Wa	
-3.428	3.541(0)	-27.3	35	T	μ=0	67Wa	
-2.960	3.577(0)	-25.8	40	T	μ=0	67Wa	
B159 BENZOIC ACID, 3-trimethylsilyl-			$C_{10}H_{14}O_2Si$		$C_6H_4(CO_2H)[Si(CH_3)_3]$		
1.890	4.142(0)	-12.4	15	T	μ=0	70Wc	
2.100	4.116(0)	-11.7	20	T	μ=0	70Wc	
2.320	4.089(0)	-10.9	25	T	μ=0	70Wc	
2.530	4.060(0)	-10.2	30	T	μ=0	70Wc	
2.750	4.029(0)	-9.5	35	T	μ=0	70Wc	
2.960	3.996(0)	-8.8	40	T	μ=0	70Wc	
B160 BENZOIC ACID, 4-trimethylsilyl-			$C_{10}H_{14}O_2Si$		$C_6H_4(CO_2H)[Si(CH_3)_3]$		
3.060	4.270(0)	-8.9	15	T	μ=0	70Wc	
3.060	4.230(0)	-8.9	20	T	μ=0	70Wc	
3.060	4.192(0)	-8.9	25	T	μ=0	70Wc	
3.060	4.155(0)	-8.9	30	T	μ=0	70Wc	
3.060	4.119(0)	-8.9	35	T	μ=0	70Wc	
3.060	4.084(0)	-8.9	40	T	μ=0	70Wc	

B161 BENZOPHENONE, 4-amino- $C_{13}H_{11}ON$

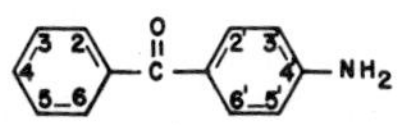

ΔH, kcal /mole	pK	ΔS, cal /mole°K	T, °C	M	Conditions	Ref.	Re-marks
B161, cont.							
4.541	2.15(25°,+1)	5.39	10-40	T	--	50Sa	c
B162 5,6-BENZOQUINOLINE			$C_{13}H_9N$		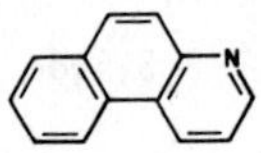		
4.54	5.00(+1)	-6.45	20	T	--	60Pd	
B163 7,8-BENZOQUINOLINE			$C_{13}H_9N$		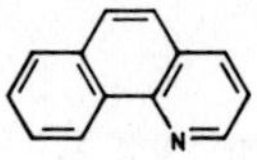		
6.09	4.15(+1)	2.7	20	T	--	60Pd	
B164 1,2,3-BENZOTRIAZOLE			$C_6H_5N_3$		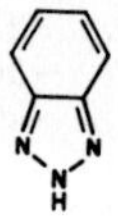		
7.47	8.38(+1)	-13.3	25	C	μ=0	68Ha	
B165 BENZOXAZOLE, 2(2'-pyridyl)-			$C_{12}H_8ON_2$		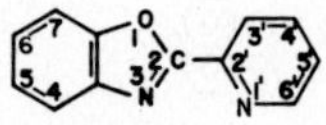		
12.2	9.27(25°,+1)	-1	15-40	T	μ=0.005 S: 50 w % Dioxane	55Ha	c
12.5	8.98(25°,+1)	0.8	0-40	T	μ=0.005	55Ha	c
B166 BETAINE			$C_5H_{12}O_2N$		$(CH_3)_3NCH_2COOH$		
-0.08	1.832(+1)	-8.7	25	C	μ=0	66Ab	
B167 BICYLCO[2.2.2]OCTANE, 4-bromo-1-carboxylic acid			$C_9H_{13}O_2Br$				
2.92	9.750(25°,0)	-34.8	0&25	T	μ=0 S: Anhydrous Methanol	67Rb	b,c,q: +1.0
B168 BICYCLO[2.2.2]OCTANE, 4-carbethoxy-			$C_{12}H_{18}O_4$		see B167		
2.15	9.933(25°,0)	-38.2	0&25	T	μ=0 S: Anhydrous Methanol	67Rb	b,c,q: +1.0

ΔH, kcal /mole	pK	ΔS, cal /mole°K	T,°C	M	Conditions	Ref.	Remarks
B169 BICYCLO[2.2.2]OCTANE, 1-carboxylic acid			$C_9H_4O_2$		see B167		
2.47	10.226(25°,0)	-38.5	0&25	T	μ=0 S: Anydrous Methanol	67Rb	b,c,q: ±1.0
B170 BICYCLO[2.2.2]OCTANE, 1-carboxylic acid-4-cyano-			$C_{10}H_{13}O_2N$		see B167		
2.01	9.617(25°,0)	-37.3	0&25	T	μ=0 S: Anhydrous Methanol	67Rb	b,c,q: ±1.0
B171 BICYCLO[2.2.2]OCTANE, 1-carboxylic acid-4-hydroxy-			$C_9H_{14}O_3$		see B167		
1.94	9.985(25°,0)	-39.2	0&25	T	μ=0 S: Anhydrous Methanol	67Rb	b,c,q: ±1.0
B172 BICYCLO[2.2.2]OCTANE, 1-carboxylic acid-4-trimethyl-amine			$C_{12}H_{22}O_2N$		see B167		
0.91	9.375(25°,+1)	-39.9	0&25	T	μ=0 S: Anhydrous Methanol	67Rb	b,c,q: ±1.0
B173 BICYCLO[2.2.2]OCTANE, 1,4-dicarboxylic acid-			$C_{10}H_{14}O_4$		see B167		
3.26	9.870(25°,0)	-33.6	0&25	T	μ=0 S: Anhydrous Methanol	67Rb	b,c,q: ±1.0
1.98	10.268(25°,-1)	-40.4	0&25	T	μ=0 S: Anhydrous Methanol	67Rb	b,c,q: ±1.0
B174 2-2'-BIIMIDAZOLYL			$C_6H_6N_4$		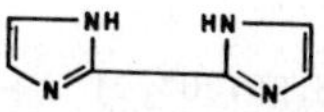		
6.08	5.01(+1)	-2.5	25	C	0.3M($NaClO_4$)	67Hc	
B175 BIPHENYL,2-amino-			$C_{12}H_{11}N$		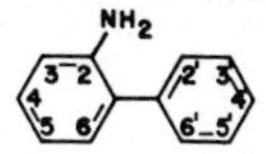		
5.7	3.03(25°,+1)	5.2	0-20	T	C=0.01 S: 50 w % Ethanol	59Eb	c

ΔH, kcal /mole	pK	ΔS, cal /mole°K	T, °C	M	Conditions	Ref.	Remarks
B176 BIPHENYL,3-amino-			$C_{12}H_{11}N$		see B175		
5.2	3.82(20°,+1)	0.2	0-20	T	C=0.01 S: 50 w % Ethanol	59Eb	c
B177 BIPHENYL, 4-amino-			$C_{12}H_{11}N$		see B175		
5.6	3.81(20°,+1)	1.5	0-20	T	C=0.01 S: 50 w % Ethanol	59Eb	c
B178 2,2'-BIPYRIDYL			$C_{10}H_8N_2$		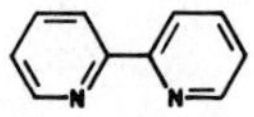		
3.37	4.352(+1)	-8.62	25	T	μ=0	61Bb	
3.5	4.34(+1)	-8	25	T	μ=0	49Ka	
3.37	4.352(+1)	-8.6	25	T	μ=0	55Na	
2.8	4.33(25°,+1)	-10	0-40	T	μ=0.005	55Ha	c
3.66	4.52(+1)	-8.2	20	C	0.1M($NaNO_3$)	63Aa	
3.55	4.44(+1)	8.4	25	C	μ=0.1M(NO_3^-)	70Eb	
4.0	4.67(+1)	-8	25	C	μ=1.0	62Ka	
4.02	4.62(+1)	-7.88	30.3	C	1M(KNO_3)	65Da	
3.67	4.37(+1)	6.55	20	T	--	60Pd	
2.0	4.47(+1)	-13	25	T	--	64La	
2.7	3.33(25°,+1)	-6	0-40	T	μ=0.05 S: 50 w % Dioxane	55Ha	c
B179 BIURET, 2,4-dimercapto-1,5-diphenyl-			$C_{14}H_{13}N_3S_2$		$H_5C_6N{=}C(SH)NHC(SH){=}NC_6H_5$		
5.238	8.44(20°,0)	-19.91	10-40	T	μ=0.1 S: 75 v % Ethanol	74Aa	c
B180 BIURET, 2,4-dimercapto-1,5-di-p-tolyl-			$C_{16}H_{17}N_3S_2$		$H_3CC_6H_4N{=}C(SH)NHC(SH){=}NC_6H_4CH_3$		
5.798	8.86(20°,0)	-19.84	10-40	T	μ=0.1 S: 75 v % Ethanol	74Aa	c
B181 BIURET, 2,4-dimercapto-1-methyl-5-p-tolyl-			$C_{10}H_{13}N_3S_2$		$H_3CC_6H_4N{=}C(SH)NHC(SH){=}NCH_3$		
6.592	9.31(20°,0)	-19.13	10-40	T	μ=0 S: 75 v % Ethanol	74Aa	c
B182 BIURET, 2,4-dimercapto-1-phenyl-5-o-tolyl-			$C_{15}H_{15}N_3S_2$		$H_5C_6N{=}C(SH)NHC(SH){=}NC_6H_4CH_3$		
5.833	8.51(20°,0)	-18.13	10-40	T	μ=0.1 S: 75 v % Ethanol	74Aa	c

ΔH, kcal /mole	pK	ΔS, cal /mole°K	T, °C	M	Conditions	Ref.	Remarks
B183 BIURET, 2,4-dimercapto-1-phenyl-5-p-tolyl-			$C_{15}H_{15}N_3S_2$		$H_5C_6N{=}C(SH)NHC(SH){=}NC_6H_4CH_3$		
5.158	8.56(20°,0)	-20.66	10-40	T	μ=0.1 S: 75 v % Ethanol	74Aa	c
B184 BORIC ACID			H_3O_3B		H_3BO_3		
3.920	9.380(0)	-29.1	10	T	μ=0	340b	
3.750	9.327(0)	-29.7	15	T	μ=0	340b	
3.570	9.280(0)	-30.3	20	T	μ=0	340b	
3.360	9.236(0)	30.0	25	T	μ=0	340b	
3.382	9.237(0)	-30.9	25	T	μ=0 ΔCp=-42.4	39Hb	1
3.191	9.230(0)	-31.5	25	T	μ=0	45Ja	
3.140	9.197(0)	-31.7	30	T	μ=0	340b	
2.900	(0)	-----	35	T	μ=0	340b	
2.630	9.132(0)	-33.4	40	T	μ=0	340b	
2.350	(0)	-----	45	T	μ=0	340b	
2.040	9.080(0)	-35.2	50	T	μ=0	340b	
3.3	9.22(0)	-31.1	25	C	C=0.01M	58Ta	
B185 BOVINE ALBUMIN							
11	10.3	-10	25	T	μ=0.5	56La	e: Tyrosyl
B186 BROMIC ACID			HO_3Br		$HBrO_3$		
1.306	1.013(30°,0)	-0.297	25-35	T	μ=0 S: Formamide	75Da	c
B187 BUTANE, 1-amino-			$C_4H_{11}N$		$CH_3CH_2CH_2CH_2NH_2$		
14.07	10.59(20°,+1)	-1.3	20-40	T	μ=0	51Eb	c
13.98	10.63(+1)	-1.74	25	C	μ=0	680b	
13.98	10.640(+1)	-1.8	25	C	μ=0	69Cc	
13.9	10.640(+1)	-2.1	25	T	μ=0	61Bb	
14.07	10.107(40°,+1)	-1.3	20-40	T	μ=0	51Ea	c
13.879	10.640(+1)	-2.1	25	T	μ=0	68Cf	
14.09	10.114(+1)	-1.3	20-40	T	μ=0.05	51Ea	c
14.04	10.127(+1)	-1.5	20-40	T	μ=0.10	51Ea	c
13.97	(+1)	-----	25	C	μ=0.2(NaCl)	62Wa	
14.09	10.141(+1)	-1.4	20-40	T	μ=0.20	51Ea	
13.9	10.95(25°,+1)	-3.46	0-25	T	μ=2	59Gc	c
B188 BUTANE, 2-amino-			$C_4H_{11}N$		$CH_3CH_2CH(NH_2)CH_3$		
14.03	10.56(+1)	-1.3	25	C	μ=0	69Cc	
B189 BUTANE, 1-amino-2-di(aminomethyl)-			$C_6H_{17}N_3$		$H_2NCH_2C(C_2H_5)(CH_2NH_2)CH_2NH_2$		
7.6	3.58(30°,+3)	-7	10-40	T	μ=0	50Ba	c
7.20	4.25(+3)	4.7	25	C	0.1M(KCl)	61Ca	
8.2	4.59(+3)	5.2	30-40	T	0.1F($BaCl_2$) 0.1F(HNO_3)	50Ja	b,c
12.25	8.59(30°,+2)	-1	10-40	T	μ=0	50Ba	c
11.95	8.98(+2)	-1.0	25	C	0.1M(KCl)	61Ca	
12.6	8.94(+2)	-0.7	30-40	T	0.1F($BaCl_2$) 0.1F(HNO_3)	50Ja	b,c
11.1	9.66(30°,+1)	8	10-40	T	μ=0	50Ba	c

ΔH, kcal /mole	pK	ΔS, cal /mole°K	T,°C	M	Conditions	Ref.	Remarks
B189, cont.							
11.20	9.78(+1)	-7.2	25	C	0.1M(KCl)	61Ca	
11.7	9.68(+1)	-6.9	30-40	T	0.1FBaCl_2 0.1FHNO_3	50Ja	b,c
B190 BUTANE, 1-amino-3-thio-			C_3H_9NS		$CH_3SCH_2CH_2NH_2$		
12.3	9.18(30°,+1)	1	10-40	T	μ=0	59Ma	c
B191 BUTANE, 1-amino-3-methyl-			$C_5H_{13}N$		$(CH_3)_2CHCH_2CH_2NH_2$		
14.03	10.64(+1)	-1.6	25	C	μ=0	69Cc	
B192 BUTANE, 1,2-diamino-			$C_4H_{12}N_2$		$H_2NCH_2CH(CH_2CH_3)NH_2$		
9.92	6.399(+2)	4.0	25	C	μ=0	66Pg	
11.45	9.388(+1)	-4.5	25	C	μ=0	66Pg	
B193 BUTANE, 1,4-diamino-			$C_4H_{12}N_2$		$H_2NCH_2CH_2CH_2CH_2NH_2$		
12.0	9.04(30°,+2)	-2	10-40	T	μ=0	50Ba	c
13.20	9.20(+2)	2.2	25	C	μ=0.02	69Cc	
13.16	9.634(+2)	0.1	25	C	μ=0.5M(KNO_3)	70Ba	
12.0	10.50(30°,+1)	-8	10-40	T	μ=0	50Ba	c
13.58	10.65(+1)	-3.2	25	C	μ=0.02	69Cc	
13.58	10.804(+1)	-3.9	25	C	μ=0.5M(KNO_3)	70Ba	
B194 BUTANE, 2,3-diamino-(meso)			$C_4H_{12}N_2$		$CH_3CH(NH_2)CH(NH_2)CH_3$		
9.4	6.92(25°,+2)	0.13	0&25	T	0.5M(KNO_3)	53Ba	b,c
9.8	9.97(25°,+1)	-12.7	0&25	T	0.5M(KNO_3)	53Ba	b,c
B195 BUTANE, 2,3-diamino-(rac)			$C_4H_{12}N_2$		$CH_3CH(NH_2)CH(NH_2)CH_3$		
10.3	6.91(25°,+2)	2.93	0&25	T	0.5M(KNO_3)	53Ba	b,c
10.3	10.00(25°,+1)	-11.2	0&25	T	0.5M(KNO_3)	53Ba	b,c
B196 BUTANE, 2,3-diamino-2,3-dimethyl-			$C_6H_{16}N_2$		$CH_3C(NH_2)(CH_3)C(NH_2)(CH_3)CH_3$		
9.2	6.56(25°,+2)	1.03	0&25	T	0.5M(KNO_3)	53Ba	b,c
8.9	10.13(25°,+1)	-16.50	0&25	T	0.5M(KNO_3)	53Ba	b,c
B197 BUTANE, 1,4-diamino-N,N,N',N'-tetraacetic acid			$C_{12}H_{20}O_8N_2$		$(HO_2CH_2C)_2N(CH_2)_4N(CH_2CO_2H)_2$		
5.81	9.05(-2)	-21.6	20	C	0.1M(KNO_3)	64Ab	
6.68	10.66(-3)	-26.0	20	C	0.1M(KNO_3)	64Ab	
B198 BUTANE 2,3-diamino-N,N,N',N'-tetraacetic acid			$C_{12}H_{20}O_8N_2$		$(HO_2CCH_2)_2NCH(CH_3)CH(CH_3)N(CH_2CO_2H)_2$		
5.6	10.74(-3)	-30.1	20	C	μ=1M(KCl)	69Sc	

ΔH, kcal /mole	pK	ΔS, cal /mole°K	T, °C	M	Conditions	Ref.	Remarks
B199 BUTANE, 1,1-dinitro-			$C_4H_8O_4N_2$		$NO_2CH(NO_2)CH_2CH_2CH_3$		
3.618	5.90	-12.33	20	T	--	61Sc	
B200 1,2,3,4-BUTANETETRACARBOXYLIC ACID			$C_8H_{10}O_8$		$HO_2CCH_2CH(CO_2H)CH(CO_2H)CH_2$-$CO_2H$		
0.2	3.43(0)	-15.0	25	T	μ=0	72Pf	
-0.3	4.58(-1)	-21.9	25	T	μ=0	72Pf	
-0.5	5.85(-2)	-28.4	25	T	μ=0	72Pf	
-1.2	7.16(-3)	-36.7	25	T	μ=0	72Pf	
B201 BUTANOIC ACID			$C_4H_8O_2$		$CH_3CH_2CH_2CO_2H$		
0.273	4.806(0)	-21.0	0	T	μ=0	34Ha	
0.106	4.804(0)	-21.6	5	T	μ=0	34Ha	
0	4.803(0)	-22.0	8	T	μ=0	34Ha	
-0.242	(0)	-----	10	C	μ=0	48Ca	
-0.073	4.803(0)	-22.2	10	T	μ=0	34Ha	
-0.266	4.805(0)	-22.9	15	T	μ=0	34Ha	
-0.542	(0)	-----	20	C	μ=0	48Ca	
-0.472	4.810(0)	-23.6	20	T	μ=0	34Ha	
-0.69	4.82(0)	-24.4	25	C	μ=0	74Md	
-0.698	4.818(0)	-24.4	25	C	μ=0	48Ca	
-0.64	4.817(0)	-24.2	25	C	μ=0	68Cd	
-0.691	4.817(0)	-24.4	25	T	μ=0 ΔCp=-45.3	67Ca 34Ha 39Hb	1
-0.73	4.82(0)	-24.5	25	C	μ=0	58Ca 61Cb	a:13.5
-0.853	(0)	-----	30	C	μ=0	48Ca	
-0.926	4.827(0)	-25.1	30	T	μ=0	34Ha	
-1.174	4.840(0)	-26.0	35	T	μ=0	34Ha	
-1.12	4.854(0)	-25.8	40	C	μ=0	68Cd	
-1.437	4.854(0)	-26.8	40	T	μ=0	34Ha	
-1.714	4.871(0)	-27.7	45	T	μ=0	34Ha	
-0.56	4.815(0)	-23.9	25	C	μ=0.05M	61Ba	
0.43	4.78(25°,0)	-20.4	25&40	T	μ=3M(NaCl)	73Sa	b,c
-0.494	4.833(35°,0)	-24	0&35	T	--	39Wa	b,c
-0.67	4.93(0)	-24.8	25	C	μ=0 S: 5 w % tert-Butanol	74Md	
-0.69	5.07(0)	-25.6	25	C	μ=0 S: 10 w % tert-Butanol	74Md	
-0.91	5.23(0)	-27.0	25	C	μ=0 S: 15 w % tert-Butanol	74Md	
-1.1	5.31(0)	-28.0	25	C	μ=0 S: 17.5 w % tert-Butanol	74Md	
-0.81	5.41(0)	-27.5	25	C	μ=0 S: 20 w % tert-Butanol	74Md	
-0.43	5.63(0)	-27.2	25	C	μ=0 S: 25 w % tert-Butanol	74Md	
-0.19	5.81(0)	-27.2	25	C	μ=0 S: 30 w % tert-Butanol	74Md	
0.07	6.06(0)	-27.5	25	C	μ=0 S: 35 w % tert-Butanol	74Md	
0.07	6.19(0)	-28.2	25	C	μ=0 S: 40 w % tert-Butanol	74Md	
B202 BUTANOIC ACID, amide, 3-oxo-N-anilino-			$C_{10}H_{12}O_2N_2$		$C_6H_5NHNHC(O)CH_2C(O)CH_3$		
4.4	11.43(25°,0)	-37	10-40	T	S: 50 % Dioxane	69Ha	c

ΔH, kcal /mole	pK	ΔS, cal /mole°K	T,°C	M	Conditions	Ref.	Remarks
B203 BUTANOIC ACID, 2-amino-			$C_4H_9O_2N$		$CH_3CH_2CHNH_2CO_2H$		
1.090	2.334(+1)	-6.7	1	T	μ=0	37Sa	
0.77	2.315(+1)	-7.7	10	C	μ=0	68Cd	
0.750	2.310(+1)	-7.9	12.5	T	μ=0	37Sa	
0.38	2.286(+1)	-9.2	25	C	μ=0	68Cd	
0.310	2.286(+1)	-9.4	25	T	μ=0	37Sa	
0.309	2.288(+1)	-9.4	25	T	μ=0 ΔCp=-38.6	39Hb	1
0.30	2.284(+1)	-9.5	25	T	μ=0	58Ea	
-0.220	2.288(+1)	-11.2	37.5	T	μ=0	37Sa	
0.04	2.289(+1)	-10.4	40	C	μ=0	68Cd	
-0.830	2.297(+1)	-13.1	50	T	μ=0	37Sa	
1.0	2.11(+1)	-6.30	25	C	μ=0.05(KCl)	72Gc	
0.42	2.34(+1)	-9.30	25	C	μ=0.2M(KCl)	73Ga	
10.750	10.530(0)	-9.0	1.0	T	μ=0	37Sa	
10.820	10.180(0)	-8.7	12.5	T	μ=0	37Sa	
10.86	9.830(0)	-8.6	25	C	μ=0	69Cc	
10.770	9.830(0)	-8.8	25	T	μ=0	37Sa	
10.580	9.518(0)	-9.4	37.5	T	μ=0	37Sa	
10.260	9.234(0)	-10.5	50	T	μ=0	37Sa	
10.3	9.62(0)	-9.47	25	C	μ=0.05(KCl)	72Gc	
10.56	9.852(25°,0)	-9.7	25-50	T	μ=0.1M(KCl)	70Ha	c,e:NH_3^+
10.73	9.56(0)	-7.7	25	C	μ=0.16(KNO_3)	70Md	
11.31	9.61(0)	-6.04	25	C	μ=0.2M(KCl)	73Ga	
10.840	9.830(0)	-8.46	25	C	--	67Ac	
B204 BUTANOIC ACID, 3-amino-			$C_4H_9O_2N$		$CH_3CH(NH_2)CH_2CO_2H$		
11.02	10.14(0)	9.4	25	C	μ=0.16	70Lb	e:NH_3^+
B205 BUTANOIC ACID, 4-amino-			$C_4H_9O_2N$		$H_2NCH_2CH_2CH_2CO_2H$		
0.75	4.057(+1)	-15.9	10	C	μ=0	68Cd	
0.39	4.031(+1)	-17.2	25	C	μ=0	68Cd	
0.405	4.0312(+1)	-17.0	25	T	μ=0	54Kb	
0.40	4.03(+1)	-17	25	T	μ=0	63Ea	
0.01	4.027(+1)	-18.5	40	C	μ=0	68Cd	
12.45	10.556(0)	-6.5	25	C	μ=0	69Cc	
12.070	10.5557(0)	-7.8	25	T	μ=0	54Kb	
11.72	10.565(25°,0)	-9.0	25-50	T	μ=0.1M(KCl)	70Ha	c,e:NH_3^+
11.950	10.555(0)	-8.05	25	C	--	67Ac	
B206 BUTANOIC ACID, 2-amino-N-acetyl-			$C_6H_{11}O_3N$		$CH_3CH_2CH(NHCOCH_3)CO_2H$		
-0.773	3.7158(+1)	-19.59	25	T	μ=0	56Kc	
B207 BUTANOIC ACID, 2-amino-N-carbamoyl-			$C_5H_{10}O_3N_2$		$H_2NCONHCH(CH_2CH_3)CO_2H$		
-0.493	3.8856(+1)	-19.43	25	T	μ=0	56Kb	
B208 BUTANOIC ACID, 4-amino-N-carbamoyl-			$C_5H_{10}O_3N_2$		$H_2NCONH(CH_2)_3CO_2H$		
-0.115	4.6831(+1)	-21.83	25	T	μ=0	56Kb	
B209 BUTANOIC ACID, 2-amino-N-glycyl-			$C_6H_{12}O_3N_2$		$H_2NCH_2CONHCH(CH_2CH_3)CO_2H$		

ΔH, kcal /mole	pK	ΔS, cal /mole°K	T, °C	M	Conditions	Ref.	Remarks
B209, cont.							
-0.672	3.1546(+1)	-16.69	25	T	μ=0	57Ka	
10.750	8.331(0)	-2.0	25	T	μ=0	75Kb	e:NH_3^+
7.7	6.62(+1)	-5.1	25	C	μ=0.1M(KNO_3) HA=[Cu(II)$NH_2CH_2CONHCH$-(CH_2CH_3)CO_2]$^+$	72Bi	
B210	BUTANOIC ACID, 4-amino-N-glycyl-		$C_6H_{12}O_3N_2$		$NH_2CH_2C(O)NHCH_2CH_2CH_2CO_2H$		
0.0	4.22(+1)	-19.3	25	C	μ=0.1M(KNO_3)	72Bi 71La	q:+0.2
10.1	8.12(0)	-3.2	25	C	μ=0.1M(KNO_3)	72Bi 71La	
B211	BUTANOIC ACID, 2-amino-4-hydroxy-		$C_4H_9O_3N$		$HOCH_2CH_2CH(NH_2)CO_2H$		
-0.3	2.265(+1)	-11.4	25	T	μ=0.10M(KCl)	71Bg	
10.9	9.257(0)	-5.9	25	T	μ=0.10M(KCl)	71Bg	
10.23	9.43(0)	8.8	25	C	μ=0.16	70Lb	e:NH_3^+
B212	BUTANOIC ACID, 4-amino-3-hydroxy-		$C_4H_9O_3N$		$NH_2CH_2CH(OH)CH_2CO_2H$		
0.78	3.834(+1)	-14.9	25	T	μ=0.1M(KCl)	75Bd	e:CO_2
10.89	9.487(0)	-6.8	25	T	μ=0.1M(KCl)	75Bd	e:NH_3^+
10.88	9.73(0)	8.0	25	C	μ=0.16	70Lb	e:NH_3^+
B213	BUTANOIC ACID, 2-amino-, methyl ester		$C_5H_{11}O_2N$		$CH_3CH_2CH(NH_2)CO_2CH_3$		
11.45	7.640(25°,0)	3.6	25-50	T	μ=0.1M(KCl)	70Ha	c,e:NH_3^+
B214	BUTANOIC ACID, 4-amino-, methyl ester		$C_5H_{11}O_2N$		$H_2NCH_2CH_2CH_2CO_2CH_3$		
13.19	9.839(25°,0)	-0.8	25-50	T	μ=0.1M(KCl)	70Ha	c,e:NH_3^+
B215	BUTANOIC ACID, 2-bromo-		$C_4H_7O_2Br$		$CH_3CH_2CH(Br)CO_2H$		
-1.263	2.939(35°,0)	-18	0&35	T	--	39Wa	b,c
B216	BUTANOIC ACID, 2,2-dimethyl-		$C_6H_{12}O_2$		$CH_3CH_2C(CH_3)_2COOH$		
-0.62	4.93(0)	-24	25	C	μ=0	70Ce	
B217	BUTANOIC ACID, 2-ethyl-		$C_6H_{12}O_2$		$(C_2H_5)_2CHCO_2H$		
-1.97	4.751(0)	-28.4	25	C	μ=0	70Ce	
-1.82	4.74(0)	-27.7	25	C	μ=0	74Md	
-1.86	4.74(0)	-27.6	25	C	μ=0	68Cd	
-2.34	4.813(0)	-29.5	40	C	μ=0	70Ce	
-1.74	4.89(0)	-28.2	25	C	μ=0 S: 5 w % tert-Butanol	74Md	
-1.67	5.11(0)	-29.1	25	C	μ=0 S: 10 w % tert-Butanol	74Md	

B217, cont.

ΔH, kcal /mole	pK	ΔS, cal /mole°K	T, °C	M	Conditions	Ref.	Remarks
-2.46	5.34(0)	-32.7	25	C	μ=0 S: 15 w % tert-Butanol	74Md	
-3.08	5.48(0)	-35.4	25	C	μ=0 S: 17.5 w % tert-Butanol	74Md	
-2.46	5.60(0)	-33.9	25	C	μ=0 S: 20 w % tert-Butanol	74Md	
-1.22	5.90(0)	-31.1	25	C	μ=0 S: 25 w % tert-Butanol	74Md	
-0.17	6.19(0)	-28.9	25	C	μ=0 S: 30 w % tert-Butanol	74Md	
0.41	6.46(0)	-28.2	25	C	μ=0 S: 35 w % tert-Butanol	74Md	
0.62	6.63(0)	-28.2	25	C	μ=0 S: 40 w % tert-Butanol	74Md	

B218 BUTANOIC ACID, 3-hydroxy- $C_4H_8O_3$ $CH_3CHOHCH_2CO_2H$

ΔH, kcal /mole	pK	ΔS, cal /mole°K	T, °C	M	Conditions	Ref.	Remarks
2.022	0.663(0)	3.48	20	T	μ=0	55Da	

B222 BUTANOIC ACID, 2-methyl- $C_5H_{10}O_2$ $CH_3CH_2CH(CH_3)CO_2H$

ΔH, kcal /mole	pK	ΔS, cal /mole°K	T, °C	M	Conditions	Ref.	Remarks
-1.24	4.761(0)	-26.1	25	C	μ=0	70Ce	

B223 BUTANOIC ACID, 3-methyl- $C_5H_{10}O_2$ $CH_3CH(CH_3)CH_2CO_2H$

ΔH, kcal /mole	pK	ΔS, cal /mole°K	T, °C	M	Conditions	Ref.	Remarks
-0.80	4.742(0)	-24.5	10	C	μ=0	70Ce	
-0.767	(0)	-----	10	C	μ=0	48Ca	
-1.039	(0)	-----	20	C	μ=0	48Ca	
-1.1	4.77(0)	-25.8	25	C	μ=0	74Md	
-1.168	4.770(0)	-25.7	25	C	μ=0	68Cd	
-1.15	4.777(0)	-25.7	25	C	μ=0	70Ce 68Cd	
-1.219	4.781(0)	-26.0	25	T	μ=0	52Ea	
-1.302	(0)	-----	30	C	μ=0	68Cd	
-1.48	4.831(0)	-26.8	40	C	μ=0	70Ce	
-----	(0)	-----	10-40	C	μ=0 ΔCp=-23	70Ce	
-0.809	4.851(35°,0)	-25	0&35	T	--	39Wa	b,c
-1.1	4.94(0)	-26.3	25	C	μ=0 S: 5 w % tert-Butanol	74Md	
-1.3	5.06(0)	-27.5	25	C	μ=0 S: 10 w % tert-Butanol	74Md	
-1.5	5.22(0)	-28.9	25	C	μ=0 S: 15 w % tert-Butanol	74Md	
-2.0	5.31(0)	-31.0	25	C	μ=0 S: 17.5 w % tert-Butanol	74Md	
-1.6	5.47(0)	-30.6	25	C	μ=0 S: 20 w % tert-Butanol	74Md	
-0.62	5.75(0)	-28.4	25	C	μ=0 S: 25 w % tert-Butanol	74Md	
-0.024	5.96(0)	-27.2	25	C	μ=0 S: 30 w % tert-Butanol	74Md	
0.48	6.14(0)	-26.5	25	C	μ=0 S: 35 w % tert-Butanol	74Md	
0.38	6.34(0)	-27.7	25	C	μ=0 S: 40 w % tert-Butanol	74Md	

B224 BUTANOIC ACID, 2-oxo- $C_4H_6O_3$ $CH_3CH_2C(O)CO_2H$

ΔH, kcal /mole	pK	ΔS, cal /mole°K	T, °C	M	Conditions	Ref.	Remarks
2.82	2.50(0)	-2.0	25	C	μ=0	67Ob	
2.932	2.50(0)	-1.61	25	C	μ=0.05	67Ob	

ΔH, kcal /mole	pK	ΔS, cal /mole°K	T, °C	M	Conditions	Ref.	Re-marks
B225 1-BUTANOL, 2-amino-			$C_4H_{11}ON$		$CH_3CH_2CH(NH_2)CH_2OH$		
12.82	9.53(0)	-0.60	25	C	μ=0	61Ta	
12.460	9.516(25°,0)	-1.8	0-60	T	μ=0	68Tb	c
B228 1-BUTANOL, 2-amino-2-ethyl-			$C_6H_{15}ON$		$CH_3CH_2C(C_2H_5)(NH_2)CH_2OH$		
9.66	9.82(0)	-12.6	25	C	μ=0	69Cc	
B226 2-BUTANONE, 3-amino-N-allyl-3-methyl-, oxime			$C_8H_{16}ON_2$		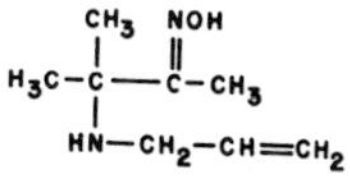		
10.8	(+1)	-----	25	C	μ=0	64Wa	
B227 2-BUTANONE, 3-amino-N-n-butyl-3-methyl, oxime			$C_9H_{20}ON_2$		see B226		
11.0	9.09(+1)	-5	25	C	μ=0	64Wa	
B229 2-BUTANONE, 3-amino-N-ethyl-3-methyl, oxime			$C_7H_{16}ON_2$		see B226		
10.6	9.23(+1)	-7	25	C	μ=0	64Wa	
B230 2-BUTANONE, 3-amino-N-isopropyl-3-methyl-, oxime			$C_8H_{18}ON_2$		see B226		
10.6	9.09(+1)	-6	25	C	μ=0	64Wa	
B231 2-BUTANONE, 3-amino-N-methyl-3-methyl-, oxime			$C_6H_{14}ON_2$		see B226		
10.3	9.23(+1)	-8	25	C	μ=0.002	64Wa	
B232 2-BUTANONE, 3-amino-3-methyl-, oxime			$C_5H_{12}ON_2$		see B226		
12.7	9.09(+1)	1	25	C	μ=0	64Wa	
B233 2-BUTANONE, 3-amino-N-n-propyl-3-methyl-, oxime			$C_8H_{18}ON_2$		see B226		
10.8	9.09(+1)	-5	25	C	μ=0	64Wa	
B234 2-BUTENOIC ACID (trans)			$C_4H_6O_2$		$CH_3CH{:}CHCO_2H$		
0.280	4.676(35°,0)	-20	0&35	T	--	39Wa	b,c
B235 BUTTER YELLOW INDICATOR							
-4.30	3.226(25°,0)	-29	18-35	T	μ=0	53Ka	c

ΔH, kcal /mole	pK	ΔS, cal /mole°K	T, °C	M	Conditions	Ref.	Remarks
B236 2-BUTYNOIC ACID			$C_4H_4O_2$		$CH_3C\vdots CCO_2H$		
0.827	2.618(0)	-9.10	10	T	μ=0 ΔCp=-61.54	69Gc	
0.516	2.620(0)	-10.18	15	T	μ=0 ΔCp=-62.62	69Gc	
0.200	2.611(0)	-11.27	20	T	μ=0 ΔCp=-63.71	69Gc	
-0.121	2.620(0)	-12.36	25	T	μ=0 ΔCp=-64.80	69Gc	
-0.447	2.618(0)	-13.44	30	T	μ=0 ΔCp=-65.88	69Gc	
-0.780	2.621(0)	-14.53	35	T	μ=0 ΔCp=-66.97	69Gc	
-1.119	2.631(0)	-15.62	40	T	μ=0 ΔCp=-68.06	69Gc	
-1.460	2.647(0)	-16.70	45	T	μ=0 ΔCp=-69.14	69Gc	

ΔH, kcal /mole	pK	ΔS, cal /mole°K	T,°C	M	Conditions	Ref.	Re-marks
			C				
C1 CAMPHORIC ACID			$C_{10}H_{16}O_4$		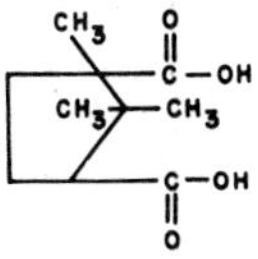		
-0.550	4.595(35°,0)	-23	0&35	T	--	39Wa	b,c
C2 CARBAZIDE, 3-thio-S-methyl			$C_2H_8N_4S$		$CH_3SC(:NNH_2)NHNH_2$		
9.9	7.563(+1)	-1.4	25	T	μ=0.1M(KCl)	72Bh	
C3 CARBONIC ACID			CH_2O_3		H_2CO_3		
0.580	3.807(0)	-15.3	0	T	μ=0	54Wa	
0.0	3.754(0)	-17.2	17	T	μ=0	54Wa	
-0.412	3.764(0)	-18.5	27	T	μ=0	54Wa	
-0.880	3.798(0)	-20.2	37	T	μ=0	54Wa	
-1.710	3.602(0)	-22.7	0	T	--	58Sa	
-0.800	3.620(0)	-19.3	23.0	T	--	58Sa	
-0.890	3.770(0)	-20.2	25.5	T	--	58Sa	
-0.960	3.60(0)	-19.6	27	T	--	41Ra	
-0.650	3.824(0)	-19.6	30.2	T	--	58Sa	
-0.480	3.824(0)	-19.0	37.5	T	--	58Sa	
3.600	10.329(-1)	-35.16	25	T	μ=0 ΔCp=-65 $\Delta H=13.2786-1.0884 \times 10^{-4}T^2$ 0<T °C<50	41Hb	
3.500	10.24(-1)	-35.2	25	C	μ=0	37Pa	
4.484	6.583	13.7	0	T	μ=0 R: $CO_2 + H_2O = H^+ + HCO_3^-$	35Sa	
2.952	6.429	-19.1	15	T	μ=0 R: $CO_2 + H_2O = H^+ + HCO_3^-$	35Sa	
1.843	6.364	-22.9	25	C	μ=0 R: $CO_2 + H_2O = H^+ + HCO_3^-$	37Pa	
2.050	6.460	-22.7	25	T	μ=0 R: $CO_2 + H_2O = H^+ + HCO_3^-$	28Sa	
2.075	6.365	-22.2	25	T	μ=0 R: $CO_2 + H_2O = H^+ + HCO_3^-$	35Sa	
1.109	7.317	-29.9	38	T	μ=0 R: $CO_2 + H_2O = H^+ + HCO_3^-$	35Sa	
2.000	6.1(37°)	-21.5	22-37	T	μ=0.3M R: $CO_2 + H_2O = H^+ + HCO_3^-$	56Bb	c
C4 CARBONIC ACID, deuterated			CD_2O_3		D_2CO_3		
4.212	11.076(-1)	-36.6	25	T	μ=0 S: Deuterium Oxide ΔCp=-61.4	69Pa	
C5 CARBONIC ACID, diethyl ester			$C_5H_{10}O_3$		$C_2H_5OC(:O)OC_2H_5$		
16.4	-----	-----	25	C	C$\sim 10^{-3}$M S: Fluorosulfuric acid	74Ab	
C6 CARBONIC ACID, trithio-			CH_2S_3		$SC(SH)_2$		
2.86	2.64(0)	-2.51	20	T	--	63Ga	

ΔH, kcal /mole	pK	ΔS, cal /mole°K	T, °C	M	Conditions	Ref.	Remarks
C7 CASEIN, Iodinated							
1.50	-----	-----	20&30	T	μ=0.15	72Ra	b,c,e: CO_2H
6.80	-----	-----	20&30	T	μ=0.15	72Ra	b,c,e: Imidazole
12.00	-----	-----	20&30	T	μ=0.15	72Ra	b,c,e: Amino
13.60	-----	-----	20&30	T	μ=0.15	72Ra	b,c,e: Guanidinium
C8 CHLORIC ACID, per			HO_4Cl		$HClO_4$		
-1.8	-1.58(0)	1.2	25	T	--	60Hb	
C9 CHLOROPHOSPHONIC ACID, diethyl ester			$C_4H_{10}O_3ClP$		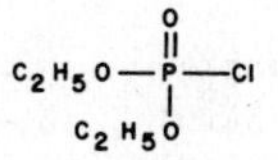		
15.5	-----	-----	25	C	C∿10^{-3}M S: Fluorosulfuric acid	74Ab	
C10 CHLOROUS ACID			HO_2Cl		$HClO_2$		
-4.100	2.021(0)	-22.7	25	C	μ=0	37Pa	
C11 CHROMIC ACID			H_2O_4Cr		H_2CrO_4		
-8.5	-0.55(0)	-26	25	T	μ=1	69La	
-0.7	6.50(-1)	-32.1	25	C	C=0.001M - 0.05M	69La 58Hb	
-1.08	5.89(-1)	-30.6	25	C	μ=3M($NaClO_4$)	72Ab 70Aa	q:±0.09
C12 CHROMIUM, trioxochloro			HO_3ClCr		$HCrO_3Cl$		
-1.13	1.05(0)	-8.6	25	T	μ=1	66Tb	
C13 α-CHYMOTRYPSIN							
50.6	18.49(20°)	87	12-20	T	Buffer solution: (0.2M NaCl + 0.02M Tris-HCl) Low-temp case (T_1) R: $A_bH_2 = A_g + 2H^+$	71Ka	c
2.2	17.66(30°)	-76	28-36	T	Buffer solution: (0.2M NaCl + 0.02M Tris-HCl) High-temp case (T_2) R: $A_fH_2 = A_g + 2H^+$	71Ka	c
22.5	8.14(25°)	38	20-28	T	Buffer solution: (0.2M NaCl + 0.02M Tris-HCl) R: $A_bH_2 = A_jH + H$	71Kb	c

ΔH, kcal /mole	pK	ΔS, cal /mole°K	T, °C	M	Conditions	Ref.	Remarks
C14 CINNAMIC ACID			$C_9H_8O_2$		$C_6H_5CH:CHCO_2H$		
0.600	4.404(0)	-18.1	25	T	--	39Ea	Form, cis or Trans, not specified
0.573	4.410(35°,0)	-18	0&35	T	--	39Wa	b,c Form, cis or Trans, not specified
C15 CINNAMIC ACID, 2-bromo			$C_9H_7O_2Br$		$C_6H_5CH:C(Br)CO_2H$		
-2.990	1.977(0)	-19.1	25	T	--	39Ea	Form, cis or Trans, not specified
C16 CINNAMIC ACID, 2-hydroxy-			$C_9H_8O_3$		$C_6H_5CH:C(OH)CO_2H$		
0.181	4.616(35°,0)	-20	0&35	T	--	39Wa	b,c Form, cis or Trans, not specified
C17 CITRIC ACID			$C_6H_8O_7$		$HOC(CH_2CO_2H)_2CO_2H$		
1.760	3.220(0)	-8.3	0	T	μ=0	49Bb	
1.612	3.200(0)	-8.8	5	T	μ=0	49Bb	
1.462	3.176(0)	-9.4	10	T	μ=0	49Bb	
1.310	3.160(0)	-9.9	15	T	μ=0	49Bb	
1.155	3.142(0)	-10.4	20	T	μ=0	49Bb	
0.997	3.128(0)	-11.0	25	T	μ=0	49Bb	
0.836	3.116(0)	-11.5	30	T	μ=0	49Bb	
0.673	3.109(0)	-12.0	35	T	μ=0	49Bb	
0.507	3.099(0)	-12.6	40	T	μ=0	49Bb	
0.338	3.097(0)	-13.1	45	T	μ=0	49Bb	
0.167	3.095(0)	-13.6	50	T	μ=0	49Bb	
1.654	4.837(-1)	-16.1	0	T	μ=0	49Bb	
1.447	4.813(-1)	-16.8	5	T	μ=0	49Bb	
1.237	4.797(-1)	-17.6	10	T	μ=0	49Bb	
1.022	4.782(-1)	-18.3	15	T	μ=0	49Bb	
0.804	4.769(-1)	-19.1	20	T	μ=0	49Bb	
0.582	4.761(-1)	-19.8	25	T	μ=0	49Bb	
0.357	4.755(-1)	-20.6	30	T	μ=0	49Bb	
0.128	4.751(-1)	-21.3	35	T	μ=0	49Bb	
-0.105	4.750(-1)	-22.1	40	T	μ=0	49Bb	
-0.342	4.754(-1)	-22.8	45	T	μ=0	49Bb	
-0.583	4.757(-1)	-23.6	50	T	μ=0	49Bb	
0.660	6.393(-2)	-26.8	0	T	μ=0	49Bb	
0.378	6.386(-2)	-27.9	5	T	μ=0	49Bb	
0.090	6.383(-2)	-28.9	10	T	μ=0	49Bb	
-0.202	6.384(-2)	-29.9	15	T	μ=0	49Bb	

ΔH, kcal /mole	pK	ΔS, cal /mole$^\circ K$	T,$^\circ$C	M	Conditions	Ref.	Re- marks

C17, cont.

ΔH, kcal /mole	pK	ΔS, cal /mole$^\circ K$	T,$^\circ$C	M	Conditions	Ref.	Remarks
−0.500	6.388(−$\underline{2}$)	−30.9	20	T	μ=0	49Bb	
−0.803	6.396(−$\underline{2}$)	−32.0	25	T	μ=0	49Bb	
−1.111	6.406(−$\underline{2}$)	−33.0	30	T	μ=0	49Bb	
−1.424	6.423(−$\underline{2}$)	−34.0	35	T	μ=0	49Bb	
−1.742	6.439(−$\underline{2}$)	−35.0	40	T	μ=0	49Bb	
−2.065	6.462(−$\underline{2}$)	−36.0	45	T	μ=0	49Bb	
−2.394	6.484(−$\underline{2}$)	−37.1	50	T	μ=0	49Bb	

C18 CONIINE $C_8H_{17}N$ $C_5H_9NH(C_3H_7)$

14.44	11.24(+$\underline{1}$)	−3.05	25	C	μ=0.5M(KNO_3)	74Be	

C19 18-CROWN 4, diaza- $C_{12}H_{26}O_4N_2$ $HN(CH_2CH_2OCH_2CH_2OCH_2CH_2)_2NH$

9.5	7.94(+$\underline{2}$)	−4.4	25	C	μ=0.1[$(CH_3)_4NCl$]	75Aa	
8.6	9.08(+$\underline{1}$)	−12.7	25	C	μ=0.1[$(CH_3)_4NCl$]	75Aa	

C20 CRYPTATE, diaza-Hexaoxa $C_{18}H_{35}O_6N_2$ $N(CH_2CH_2OCH_2CH_2OCH_2CH_2)_3N$

4.5	7.31(+$\underline{2}$)	−18.3	25	C	μ=0.1[$(CH_3)_4NCl$]	75Aa	
10.8	9.71(+$\underline{1}$)	−8.2	25	C	μ=0.1[$(CH_3)_4NCl$]	75Aa	

C21 CYANIC ACID CHON HOCN

1	3.57(26°,$\underline{0}$)	−10	20-33	T	μ=0	66Bh	c
6.390	1.74($\underline{0}$)	6.16	0	T	−−	58Cb	
3.690	2.69($\underline{0}$)	−3.18	18	T	−−	58Cb	
2.480	3.47($\underline{0}$)	−7.37	25	T	−−	58Cb	
−1.430	3.31($\underline{0}$)	−20.2	45	T	−−	58Cb	

C23 CYCLOBUTANECARBOXYLIC ACID $C_5H_8O_2$ $\underline{CH_2CH_2CH_2}CHCO_2H$

−0.08	4.60($\underline{0}$)	−21.3	10	C	μ=0	72Cb	q:$\underline{+}$0.01
−0.68	4.785($\underline{0}$)	−24.2	25	C	μ=0	72Cb	
−1.38	4.82($\underline{0}$)	−26.4	40	C	μ=0	72Cb	
-----	($\underline{0}$)	-----	10-40	C	μ=0	ΔCp=−43	72Cb

C24 1,1-CYCLOBUTANEDICARBOXYLIC ACID $C_6H_8O_4$

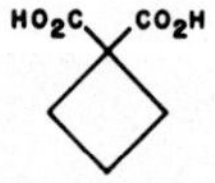

−0.259	2.922($\underline{0}$)	−14.23	25	C	0.1M($NaClO_4$, $HClO_4$)	75Ca	
−0.805	5.454(−$\underline{1}$)	−27.64	25	C	0.1M($NaClO_4$, $HClO_4$)	75Ca	

C25 3-CYCLOBUTENE-1,2-DIONE, 3,4-dihydroxy- $C_4H_2O_4$

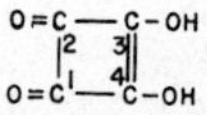

−1.493	0.541($\underline{0}$)	−7.48	25	T	μ=0	71Sb	q:$\underline{+}$0.12
−3.028	3.480(−$\underline{1}$)	−26.08	25	T	μ=0	70Sa	q:$\underline{+}$0.45

C26 CYCLOHEPTANE, 1-aza- $C_6H_{13}N$ $\underline{CH_2(CH_2)_4CH_2}NH$

ΔH, kcal /mole	pK	ΔS, cal /mole°K	T, °C	M	Conditions	Ref.	Remarks
C26, cont.							
13.01	11.11(+1)	-7.2	25	C	μ=0	71Ca 68Ca	
C27 CYCLOHEXANE, amino-			$C_6H_{13}N$		$C_6H_{11}(NH_2)$		
14.26	11.31(+1)	-0.72	5	C	μ=0	75Bb	
14.38	10.64(+1)	-0.5	25	C	μ=0	69Cc	
14.26	10.56(+1)	-0.49	25	C	μ=0 ΔCp=5.74 ΔH=14.161 + 0.00499T 25<T °C<125	75Bb	
14.35	9.75(+1)	-0.23	50	C	μ=0	75Bb	
14.50	9.04(+1)	0.30	75	C	μ=0	75Bb	
14.65	8.43(+1)	0.71	100	C	μ=0	75Bb	
14.84	7.89(+1)	1.18	125	C	μ=0	75Bb	
C28 CYCLOHEXANE, amino-N,N-dimethyl-			$C_8H_{12}N$		$C_6H_{11}[N(CH_3)_2]$		
10.11	10.72(+1)	-15.2	25	C	μ=0	69Cc	
C29 CYCLOHEXANE, 1,2-diamino(cis)-			$C_6H_{14}N_2$		$C_6H_{10}(NH_2)_2$		
10.1	5.94(30°,+2)	6	10-40	T	μ=0	50Ba	c
11.5	9.66(30°,+1)	-6	10-40	T	μ=0	50Ba	c
C30 CYCLOHEXANE, 1,2-diamino(trans)			$C_6H_{14}N_2$		$C_6H_{10}(NH_2)_2$		
10.2	6.20(30°,+2)	5	10-40	T	μ=0	50Ba	c
11.7	9.60(30°,+1)	-6	10-40	T	μ=0	50Ba	c
C31 CYCLOHEXANE, 1,2-diamino-N,N,N',N'-tetraacetic acid (trans)			$C_{14}H_{22}O_8N_2$		$C_6H_{10}[N(CH_2CO_2H)_2]_2$		
2.06	6.15(-2)	-21.1	20	C	0.1M(KNO_3)	63Ab	
2.5	6.16(-2)	-20	25	C	μ=0.1(KNO_3)	70Kb	
1.34	6.12(-2)	-23.51	25	T	0.1M(KNO_3)	62Mb	
6.65	12.35(-3)	-33.9	20	C	0.1M(KNO_3)	63Ab	
9.0	12.35(-3)	-26	25	C	μ=0.1(KNO_3)	70Kb	
7.35	11.58(-3)	-28.44	25	T	0.1M(KNO_3)	62Mb	
6.63	(-3)	-----	25	C	μ=0.20(KNO_3)	69Cb	
C32 CYCLOHEXANECARBOXYLIC ACID			$C_7H_{12}O_2$		$C_6H_{11}(CO_2H)$		
0.23	4.62(0)	-20.3	10	C	μ=0	72Cb	q:±0.02
-0.37	4.899(0)	-23.7	25	C	μ=0	72Cb	
-0.92	4.903(0)	-25.5	25	C	μ=0	67Ca	
-0.79	4.90(0)	-25.1	25	C	μ=0	74Md	
-0.91	5.01(0)	-25.8	40	C	μ=0	72Cb	
-----	(0)	-----	10-40	C	μ=0 ΔCp=-37	72Cb	
-0.91	5.02(0)	-26.0	25	C	μ=0 S: 5 w % tert-Butanol	74Md	
-0.91	5.22(0)	-27.0	25	C	μ=0 S: 10 w % tert-Butanol	74Md	
-1.7	5.48(0)	-30.8	25	C	μ=0 S: 15 w % tert-Butanol	74Md	
-2.1	5.61(0)	-32.7	25	C	μ=0 S: 17.5 w % tert-Butanol	74Md	

ΔH, kcal /mole	pK	ΔS, cal /mole°K	T, °C	M	Conditions	Ref.	Remarks
C32, cont.							
-2.1	5.73(0)	-31.3	25	C	μ=0 S: 20 w % tert-Butanol	74Md	
0.0	6.01(0)	-27.5	25	C	μ=0 S: 25 w % tert-Butanol	74Md	
0.76	6.26	-26.0	25	C	μ=0 S: 30 w % tert-Butanol	74Md	
1.2	6.41(0)	-25.3	25	C	μ=0 S: 35 w % tert-Butanol	74Md	
1.0	6.62(0)	-26.8	25	C	μ=0 S: 40 w % tert-Butanol	74Md	
C33 1,1-CYCLOHEXANEDICARBOXYLIC ACID			$C_8H_{12}O_4$		$C_6H_{10}(CO_2H)_2$		
-1.045	3.270(0)	-18.46	25	C	0.1M($NaClO_4$, $HClO_4$)	75Ca	
0.079	5.683(-1)	-25.72	25	C	0.1M($NaClO_4$, $HClO_4$)	75Ca	
C34 1,2-CYCLOHEXANEDICARBOXYLIC ACID (cis)			$C_8H_{12}O_4$		$C_6H_{10}(CO_2H)$		
1.1	4.34(0)	-16	25	C	μ=0	63Ea	
-0.30	6.76(-1)	-31	25	C	μ=0	63Ea	
C35 1,2-CYCLOHEXANEDICARBOXYLIC ACID (trans)			$C_8H_{12}O_4$		$C_6H_{10}(CO_2H)_2$		
-1.9	4.18(0)	-25	25	C	μ=0	63Ea	
-0.24	5.93(-1)	-27	25	C	μ=0	63Ea	
C36 CYCLOOCTANE, 1-aza			$C_7H_{15}N$		$CH_2(CH_2)_5CH_2NH$ (ring)		
12.857	11.1(+1)	-7.7	25	C	μ=0	71Ca	
C37 CYCLOPENTANE, amino-			$C_5H_{11}N$		$C_5H_9(NH_2)$		
14.30	10.65(+1)	-0.8	25	C	μ=0	69Cc	
C38 CYCLOPENTANE, 1,2-diamino-N,N,N',N'-tetraacetic acid			$C_{13}H_{20}O_8N_2$		$C_5H_8[N(CH_2CO_2H)_2]_2$		
6.44	10.20(-3)	-24.7	20	C	μ=1M(KCl)	69Sc	
C39 1,2-CYCLOPENTANEDIONE, dioxime			$C_5H_8O_2N_2$		$C_5H_6(:NOH)_2$		
3.7	0.717(+1)	9.0	20-30	T	μ=1M HA=[Co(III)-(Cpdox·H)(Cpdox·H_2)(NO_2)-(NO_2H)]$^+$ where Cpdox·H_2 = 1,2-Cyclopentanedione-dioxime	72Vb	c,q: ±0.4
C40 CYCLOPENTANECARBOXYLIC ACID			$C_6H_{10}O_2$		$C_5H_9(CO_2H)$		
0.04	4.67(0)	-21.2	10	C	μ=0	72Cb	q:±0.01
-0.52	4.905(0)	-24.2	25	C	μ=0	72Cb	

ΔH, kcal /mole	pK	ΔS, cal /mole°K	T,°C	M	Conditions	Ref.	Remarks
C40, cont.							
-1.23	4.96(0)	-26.6	40	C	μ=0	72Cb	
-----	(0)	-----	10-40	C	μ=0 ΔCp=-42	72Cb	
C41 1,1-CYCLOPENTANEDICARBOXYLIC ACID			$C_7H_{10}O_4$		$C_5H_8(CO_2H)_2$		
-0.063	3.081(0)	-14.30	25	C	0.1M($NaClO_4$, $HClO_4$)	75Ca	
-0.902	5.769(-1)	-29.41	25	C	0.1M($NaClO_4$, $HClO_4$)	75Ca	
C42 CYCLOPROPANE, amino-			C_3H_7N		$C_3H_5(NH_2)$		
11.72	9.10(+1)	-2.3	25	C	μ=0	69Cc	
C43 CYCLOPROPANECARBOXYLIC ACID			$C_4H_6O_2$		$C_3H_5(CO_2H)$		
0.52	4.93(0)	-20.7	10	C	μ=0	72Cb	
-0.01	4.827(0)	-22.1	25	C	μ=0	72Cb	q:±0.01
-0.67	4.92(0)	-24.6	40	C	μ=0	72Cb	
-----	(0)	-----	10-40	C	μ=0 ΔCp=-40	72Cb	
C44 1,1-CYCLOPROPANEDICARBOXYLIC ACID			$C_5H_6O_4$		$C_3H_4(CO_2H)_2$		
-0.310	1.632(0)	-8.50	25	C	0.1M($NaClO_4$, $HClO_4$)	75Ca	
0.385	7.197(-1)	-31.62	25	C	0.1M($NaClO_4$, $HClO_4$)	75Ca	
C45 1,2-CYCLOPROPANEDICARBOXYLIC ACID, 3,3-dimethyl-(cis)-			$C_7H_{10}O_4$		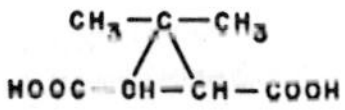		
-1	2.34(0)	-14	25	T	μ=0	63Ea	
-1	8.31(-1)	-41	25	T	μ=0	63Ea	
C46 1,2-CYCLOPROPANEDICARBOXYLIC ACID, 3,3-dimethyl-(trans)-			$C_7H_{10}O_4$		see C45		
-2	3.92(0)	-25	25	T	μ=0	63Ea	
-2	5.32(-1)	-31	25	T	μ=0	63Ea	
C47 CYSTEINE			$C_3H_7O_2NS$		$HSCH_2CH(NH_2)CO_2H$		
0.071	2.44(+1)	-10.9	25	C	μ=3.0M($NaClO_4$)	72Ga	e:CO_2H, q:±0.14
8.6	8.39(0)	-9.5	25	C	μ=0	64Wb	For a correction involving this article, see ref. 60Ka
8.4	8.37(0)	-10.1	25	T	μ=0	69Cd	e:SH
7.5	8.54(0)	-14	25	T	μ=0	69Ce	e:SH

ΔH, kcal /mole	pK	ΔS, cal /mole°K	T,°C	M	Conditions	Ref.	Re-marks

C47, cont.

ΔH, kcal /mole	pK	ΔS, cal /mole°K	T,°C	M	Conditions	Ref.	Re-marks
10.5	8.86(0)	-5	25	T	μ=0	69Ce	e:NH_3^+
8.0	7.98(45°,0)	-11.4	50	T	μ=0	69Cd	
10.2	8.28(0)	-6	50	T	μ=0	69Ce	e:NH_3^+
6.0	8.15(0)	-19	50	T	μ=0	69Ce	e:SH
7.6	7.52(0)	-12.5	75	T	μ=0	69Cd	
7.71	(0)	-----	20	C	μ≃0.01	71Ma	
9.27	8.784(0)	-9.1	25	C	μ=3.0M($NaClO_4$)	72Ga	e:NH_3^+
8.4	10.70(-1)	-20.8	25	T	μ=0	69Cd	e:NH_3^+
8.1	10.76(-1)	-21.9	25	C	μ=0	64Wb	
6.4	10.21(-1)	-25	25	T	μ=0	69Ce	e:SH
9.4	10.53(-1)	-17	25	T	μ=0	69Ce	e:NH_3^+
7.0	10.34(45°,-1)	-25.3	50	T	μ=0	69Cd	
8.5	10.03(-1)	-20	50	T	μ=0	69Ce	e:NH_3^+
4.3	9.89(-1)	-32	50	T	μ=0	69Ce	e:SH
5.8	9.93(-1)	-29.1	75	T	μ=0	69Cd	
8.58	(-1)	-----	20	C	μ≃0.01	71Ma	
9.66	10.709(-1)	-16.6	25	C	μ=3.0M($NaClO_4$)	72Ga	e:SH

C48 CYSTEINE(L), N-benzenesulphonyl $C_9H_{11}O_4NS$ $C_6H_5SO_2NHCH(CO_2H)CH_2SH$

ΔH, kcal /mole	pK	ΔS, cal /mole°K	T,°C	M	Conditions	Ref.	Re-marks
14.8	4.44(24°,0)	70.2	21-32	T	μ=0.50M S: 50 v % Dioxane	74Ga	
-9.8	11.15(24°,-1)	17.9	21-32	T	μ=0.50M S: 50 v % Dioxane	74Ga	

C49 CYSTEINE(L), S-methyl- $C_4H_9O_2NS$ $H_3CSCH_2CH(NH_2)COOH$

ΔH, kcal /mole	pK	ΔS, cal /mole°K	T,°C	M	Conditions	Ref.	Re-marks
10.1	8.97(0)	-7.2	25	C	μ=0	64Wb	

C50 CYSTEINE(L), S-methyl-, methyl ester $C_5H_{11}O_2NS$ $CH_3SCH_2CH(NH_2)CO_2CH_3$

ΔH, kcal /mole	pK	ΔS, cal /mole°K	T,°C	M	Conditions	Ref.	Re-marks
10.99	6.70(25°,0)	6.2	25-50	T	μ=0.1M	67Hb	c,e: NH_3^+

C51 CYSTINE, diacetyl ester- $C_{10}H_{16}O_6N_2S_2$ $[CH_3C(O)C(O)OCH(NH_2)CH_2S-]_2$

ΔH, kcal /mole	pK	ΔS, cal /mole°K	T,°C	M	Conditions	Ref.	Re-marks
5.800	8.53	-20	25	T	μ=0.03-0.15	55Ca	

C52 CYSTINE, N,N'-diformyl- $C_8H_{12}O_6N_2S_2$ $[HO_2CCH(NHCHO)CH_2S-]_2$

ΔH, kcal /mole	pK	ΔS, cal /mole°K	T,°C	M	Conditions	Ref.	Re-marks
6.300	9.50	-23	25	T	μ=0.03-0.15	55Ca	

C53 CYTIDINE $C_9H_{13}O_5N_3$

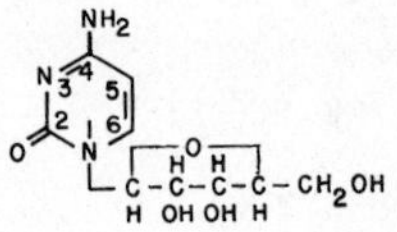

ΔH, kcal /mole	pK	ΔS, cal /mole°K	T,°C	M	Conditions	Ref.	Re-marks
5.31	4.29(+1)	-0.9	10	C	μ=0	70Cc	e:N_3
4.47	4.08(+1)	-3.9	25	C	μ=0	67Ia	
5.11	4.08(+1)	-1.5	25	C	μ=0	70Cc	e:N_3
3.7	4.54(+1)	-8.6	25	T	μ=0	66Lc	
4.83	3.92(+1)	-2.5	40	C	μ=0	70Cc	e:N_3
-----	(+1)	-----	10-40	C	μ=0 ΔCp=-16		
4.01	3.99(30°,+1)	-5.0	10-80	T	μ=0	70We	
4.88	4.07(20°,+1)	1.9	10-80	T	μ=0	70We	j:Mg^{2+} (10^{-2}M)

ΔH, kcal /mole	pK	ΔS, cal /mole°K	T, °C	M	Conditions	Ref.	Remarks
C53, cont.							
4.4	4.22(+1)	-5.0	25	T	μ=0.1	64Sd	
10.6	13.1(0)	-22.3	10	C	μ=0	70Cc	e: Ribose OH
10.7	12.24(0)	-20.2	25	C	μ=0	67Ia	e: Ribose OH
10.3	12.5(0)	-22.9	25	C	μ=0	70Cc	e: Ribose OH
10.2	12.0(0)	-22.5	40	C	μ=0	70Cc	e: Ribose OH
C54 CYTIDINE, 2'-deoxy-			$C_9H_{13}O_4N_3$		see C53		
4.300	4.3	-5.3	25	C	μ=0.1	60Ra	
C55 CYTIDINE-5'-diphosphoric acid			$C_9H_{15}O_{11}N_3P_2$		see C53		
5.5	4.46(25°)	-1.96	5-35	T	μ=0.1M(KNO_3)	73Ba	c
-1.34	7.18(-2)	-37.4	25	T	μ=0	65Pe	
1.2	6.44(25°,-2)	-25.4	5-35	T	μ=0.1M(KNO_3)	73Ba	c
C56 CYTIDINE-5'-monophosphoric acid			$C_9H_{14}O_8N_3P$		see C53		
3.64	4.39(30°)	-8.0	10-80	T	μ=0	70We	c
4.29	4.53(20°)	2.9	10-80	T	μ=0	70We	c,j: Mg^{2+} (10^{-2}M)
-1.35	6.62(-1)	-34.8	25	T	μ=0	65Pe	
C57 CYTIDINE, 5'-monophosphoric acid-2'-deoxy-			$C_9H_{14}O_7N_3P$		see C53		
4.280	4.4	-5.8	25	C	μ=0.1	60Ra	
C58 CYTIDINE, 5'-triphosphoric acid-2'-deoxy-			$C_9H_{14}O_7N_3P$		see C53		
4.280	4.4	-5.8	25	C	μ=0.1	60Ra	
C59 CYTIDYLIC ACID			$C_9H_{14}O_8N_3P$		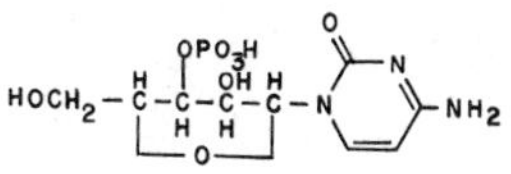		
5.2	3.75(66°)	-1.7	25-75	T	μ=0.1M(Sodium Phosphate)	66Ce	c

C60 CYTOSINE $C_4H_5ON_3$

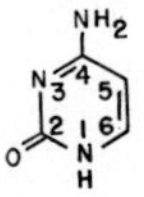

ΔH, kcal /mole	pK	ΔS, cal /mole°K	T, °C	M	Conditions	Ref.	Remarks
5.25	4.79(+1)	-3.4	10	C	μ=0	70Cc	e:N_3 nitrogen
5.14	4.58(+1)	-3.7	25	C	μ=0	67Ia	e:N_3 nitrogen
4.98	4.42(+1)	-4.3	40	C	μ=0	70Cc	e:N_3 nitrogen
-----	(+1)	-----	10-40	-	μ=0 ΔCp=-9	70Cc	
5.0	4.5(+1)	-3.7	25	T	μ=0.1	64Sd	
4.470	4.5(+1)	-----	25	C	μ=0.1	60Ra	
3.85	4.91(+1)	7.1	30	T	μ=0.20	66Lc	
12.05	12.62(0)	-15.2	10	C	μ=0	70Cc	e: N_1-C_2O group
11.5	12.15(0)	-17.0	25	C	μ=0	67Ia	e: N_1-C_2O group
11.07	11.68(0)	-18.1	40	C	μ=0	70Cc	e: N_1-C_2O group
-----	(0)	-----	10-40	-	μ=0 ΔCp=-33		
11.0	11.82(0)	-17.1	25	T	μ=0.1	64Sd	
7.24	12.3(0)	-7	30	T	μ=0.20	66Lc	

ΔH, kcal /mole	pK	ΔS, cal /mole°K	T, °C	M	Conditions	Ref.	Remarks

D

D1 DECANE, 4,7-diaza-1,10-diamino- $C_8H_{22}N_4$ $H_2N(CH_2)_3NH(CH_2)_2NH(CH_2)_3NH_2$

ΔH, kcal /mole	pK	ΔS, cal /mole°K	T, °C	M	Conditions	Ref.	Remarks
8.16	5.59(+4)	1.8	25	C	μ=0.1M(NaCl)	73Hb	
9.77	5.84(+4)	6.1	25	C	μ=0.5M(KNO_3)	73Bb	
10.32	8.30(+3)	-3.37	25	C	μ=0.1M(NaCl)	73Hb	
10.71	8.53(+3)	-3.1	25	C	μ=0.5M(KNO_3)	73Bb	
12.38	9.77(+2)	-3.18	25	C	μ=0.1M(NaCl)	73Hb	
12.79	9.95(+2)	-2.6	25	C	μ=0.5M(KNO_3)	73Bb	
12.35	10.53(+1)	-6.76	25	C	μ=0.1M(NaCl)	73Hb	
12.23	10.66(+1)	-7.8	25	C	μ=0.5M(KNO_3)	73Bb	

D2 DECANE, 1,1-dinitro- $C_{10}H_{20}O_4N_2$ $C_9H_{19}CH(NO_2)_2$

ΔH, kcal /mole	pK	ΔS, cal /mole°K	T, °C	M	Conditions	Ref.	Remarks
2.50	3.597(0)	-16.46	5-60	T	--	62Sa	c

D3 DECANEDIOIC ACID, 5,6-dithio- $C_8H_{14}O_4S_2$ $[S(CH_2)_3CO_2H]_2$

ΔH, kcal /mole	pK	ΔS, cal /mole°K	T, °C	M	Conditions	Ref.	Remarks
6.300	10.35	-26	25	T	μ=0.03-0.15	55Ca	

D4 DIETHYLENETRIAMINE $C_4H_{13}N_3$ $(NH_2CH_2CH_2)_2NH$

ΔH, kcal /mole	pK	ΔS, cal /mole°K	T, °C	M	Conditions	Ref.	Remarks
7.20	4.25(+3)	4.7	25	C	0.1M(KCl)	61Ca	
8.2	4.59(+3)	5.2	30&40	T	0.1M($BaCl_2$) 0.1M(HNO_3)	50Ja	b,c
11.95	8.98(+2)	-1.0	25	C	0.1M(KCl)	61Ca	
12.6	8.94(+2)	-0.7	30&40	T	0.1M($BaCl_2$) 0.1M(HNO_3)	50Ja	b,c
11.20	9.78(+1)	-7.2	25	C	0.1M(KCl)	61Ca	
11.7	9.68(+1)	-6.9	30&40	T	0.1M($BaCl_2$) 0.1M(HNO_3)	50Ja	b,c

D5 DIMETHYL SULFATE $C_2H_6O_4S$ $(CH_3)_2SO_4$

ΔH, kcal /mole	pK	ΔS, cal /mole°K	T, °C	M	Conditions	Ref.	Remarks
5.3	(+1)	-----	25	C	C∼10^{-3}M S: Fluorosulfuric acid	74Ab	

D6 1,4-DIOXANE $C_4H_8O_2$ $OCH_2CH_2OCH_2CH_2$ (ring)

ΔH, kcal /mole	pK	ΔS, cal /mole°K	T, °C	M	Conditions	Ref.	Remarks
21.5	-----	-----	25	C	C∼10^{-3}M S: Fluorosulfuric acid	74Ab	

D7 1,3-DIOXOLAN-2-ONE, 4-methyl- $C_4H_6O_3$

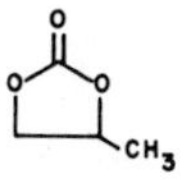

ΔH, kcal /mole	pK	ΔS, cal /mole°K	T, °C	M	Conditions	Ref.	Remarks
17.8	-----	-----	25	C	C∼10^{-3}M S: Fluorosulfuric acid	74Ab	

D8 DIPROPYLENETRIAMINE $C_6H_{17}N_3$ $(NH_2CH_2CH_2CH_2)_2NH$

ΔH, kcal /mole	pK	ΔS, cal /mole°K	T, °C	M	Conditions	Ref.	Remarks
10.47	7.72(+3)	-0.2	25	C	0.1M(KCl)	66Pd	
12.99	9.56(+2)	-0.2	25	C	0.1M(KCl)	66Pd	
12.26	10.65(+1)	-7.5	25	C	0.1M(KCl)	66Pd	

ΔH, kcal /mole	pK	ΔS, cal /mole°K	T, °C	M	Conditions	Ref.	Remarks
D9 DIPROPYLENETRIAMINE, 5-methyl-			$C_7H_{19}N_3$		$(NH_2CH_2CH_2CH_2)_2NCH_3$		
9.96	6.32(30°,+3)	4	10-40	T	μ=0	59Gb	c
12.0	9.19(30°,+2)	-2	10-40	T	μ=0	59Gb	c
11.0	10.33(30°,+1)	-11	10-40	T	μ=0	59Gb	c
D10 DJENKOLIC ACID			$C_7H_{14}O_4N_2S_2$		$CH_2[SCH_2CH(NH_2)CO_2H]_2$		
1.205	2.200(+1)	-6.02	25	T	μ=0.1M(KCl)	72Bg	e:CO_2H
10.616	8.158(0)	-1.71	25	T	μ=0.1M(KCl)	72Bg	e:NH_3^+
12.835	8.954(0)	2.09	25	T	μ=0.1M(KCl)	72Bg	e:NH_3^+
D11 DODECABORANE, 1,12-dicarboxy-			$C_2H_{12}O_4B_{12}$		$[1,12\text{-}B_{12}H_{10}(CO_2H)_2]$		
2.15	9.07(0)	-34.4	25	C	μ=0	66Hb	
2.30	10.23(-1)	-39.1	25	C	μ=0	66Hb	
D12 DODECANE, 1,12-diamino, 4,9-diaza-			$C_{10}H_{26}N_4$		$H_2N(CH_2)_3NH(CH_2)_4NH(CH_2)_3NH_2$		
11.4	7.96(+4)	2.0	25	C	μ=0.10M	74Pa	
12.4	8.85(+3)	1.2	25	C	μ=0.10M	74Pa	
12.4	10.02(+2)	-4.1	25	C	μ=0.10M	74Pa	
13.1	10.80(+1)	-5.2	25	C	μ=0.10M	74Pa	
D13 DODECANE, 5,8-diaza-2,11-diamino-			$C_{14}H_{34}N_4$		$[CH_3CH(NH_2)CH_2C(CH_3)_2NHCH_2]_2$		
7.95	4.90(+4)	4.23	25	C	μ=0.10M(NaCl)	73Hc	
11.08	7.61(+3)	2.34	25	C	μ=0.10M(NaCl)	73Hc	
12.67	10.06(+2)	-3.49	25	C	μ=0.10M(NaCl)	73Hc	
12.24	11.09(+1)	-9.68	25	C	μ=0.10M(NaCl)	73Hc	
D14 DODECANE, 5,8-diaza-5,8-hydroxyimino-4,4,9,9-tetramethyl-			$C_{14}H_{30}O_2N_4$		$CH_3C(NOH)CH_2C(CH_3)_2NHCH_2CH_2\text{-}NHC(CH_3)_2CH_2C(NOH)CH_3$		
9.42	5.89(+2)	4.5	25	C	μ=0	74Ha	
9.46	9.35(+1)	-11	25	C	μ=0	74Ha	

ΔH, kcal /mole	pK	ΔS, cal /mole°K	T, °C	M	Conditions	Ref.	Remarks

E

ΔH, kcal /mole	pK	ΔS, cal /mole°K	T, °C	M	Conditions	Ref.	Remarks
E1 EPHEDRINE(-) ENANTIOMORPH			$C_{10}H_{15}ON$		$C_6H_5CH(OH)CH(CH_3)NHCH_3$		
10.837	9.39(+1)	-6.57	25	C	μ=0	68Ra	
E2 PSEUDO-EPHEDRINE			$C_{10}H_{15}ON$		see E1		
11.037	9.53(+1)	-6.57	25	C	μ=0	68Ra	
E3 ETHANE, amino-			C_2H_7N		$H_2NCH_2CH_3$		
13.71	10.63(+1)	-2.7	25	C	μ=0	69Cc	
11.15	10.67(+1)	-11.4	25	T	μ=0	55Fa	
13.58	10.158(40°,+1)	-2.4	20-40	T	μ=0	51Ea	c
13.60	10.153(40°,+1)	-2.9	20-40	T	μ=0.05	51Ea	c
13.63	10.150(40°,+1)	-3.0	20-40	T	μ=0.10	51Ea	c
13.76	10.134(40°,+1)	-3.1	20-40	T	μ=0.20	51Ea	c
9	10.8(25°,+1)	19	0-25	T	μ=3	55Ma	c
16.16	12.15(+1)	-1.4	25	T	μ=0 S: Ethylene Glycol ΔCp=-2	70Kc	
E4 ETHANE, 1-amino-2-methoxy-			C_3H_9ON		$CH_3OCH_2CH_2NH_2$		
11.7	9.28(30°,+1)	-4	10-40	T	μ=0	59La	c
E5 ETHANE, 1-amino-2(methylthio)-			C_3H_9NS		$H_2NCH_2CH_2SCH_3$		
12.3	9.18(30°,+1)	1	10-40	T	μ=0	59Ma	c
E6 ETHANE, 1,2-diamino-			$C_2H_8N_2$		$H_2NCH_2CH_2NH_2$		
11.03	7.13(+2)	4.4	25	C	μ=0	66Pe	
11.08	6.859(+2)	5.8	25	C	μ=0	66Pg	
10.83	(+2)	-----	25	C	μ=0	61Ta	e:NH_3^+
10.870	6.848(+2)	5.1	25	T	μ=0	52Eb	
10.3	6.79(+2)	-3	10-40	T	μ=0	50Ba 59Ma	c
10.940	(+2)	-----	25	C	0.1M(KCl)	54Db	
10.89	7.32(+2)	3.0	25	C	0.3M($NaClO_4$)	67Hc	
11.1	7.24(25°,+2)	4.0	15-40	T	μ=0.3M($NaClO_4$)	71Mb	
10.90	7.28(+2)	3.3	25	C	μ=0.5M(KNO_3)	66Va 67Va	
10.60	7.44(+2)	1.5	25	C	1M(KCl)	61Ca	
11.33	7.536(+2)	3.5	25	C	μ=1M(KNO_3)	72Cd	
11.94	9.910(+1)	-5.3	25	C	μ=0	66Pe	
11.91	9.922(+1)	-5.5	25	C	μ=0	66Pg	
12.38	(+1)	-----	25	C	μ=0	61Ta	e:NH_3^+
11.820	9.928(+1)	-5.8	25	T	μ=0	52Eb	
11.5	9.81(+1)	7	10-40	T	μ=0	50Ba	c
11.910	(+1)	-----	25	C	0.1M(KCl)	54Db	
12.00	9.90(+1)	-5.0	25	C	0.3M($NaClO_4$)	67Hc	
12.1	10.03(25°,+1)	-5.4	15-40	T	μ=0.3M($NaClO_4$)	71Mb	c
12.18	9.98(+1)	-4.8	25	C	μ=0.5M(KNO_3)	66Va 67Va	
12.20	10.19(+1)	-5.7	25	C	1M(KCl)	61Ca	
12.18	10.225(+1)	-5.9	25	C	μ=1M(KNO_3)	72Cd	
3.50	3.26(+2)	-3.18	25.3	C	μ=0.02 HA=[Re(1,2-Diaminoethane)$_2$(O)(OH)]$^{2+}$	63Mb	

ΔH, kcal /mole	pK	ΔS, cal /mole°K	T,°C	M	Conditions	Ref.	Remarks
E6, cont.							
-3.7	0.064(30°,+2)	-11	26-60	T	μ=2.0M HA=trans $[Co(III)(1,2\text{-Diaminoethane})_2(CH_3CO_2H)(CH_3CO_2)]^{2+}$	72Da	
-1.7	0.813(30°,+2)	-9	30-50	T	μ=2.0M HA=cis $[Co(III)(1,2\text{-Diaminoethane})_2(CH_3CO_2H)(CH_3CO_2)]^{2+}$	72Da	
5.5	-0.05(25°,+4)	18	15-25	T	μ=0 HA=$[(Co(III))_2(1,2\text{-Diaminoethane})_4(NH_2)(H_2O)]^{4+}$	64Ed	c

E7 ETHANE, 1,2-diamino-N,N'-diacetic acid — $C_6H_{12}O_4N_2$ — $HO_2CCH_2NHCH_2CH_2NHCH_2CO_2H$

ΔH, kcal /mole	pK	ΔS, cal /mole°K	T,°C	M	Conditions	Ref.	Remarks
7.31	6.550(0)	-5.45	25	C	μ=0.10M(KNO_3)	70Da	
6.4	6.59(25°,0)	-9	20-40	T	μ=0.1M(KNO_3)	69Sb	
8.89	6.716(0)	-0.72	25	C	μ=1.00($NaClO_4$)	74Gb	
7.45	9.62(-1)	-19.0	25	C	μ=0.10M(KNO_3)	70Da	
6.7	9.58(25°,-1)	-22	20-40	T	μ=0.1M(KNO_3)	69Sb	c
8.91	9.695(-1)	-14.6	25	C	μ=1.00($NaClO_4$)	74Gb	

E8 ETHANE, 1,2-diamino-N,N'-di(2-aminoethyl)- — $C_6H_{18}N_4$ — $H_2NCH_2CH_2NHCH_2CH_2NHCH_2CH_2NH_2$

ΔH, kcal /mole	pK	ΔS, cal /mole°K	T,°C	M	Conditions	Ref.	Remarks
6.83	3.24(+4)	8.1	25	C	0.1M(KCl)	63Pb	
5.6	3.76(+4)	0.7	40	T	0.1F($BaCl_2$) 0.1F(HNO_3)	50Ja	
9.53	6.54(+3)	2.0	25	C	0.1M(KCl)	63Pb	
9.6	6.79(+3)	-0.4	40	T	0.1F($BaCl_2$) 0.1F(HNO_3)	50Ja	
11.27	9.06(+2)	-3.7	25	C	0.1M(KCl)	63Pb	
9.6	9.14(+2)	-11.2	40	T	0.1F($BaCl_2$) 0.1F(HNO_3)	50Ja	
11.01	9.75(+1)	-7.8	25	C	0.1M(KCl)	63Pb	
10.0	9.76(+1)	-12.7	40	T	0.1F($BaCl_2$) 0.1F(HNO_3)	50Ja	

E9 ETHANE, 1,2-diamino-N,N'-di-n-butyl- — $C_{10}H_{24}N_2$ — $CH_3CH_2CH_2CH_2NHCH_2CH_2NHCH_2CH_2CH_2CH_3$

ΔH, kcal /mole	pK	ΔS, cal /mole°K	T,°C	M	Conditions	Ref.	Remarks
10.7	7.46(25°,+2)	1.74	0&25	T	0.5M(KNO_3)	53Ba	b,c
11.0	10.19(25°,+1)	-9.73	0&25	T	0.5M(KNO_3)	53Ba	b,c

E10 ETHANE, 1,2-diamino-N,N-diethyl- — $C_6H_{16}N_2$ — $H_2NCH_2CH_2N(CH_2CH_3)_2$

ΔH, kcal /mole	pK	ΔS, cal /mole°K	T,°C	M	Conditions	Ref.	Remarks
9.4	6.92(25°,+2)	-0.1	15-40	T	μ=0.3M($NaClO_4$)	72Ma	c
8.9	9.98(25°,+1)	-15.7	15-40	T	μ=0.3M($NaClO_4$)	72Ma	c

E11 ETHANE, 1,2-diamino-N,N'-diethyl- — $C_6H_{16}N_2$ — $CH_3CH_2NHCH_2CH_2NHCH_2CH_3$

ΔH, kcal /mole	pK	ΔS, cal /mole°K	T,°C	M	Conditions	Ref.	Remarks
12.4	7.70(25°,+2)	6.37	0&25	T	0.5M(KNO_3)	53Ba	b,c
8.9	10.46(25°,+1)	-18.0	0&25	T	0.5M(KNO_3)	53Ba	b,c

E12 ETHANE, 1,2-diamino-N,N-dimethyl- — $C_4H_{12}N_2$ — $NH_2CH_2CH_2N(CH_3)_2$

ΔH, kcal /mole	pK	ΔS, cal /mole°K	T,°C	M	Conditions	Ref.	Remarks
E12, cont.							
8.42	6.694(+2)	-2.4	25	C	μ=0.5M(KNO_3)	71Pa	
8.59	6.925(+2)	-2.8	25	C	μ=1M(KNO_3)	72Cd	
10.42	9.686(+1)	-9.3	25	C	μ=0.5M(KNO_3)	71Pa	
10.19	9.834(+1)	-10.2	25	C	μ=1M(KNO_3)	72Cd	

E13 ETHANE, 1,2-diamino-N,N'-dimethyl- $C_4H_{12}N_2$ $CH_3NHCH_2CH_2NHCH_3$

ΔH, kcal /mole	pK	ΔS, cal /mole°K	T,°C	M	Conditions	Ref.	Remarks
12.32	7.47(+2)	7.17	25	T	μ=0	68Hb	
9.67	7.303(+2)	-0.9	25	C	μ=0.5M(KNO_3)	71Pa	
12.4	7.47(25°,+2)	7.41	0&25	T	0.5M(KNO_3)	53Ba	b,c
10.17	7.525(+2)	-0.3	25	C	μ=1M(KNO_3)	72Cd	
10.74	10.166(+1)	-10.5	25	C	μ=0.5M(KNO_3)	71Pa	
8.9	10.29(25°,+1)	-17.24	0&25	T	0.5M(KNO)	53Ba	b,c
10.51	10.258(+1)	-11.7	25	C	μ=1M(KNO_3)	72Cd	

E14 ETHANE, 1,2-diamino-N,N'-dimethyl-N,N'-diacetic acid- $C_8H_{16}O_4N_2$ $HO_2CCH_2N(CH_3)CH_2CH_2N(CH_3)CH_2$-$CO_2H$

ΔH, kcal /mole	pK	ΔS, cal /mole°K	T,°C	M	Conditions	Ref.	Remarks
3.68	6.294(0)	-15.3	0	T	μ=0	56Oa	
3.68	6.169(0)	-15.2	10	T	μ=0	56Oa	
3.69	6.047(0)	-15.1	20	T	μ=0	56Oa	
3.70	5.926(0)	-14.9	30	T	μ=0	56Oa	
3.72	5.803(0)	-14.7	40	T	μ=0	56Oa	
6.78	10.446(-1)	-23.1	0	T	μ=0	56Oa	
6.82	10.268(-1)	-22.9	10	T	μ=0	56Oa	
6.83	10.068(-1)	-22.8	20	T	μ=0	56Oa	
6.85	9.882(-1)	-22.6	30	T	μ=0	56Oa	
6.88	9.684(-1)	-22.3	40	T	μ=0	56Oa	

E15 ETHANE, 1,2-diamino-1,2-diphenyl-(meso)- $C_{14}H_{16}N_2$ $C_6H_5CH(NH_2)CH(NH_2)C_6H_5$

ΔH, kcal /mole	pK	ΔS, cal /mole°K	T,°C	M	Conditions	Ref.	Remarks
11.6	4.78(25°,+2)	17.10	0&25	T	0.005M[$Ba(ClO_4)_2$] S: 50 w % Dioxane	53Ba	b,c
11.0	7.85(25°,+1)	1.01	0&25	T	0.005M[$Ba(ClO_4)_2$] S: 50 w % Dioxane	53Ba	b,c

E16 ETHANE, 1,2-diamino-1,2-diphenyl-(racemic)- $C_{14}H_{16}N_2$ $C_6H_5CH(NH_2)CH(NH_2)C_6H_5$

ΔH, kcal /mole	pK	ΔS, cal /mole°K	T,°C	M	Conditions	Ref.	Remarks
9.7	3.95(25°,+2)	14.46	0&25	T	0.005M[$Ba(ClO_4)_2$] S: 50 w % Dioxane	53Ba	b,c
11.3	8.09(25°,+1)	0.872	0&25	T	0.005M[$Ba(ClO_4)_2$] S: 50 w % Dioxane	53Ba	b,c

E17 ETHANE, 1,2-diamino-N,N'-di-n-propyl- $C_8H_{20}N_2$ $CH_3CH_2CH_2NHCH_2CH_2NHCH_2CH_2CH_3$

ΔH, kcal /mole	pK	ΔS, cal /mole°K	T,°C	M	Conditions	Ref.	Remarks
9.1	7.53(25°,+2)	-4.02	0&25	T	0.5M(KNO_3)	53Ba	b,c
10.4	10.27(25°,+1)	-12.07	0&25	T	0.5M(KNO_3)	53Ba	b,c

E18 ETHANE, 1,2-diamino-N,N'-di-i-propyl- $C_8H_{20}N_2$ $CH_3CH(CH_3)NHCH_2CH_2NHCH(CH_3)CH_3$

ΔH, kcal /mole	pK	ΔS, cal /mole°K	T,°C	M	Conditions	Ref.	Remarks
10.0	7.59(25°,+2)	-1.21	0&25	T	0.5M(KNO_3)	53Ba	b,c
10.7	10.40(25°,+1)	-11.74	0&25	T	0.5M(KNO_3)	53Ba	b,c

ΔH, kcal /mole	pK	ΔS, cal /mole°K	T,°C	M	Conditions	Ref.	Remarks
E19 ETHANE, 1,2-diamino-N-(2-hydroxyethyl)-			$C_4H_{12}ON_2$		$H_2NCH_2CH_2NH(CH_2CH_2OH)$		
10.07	6.84(+2)	2.5	25	C	μ=0.5M($NaClO_4$)	75Ba	
11.13	9.74(+1)	-7.2	25	C	μ=0.5M($NaClO_4$)	75Ba	
E20 ETHANE, 1,2-diamino-N-hydroxy-N,N',N'-triacetic acid			$C_8H_{14}O_7N_2$		$(HO_2CCH_2)N(OH)CH_2CH_2N(CH_2CO_2H)_2$		
-1.12	2.63	-15.8	25	C	μ=0.1(KCl)	68Fa	
3.09	5.37	-14.2	25	C	μ=0.1(KCl)	68Fa	
6.65	9.80	-22.5	25	C	μ=0.1(KCl)	68Fa	
E21 ETHANE, 1,2-diamino-N-hydroxyethyl-N,N,N'-triacetic acid			$C_{10}H_{18}O_7N_2$		$(CH_2CO_2H)_2NCH_2CH_2N(CH_2CH_2OH)$-$CH_2CO_2H$		
-0.34	2.39(25°)	-12.1	15-40	T	μ=0.1(KNO_3)	61Ma	c
2.83	5.37(25°)	-15.1	15-40	T	μ=0.1(KNO_3)	61Ma	c
7.29	9.93(25°)	-21.0	15-40	T	μ=0.1(KNO_3)	61Ma	c
E22 ETHANE, 1,2-diamino-N-methyl-			$C_3H_{10}N_2$		$NH_2CH_2CH_2NHCH_3$		
10.1	6.63(30°,+2)	-3	10-40	T	μ=0	59Ma	c
10.34	7.26(+2)	1.5	25	C	μ=0.5M(KNO_3)	66Va 67Va	
10.85	7.473(+2)	2.2	25	C	μ=1M(KNO_3)	72Cd	
11.1	9.90(30°,+1)	9	0-40	T	μ=0	59Ma	c
11.25	10.14(+1)	-8.7	25	C	μ=0.5M(KNO_3)	66Va 67Va	
11.22	10.280(+1)	-9.4	25	C	μ=1M(KNO_3)	72Cd	
E23 ETHANE, 1,2-diamino-1-phenyl-			$C_8H_{12}N_2$		$H_2NCH_2CH(C_6H_5)NH_2$		
10.62	5.757(+2)	9.3	25	C	μ=0	66Pg	
10.4	6.26(25°,+2)	6.3	15-40	T	μ=0.3M($NaClO_4$)	72Ma	c
11.10	8.55(+1)	-1.9	25	C	μ=0	66Pg	
10.7	9.22(25°,+1)	-6.1	15-40	T	μ=0.3M($NaClO_4$)	72Ma	c
E24 ETHANE, 1,2-diamino-N,N,N',N'-tetraacetic acid			$C_{10}H_{16}O_8N_2$		$(CH_2CO_2H)_2NCH_2CH_2N(CH_2CO_2H)_2$		
-0.490	1.55(+1)	-8.8	20	C	μ=0.1	58Ta	
-0.364	2.21(0)	-11.34	25	C	μ=0	73Vb	
-0.180	1.99(0)	-9.8	20	C	μ=0.1	58Ta	
-0.180	(0)	-----	25	C	μ=0.44	65Ka 62Ya	
-1.486	3.11(-1)	-19.22	C	25	μ=0	73Vb	
-1.430	2.67(-1)	-17.1	20	C	μ=0.1	58Ta	
-1.43	(-1)	-----	25	C	μ=0.44	65Ka 62Ya	
4.030	6.935(-2)	-18.25	25	C	μ=0	74Vb	
3.69	6.236(-2)	-16.4	30	T	μ=0	56Oa 53Ca	
3.61	6.273(-2)	-16.7	25	C	μ=0.03-0.10	69Cc	
4.34	6.16(-2)	-13.4	20	C	0.1(KNO_3)	63Ab	
4.390	6.16(-2)	-13.2	20	C	μ=0.1	58Ta	
4.35	6.16(-2)	-13.6	25	C	μ=0.1(KNO_3)	70Kb	
4.76	6.17(-2)	-12.3	25	T	μ=0.1(KNO_2)	66Mc	
4.4	6.16(25°,-2)	-14	20-40	T	μ=0.1M(KNO_3)	69Sb	c
4.340	(-2)	-----	25	C	μ=0.2($LiNO_3$)	74Vb	
4.750	(-2)	-----	25	C	μ=0.2(KNO_3)	74Vb	
4.390	(-2)	-----	25	C	μ=0.2(NaCl)	74Vb	

ΔH, kcal /mole	pK	ΔS, cal /mole°K	T, °C	M	Conditions	Ref.	Remarks
E24, cont.							
4.600	(-2)	-----	25	C	μ=0.2($NaNO_3$)	74Vb	
4.460	(-2)	-----	25	C	μ=0.2($NaClO_4$)	74Vb	
4.560	(-2)	-----	25	C	μ=0.5($LiNO_3$)	74Vb	
5.310	(-2)	-----	25	C	μ=0.5(KNO_3)	74Vb	
4.930	(-2)	-----	25	C	μ=0.5(NaCl)	74Vb	
5.070	(-2)	-----	25	C	μ=0.5($NaNO_3$)	74Vb	
5.060	(-2)	-----	25	C	μ=0.5($NaClO_4$)	74Vb	
5.275	(-2)	-----	25	C	μ=1.0($LiNO_3$)	74Vb	
5.760	(-2)	-----	25	C	μ=1.0(KNO_3)	74Vb	
5.620	(-2)	-----	25	C	μ=1.0(NaCl)	74Vb	
5.730	(-2)	-----	25	C	μ=1.0($NaNO_3$)	74Vb	
5.770	(-2)	-----	25	C	μ=1.0($NaClO_4$)	74Vb	
5.070	11.245(-3)	-35.05	25	C	μ=0	74Vb	
5.34	10.883(-3)	-32.2	30	T	μ=0	56Oa 53Ca	
4.35	10.948(-3)	-35.6	25	C	μ=0.04-0.17	69Cc	
5.690	10.26(-3)	-27.5	20	C	μ=0.1	58Ta	
5.67	10.26(-3)	-27.6	20	C	0.1(KNO_3)	63Ab	
5.59	10.26(-3)	-28.2	25	C	μ=0.1(KNO_3)	70Kb	
7.86	10.21(-3)	-20.3	25	T	μ=0.1(KNO_3)	66Mc	
4.9	10.20(25°,-3)	-30	20-40	T	μ=0.1M(KNO_3)	69Sb	c
5.603	(-3)	-----	25	C	μ=0.2(KNO_3)	74Vb	
5.875	(-3)	-----	25	C	μ=0.5(KNO_3)	74Vb	
6.327	(-3)	-----	25	C	μ=1.0(KNO_3)	74Vb	
10.0	-----	-----	25	C	μ=0.45	62Ya	
					R: $H_2A = A + 2H^+$		
1.2	3.08(-1)	-10	25	C	μ=0.10M(KNO_3) HA=[Mn(II)(EDTA)]$^-$ EDTA=1,2-diamino-N,N,N', N'-tetraacetic acid	69Bg	q:±0.3
1.9	3.0(-1)	-7	25	C	μ=0.1M(KNO_3) HA=[Co(II)(EDTA)]$^-$ EDTA=1,2-diamino-N,N,N', N'-tetraacetic acid	69Bg	q:±0.3
1.8	3.22(-1)	-9	25	C	μ=0.10M(KNO_3) HA=[Ni(II)(EDTA)]$^-$ EDTA=1,2-diamino-N,N,N'- N'-tetraacetic acid	69Bg	q:±0.3
2.0	3.00(-1)	-7	25	C	μ=0.10M(KNO_3) HA=[Ni(II)(EDTA)]$^-$ EDTA=1,2-diamino-N,N,N', N'-tetraacetic acid	69Bg	q:±0.3
2.2	3.00(-1)	-6	25	C	μ=0.10M(KNO_3) HA=[Zn(II)(EDTA)]$^-$ EDTA=1,2-diamino-N,N,N', N'-tetraacetic acid	69Bg	q:±0.3
0.4	2.93(-1)	-12	25	C	μ=0.10M(KNO_3) HA=[Cd(II)(EDTA)]$^-$ EDTA=1,2-diamino-N,N,N', N'-tetraacetic acid	69Bg	q:±0.3
0.7	3.08(-1)	-12	25	C	μ=0.10M(KNO_3) HA=[Hg(II)(EDTA)]$^-$ EDTA=1,2-diamino-N,N,N', N'-tetraacetic acid	69Bg	q:±0.3
3.9	2.96(0)	0	20	C	μ=0.1 HA=[Al(H_2O)(EDTA)] EDTA=1,2-diamino-N,N,N', N'-tetraacetic acid	59Ra	
5.0	6.12(-1)	-10.9	20	C	μ=0.1 HA=[Al(H_2O)(EDTA)]$^{-1}$ EDTA=1,2-diamino-N,N,N', N'-tetraacetic acid	59Ra	

ΔH, kcal /mole	pK	ΔS, cal /mole°K	T,°C	M	Conditions	Ref.	Remarks
E25 ETHANE, 1,2-diamino-N,N,N',N'-tetra(2-aminoethyl)			$C_{10}H_{28}N_6$		$(NH_2CH_2CH_2)_2NCH_2CH_2N(CH_2CH_2NH_2)_2$		
4.50	1.33(+5)	9.0	25	C	μ=0.1(KNO_3)	71Pc 63Pa	
11.30	(+4)	-----	25	C	μ=0.1(KCl)	64Sa	
12.00	8.44(+4)	1.8	25	C	μ=0.1(KNO_3)	71Pc 64Sa	
13.15	9.00(+3)	3.1	25	C	μ=0.1(KNO_3)	71Pc 64Sa	
11.45	9.60(+2)	-5.3	25	C	μ=0.1(KNO_3)	71Pc 64Sa	
11.30	10.10(+1)	-8.1	25	C	μ=0.1(KNO_3)	71Pc 64Sa	
22.75	-----	-----	25	C	μ=0.1(KCl) R: $H_4A^{4+} = H_2A^{2+} + 2H^+$	64Sa	
35.90	-----	-----	25	C	μ=0.1(KCl) R: $H_4A^{4+} = HA^+ + 3H^+$	64Sa	
47.00	-----	-----	25	C	μ=0.1(KCl) R: $H_4A^{4+} = A + 4H^+$	64Sa	
10.00	6.62(+3)	3.3	25	C	μ=0.1(KNO_3) HA=[Ni(II)(Penten)]$^{3+}$ Penten=1,2-diamino-N,N,N',N'-tetra(2-aminoethyl) ethane	71Pc	
11.60	8.12(+3)	1.8	25	C	μ=0.1(KNO_3) HA=[Cu(II)(Penten)]$^{3+}$ Penten=1,2-diamino-N,N,N',N'-tetra(2-aminoethyl) ethane	71Pc	
11.45	8.00(+3)	1.9	25	C	μ=0.1(KNO_3) HA=[Zn(II)(Penten)]$^{3+}$ Penten=1,2-diamino-N,N,N',N'-tetra(2-aminoethyl) ethane	71Pc	
E26 ETHANE, 1,2-diamino-N,N,N',N'-tetra(2-hydroxyethyl)			$C_{10}H_{24}O_4N_2$		$(HOCH_2CH_2)_2NCH_2CH_2N(CH_2CH_2OH)_2$		
6.5	4.55(25°,+2)	1	10-55	T	μ=0.5M($NaClO_4$)	69Pc	c
7.8	8.54(25°,+1)	-13	10-55	T	μ=0.5M($NaClO_4$)	69Pc	c
E27 ETHANE, 1,2-diamino-N,N,N',N'-tetramethyl-			$C_6H_{16}N_2$		$(CH_3)_2NCH_2CH_2N(CH_3)_2$		
6.64	6.130(+2)	-5.8	25	C	μ=0.5M(KNO_3)	71Pa	
7.15	6.357(+2)	-5.0	25	C	μ=1M(KNO_3)	72Cd	
7.40	9.281(+1)	-17.6	25	C	μ=0.5M(KNO_3)	71Pa	
7.47	9.450(+1)	-13.1	25	C	μ=1M(KNO_3)	72Cd	
E28 ETHANE, 1,2-diamino-N,N,N'-trimethyl-			$C_5H_{14}N_2$		$CH_3NHCH_2CH_2N(CH_3)_2$		
7.64	6.827(+2)	-5.6	25	C	μ=1M(KNO_3)	72Cd	
9.72	9.979(+1)	-13.0	25	C	μ=1M(KNO_3)	72Cd	
E29 ETHANE, 1,1-dinitro-			$C_2H_4O_4N_2$		$CH_3CH(NO_2)_2$		
3.540	5.21(0)	-11.82	20	T	--	61Sc	

ΔH, kcal /mole	pK	ΔS, cal /mole°K	T, °C	M	Conditions	Ref.	Remarks
E30 ETHANE, nitro-			$C_2H_5O_2N$		$CH_3CH_2NO_2$		
2.42	8.57(0)	-31.1	25	C	μ=0	74Ma	
1.710	8.452(25°,0)	-32.9	18-30	T	μ<0.007	73Mc 43Ta	c
2.4	8.80(0)	-32.2	25	C	S: 0.05 mole fraction Ethanol	74Ma	
2.15	9.04(0)	-34.2	25	C	S: 0.10 mole fraction Ethanol	74Ma	
1.85	9.18(0)	-35.8	25	C	S: 0.125 mole fraction Ethanol	74Ma	
1.75	9.30(0)	-36.7	25	C	S: 0.15 mole fraction Ethanol	74Ma	
1.65	9.40(0)	-37.5	25	C	S: 0.175 mole fraction Ethanol	74Ma	
1.55	9.52(0)	-38.4	25	C	S: 0.20 mole fraction Ethanol	74Ma	
1.45	10.03(0)	-41.0	25	C	S: 0.30 mole fraction Ethanol	74Ma	
E31 1,2-ETHANEDIOL			$C_2H_6O_2$		$HOCH_2CH_2OH$		
12.14	16.47	-31.7	5	T	μ=0 S: Ethylene Glycol R: $2HA = H_2A^+ + A^-$ ΔCp=-27.5	70Kd	
12.00	16.30	-32.2	10	T	μ=0 S: Ethylene Glycol R: $2HA = H_2A^+ + A^-$ ΔCp=-28.0	70Kd	
11.86	16.14	-32.7	15	T	μ=0 S: Ethylene Glycol R: $2HA = H_2A^+ + A^-$ ΔCp=-28.5	70Kd	
11.71	15.99	-33.2	20	T	μ=0 S: Ethylene Glycol R: $2HA = H_2A^+ + A^-$ ΔCp=-29.0	70Kd	
11.57	15.84	-33.7	25	T	μ=0 S: Ethylene Glycol R: $2HA = H_2A^+ + A^-$ ΔCp=-29.4	70Kd	
11.42	15.71	-34.2	30	T	μ=0 S: Ethylene Glycol R: $2HA = H_2A^+ + A^-$ ΔCp=-29.9	70Kd	
11.27	15.57	-34.7	35	T	μ=0 S: Ethylene Glycol R: $2HA = H_2A^+ + A^-$ ΔCp=-30.4	70Kd	
11.11	15.44	-35.2	40	T	μ=0 S: Ethylene Glycol R: $2HA = H_2A^+ + A^-$ ΔCp=-30.9	70Kd	
10.96	15.32	-35.7	45	T	μ=0 S: Ethylene Glycol R: $2HA = H_2A^+ + A^-$ ΔCp=-31.4	70Kd	
E32 1,2-ETHANEDIOL, di(2'-aminoethyl)ether			$C_6H_{16}O_2N_2$		$H_2NCH_2CH_2OCH_2CH_2OCH_2CH_2NH_2$		
12.1	8.91(+2)	0.7	25	C	μ=0.1[$(CH_3)_4NCl$]	75Aa	
11.9	9.71(+1)	-5.4	25	C	μ=0.1[$(CH_3)_4NCl$]	75Aa	
E33 1,2-ETHANEDIOL, dimethyl ether			$C_4H_{10}O_2$		$CH_3OCH_2CH_2OCH_3$		
28.0	(0)	-----	25	C	C∿10^{-3}M S: Fluorosulfuric acid	74Ab	
E34 ETHANEPHOSPHONIC ACID, 2-amino-			$C_2H_8O_3NP$		$NH_2CH_2CH_2PO_3H_2$		
-0.452	5.84(0)	-28.2	25	T	μ=0	75Ga	

ΔH, kcal /mole	pK	ΔS, cal /mole°K	T,°C	M	Conditions	Ref.	Remarks
E34, cont.							
11.04	10.64(-1)	-11.7	25	T	μ=0	75Ga	
E35 ETHANEPHOSPHONIC ACID, diethyl ester			$C_6H_{15}O_3P$		$CH_3CH_2OP(O)(C_2H_5)OC_2H_5$		
23.0	(0)	-----	25	C	C∼10^{-3}M S: Fluorosulfuric acid	74Ab	
E36 ETHANETHIOL			C_2H_6S		CH_3CH_2SH		
6.42	10.61(0)	-27.0	25	C	μ=0.015	64Ia	
E37 ETHANETHIOL, 2-acetylamino-			C_4H_9ONS		$CH_3CONHCH_2CH_2SH$		
6.26	9.92	-24.4	25	C	μ=0.015	64Ta	
E38 ETHANETHIOL, 2-amino-			C_2H_7NS		$H_2NCH_2CH_2SH$		
7.43	8.23(+1)	-12.7	25	C	μ=0.01	64Ia	
E39 ETHANETHIOL, 1,1-dimethyl-			$C_4H_{10}S$		$CH_3C(CH_3)_2SH$		
5.30	11.22	-33.6	25	C	μ=0.01-0.10	64Ia	
E40 ETHANETHIOL, 1-methyl-			C_3H_8S		$CH_3CH(CH_3)SH$		
5.38	10.86	-31.6	25	C	μ=0.10	64Ia	
E41 ETHANOL, 2-amino-			C_2H_7ON		$H_2NCH_2CH_2OH$		
12.107	10.306	-2.84	0	T	μ=0	51Bc	
12.102	10.132	-2.86	5	T	μ=0	51Bc	
12.096	9.964	-2.88	10	T	μ=0	51Bc	
12.090	9.803	-2.90	15	T	μ=0	51Bc	
12.085	9.646	-2.92	20	T	μ=0	51Bc	
12.10	9.45(+1)	-2.65	25	C	μ=0	61Ta	
11.54	9.498(+1)	-4.7	25	C	μ=0	70Ea	
12.04	9.498(+1)	-3.1	25	C	μ=0	69Cc	
12.07	9.50(+1)	-2.94	25	C	μ=0	68Ob	
12.079	9.495(+1)	-2.94	25	T	μ=0	51Bc	
12.073	9.349(+1)	-2.96	30	T	μ=0	51Bc	
12.068	9.208(+1)	-2.98	35	T	μ=0	51Bc	
12.061	9.071(+1)	-3.00	40	T	μ=0	51Bc	
12.055	8.939(+1)	-3.02	45	T	μ=0	51Bc	
12.049	8.811(+1)	-3.04	50	T	μ=0	51Bc	
11.965	(+1)	-----	10	C	C=0.06	49La	
12.050	(+1)	-----	20	C	C=0.06	49La	
12.070	9.45(+1)	-2.7	25	C	C=0.06	49La	
12.090	(+1)	-----	30	C	C=0.06	49La	
15.27	11.34(+1)	-0.7	25	T	μ=0 S: Ethylene glycol ΔCp=-19	70Kc	
E42 ETHANOL, 2-amino-1-phosphoric acid-			$C_2H_8O_5NP$		$NH_2CH_2CH(OH)OP(O)(OH)_2$		
0.335	5.836(0)	-25.5	5	T	μ=0	55Cb	

ΔH, kcal /mole	pK	ΔS, cal /mole°K	T,°C	M	Conditions	Ref.	Remarks
E42, cont.							
0.143	5.832(0)	-26.2	10	T	μ=0	55Cb	
-0.052	5.832(0)	-26.9	15	T	μ=0	55Cb	
-0.250	5.834(0)	-27.5	20	T	μ=0	55Cb	
-0.452	5.838(0)	-28.2	25	T	μ=0	55Cb	
-0.658	5.845(0)	-28.9	30	T	μ=0	55Cb	
-0.866	5.854(0)	-29.6	35	T	μ=0	55Cb	
-0.951	5.858(0)	-29.9	37	T	μ=0	55Cb	
-1.079	5.865(0)	-30.3	40	T	μ=0	55Cb	
-1.294	5.878(0)	-31.0	45	T	μ=0	55Cb	
-1.513	5.893(0)	-31.6	50	T	μ=0	55Cb	

E43 ETHANOL, 2-dimethylamino- $C_4H_{11}ON$ $(CH_3)_2NCH_2CH_2OH$

8.74	9.26(+1)	-13.1	25	C	μ=0	69Cc	

E44 ETHANOL, 2-mercapto- C_2H_6OS $HSCH_2CH_2OH$

6.21	9.72(0)	-23.6	25	C	μ=0.015	64Ia	
6.5	9.54(0)	-21.3	20	C	μ=0.1(KNO_3)	65Sb	
6.15	9.49(0)	-22.77	25	C	μ=0.5M(KNO_3)	74Da	

E45 ETHANOL, 2-methylamino- C_3H_9ON $CH_3NHCH_2CH_2OH$

11.06	9.88(+1)	-8.1	25	C	μ=0	69Cc	

E46 ETHER, di(2-aminoethyl)- $C_4H_{12}ON_2$ $(NH_2CH_2CH_2)_2O$

13.2	8.62(30°,+2)	4	10-40	T	μ=0	59La	c
11.7	9.59(30°,+1)	-5	10-40	T	μ=0	59La	c

E47 ETHER, diethyl, 2,2'-diamino-N,N,N',N'-tetraacetic acid $C_{12}H_{20}O_9N_2$ $(HO_2CCH_2)_2N(CH_2)_2O(CH_2)_2N(CH_2CO_2H)_2$

7.25	8.83(-2)	-15.7	20	C	0.1M(KNO_3)	64Ab	
6.23	9.47(-3)	-22.1	20	C	0.1M(KNO_3)	64Ab	

ΔH, kcal /mole	pK	ΔS, cal /mole°K	T,°C	M	Conditions	Ref.	Remarks
			F				
F1 FACTOR B							
9.20	10.94(25°,+1)	-19	15-35	T	μ=0 HA=Factor B (corrinoids)	67Ha	c
4.7	4.49	-5	25	T	μ=0.11 HA=Factor B-histamine complex (amino)	66Ha	
11.7	11.00	-11	25	T	μ=0.11 HA=Factor B-histamine complex (amino)	66Ha	
12.1	11.39	-12	25	T	μ=0.16 HA=Factor B-imidazole complex	66Ha	
F2 FACTOR V_{1A} (corrinoids)							
6.18	11.01(25°,0)	-30	25-35	T	μ=0	67Ha	c
F3 FERRIC HORSERADISH PEROXIDASE							
5.7	10.8(25°,0)	-30	18-35	T	μ=0 HA=[Ferric Horse-radish Peroxidase (H_2O)]	69Ea	c
F4 FERRIHEMOPROTEINS							
2.20	8.49(+1)	-31.5	25	T	μ=0 Soya-bean-nodule pigments HA=[Hematin Fe(III)OH_2]$^+$	63Ta	
2.60	8.57(+1)	-30.5	25	T	μ=0 Slow component HA=[Hematin Fe(III)OH_2]$^+$	63Ta	
3.91	8.81(+1)	-27.2	25	T	μ=0 Fast component HA=[Hematin Fe(III) OH_2]$^+$	63Ta	
5.75	8.97(+1)	-21.6	25	T	μ=0 Horse blood ferrihemoglobin HA=[Hematin Fe(III) OH_2]$^+$	63Ta	
3.8	8.19(+1)	-25	25	T	μ=0 Chironomus blood ferrihemoglobin HA=[Hematin Fe(III) OH_2]$^+$	63Ta	
F5 FLUORENE, 2-amino-			$C_{13}H_{11}N$		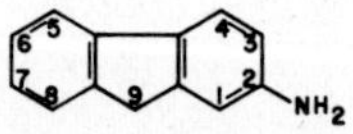		
11.4	10.34(+1)	-9	25	T	μ=0	66Ha	
5.7	4.21(20°,+1)	0.3	0-20	T	C=0.005 S: 50 w % Ethanol	59Eb	c
F6 FORMIC ACID			CH_2O_2		HCO_2H		
0.931	3.786(0)	-13.9	0	T	μ=0	34Hb	
0.755	3.772(0)	-14.5	5	T	μ=0	34Hb	
0.37	3.763(0)	-15.9	10	C	μ=0	70Ce	
0.573	3.762(0)	-15.2	10	T	μ=0	34Hb	
0.550	(0)	-----	10	T	μ=0	34Hc	
0.384	3.757(0)	-15.8	15	T	μ=0	34Hb	

ΔH, kcal /mole	pK	ΔS, cal /mole°K	T, °C	M	Conditions	Ref.	Remarks
F6, cont.							
0.189	3.753(0)	-16.5	20	T	μ=0	34Hb	
0.01	3.751(0)	-17.1	25	C	μ=0	67Ca	
-0.13	3.77(0)	-17.7	25	C	μ=0	58Ca	a:13.50
0.0	3.75(0)	-17.2	25	C	μ=0	74Md	
-0.08	3.752(0)	-17.4	25	C	μ=0	70Ce	q:±0.01
-0.023	3.752(0)	-17.6	25	T	μ=0	41Ha 62Aa	
-0.051	3.748(0)	-17.3	25	T	μ=0	45Ja	
-0.041	3.7518(0)	-17.31	25	T	μ=0	52Ea	
-0.012	3.751(0)	-17.2	25	T	μ=0 ΔCp=-40.7	34Hb 39Hb 34Hc	
-0.031	3.751(0)	-17.3	25	T	μ=0 ΔCp=-41.5	40Hb	g:34Hb
-0.221	3.752(0)	-17.9	30	T	μ=0	34Hb	
-0.436	3.758(0)	-18.5	35	T	μ=0	34Hb	
-0.57	3.766(0)	-19.0	40	C	μ=0	70Ce	
			10-40	C	μ=0 ΔCp=-31	70Ce	
-0.657	3.766(0)	-19.3	40	T	μ=0	34Hb	
-0.690	(0)	-----	40	T	μ=0	34Hc	
-0.884	3.773(0)	-20.0	45	T	μ=0	34Hb	
-1.118	3.782(0)	-20.8	50	T	μ=0	34Hb	
-1.358	3.793(0)	-21.5	55	T	μ=0	34Hb	
-1.605	3.809(0)	-22.2	60	T	μ=0	34Hb	
-0.959	3.66(25°,0)	-20.0	25-160	T	--	67Ma	c
0.0	3.86(0)	-17.7	160-180	T	--	67Ma	c
1.324	4.05(0)	-14.09	200-300	T	--	67Ma	c
-0.024	3.79(0)	-17.4	25	C	μ=0 S: 5 w % tert-Butanol	74Md	
-0.072	3.85(0)	-17.9	25	C	μ=0 S: 10 w % tert-Butanol	74Md	
-0.024	3.96(0)	-18.2	25	C	μ=0 S: 15 w % tert-Butanol	74Md	
-0.096	3.95(0)	-18.4	25	C	μ=0 S: 17.5 w % tert-Butanol	74Md	
-0.24	3.99(0)	-19.1	25	C	μ=0 S: 20 w % tert-Butanol	74Md	
-0.50	4.07(0)	-20.3	25	C	μ=0 S: 25 w % tert Butanol	74Md	
-0.69	4.20(0)	-21.5	25	C	μ=0 S: 30 w % tert-Butanol	74Md	
-0.84	4.35(0)	-22.7	25	C	μ=0 S: 35 w % tert-Butanol	74Md	
-0.98	4.48(0)	-23.9	25	C	μ=0 S: 40 w % tert-Butanol	74Md	
-1.41	11.56(0)	-57.8	15	T	S: Dimethylformamide	68Pc	
0.59	11.57(0)	-50.9	20	T	S: Dimethylformamide	68Pc	
2.63	11.55(0)	-44.0	25	T	S: Dimethylformamide	68Pc	
4.69	11.50(0)	-37.2	30	T	S: Dimethylformamide	68Pc	
6.79	11.43(0)	-30.3	35	T	S: Dimethylformamide	68Pc	
8.93	11.35(0)	-23.4	40	T	S: Dimethylformamide	68Pc	
-0.380	4.177(0)	-20.4	25	T	S: 20 w % Dioxane	45Ja	
0.359	4.181(0)	-17.92	25	T	S: 20 w % Dioxane	41Ha 62Aa	
-1.102	5.088(0)	-27.0	25	T	S: 45 w % Dioxane	45Ja	
1.067	5.092(0)	-19.72	25	T	S: 45 w % Dioxane	41Ha 62Aa	
-1.510	7.011(0)	-37.2	25	T	S: 70 w % Dioxane	45Ja	
1.474	7.017(0)	-27.16	25	T	S: 70 w % Dioxane	41Ha 62Aa	
2.509	8.801(0)	-31.84	25	T	S: 82 w % Dioxane	41Ha 62Aa	

ΔH, kcal /mole	pK	ΔS, cal /mole°K	T, °C	M	Conditions	Ref.	Re-marks
F7 FORMIC ACID, amide, N,N-dimethyl-			C_3H_7ON		$HCON(CH_3)_2$		
29.5	-----	-----	25	C	C∿10^{-3}M S: Fluorosulfuric acid	74Ab	
F8 FORMIC ACID, amide, N-methyl-			C_2H_5ON		$HCONHCH_3$		
29.6	-----	-----	25	C	C∿10^{-3}M S: Fluorosulfuric acid	74Ab	
F9 FRUCTOSE			$C_6H_{12}O_6$		$CH_2CH(OH)CH(OH)CH(OH)C(OH)$- CH_2OH (ring O linking first CH_2 and C(OH))		
9.36	12.53(0)	-24.4	10	C	μ=0	70Cd	
9.37	12.03(0)	-23.6	25	C	μ=0	70Cd	
8.2	12.27(0)	-28.6	25	C	μ=0	66Ib	
F10 FUMARIC ACID			$C_4H_4O_4$		$HO_2CCH:CHCO_2H$		
0.11	3.095(0)	-13.8	25	C	μ=0	67Ca	
-0.37	3.10(0)	-15	25	C	μ=0	63Ea	
-0.68	4.602(-1)	-23.3	25	C	μ=0	67Ca	
0.90	4.60(-1)	-16	25	C	μ=0	63Ea	
0.23	3.020(0)	-13	25	C	μ=0 HA=[Co(III)$(NH_3)_5$- OCOCH:$CHCO_2$]	74La	
F11 2-FURANCARBOXYLIC ACID			$C_5H_4O_3$		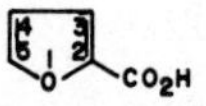		
-2.0	3.164(0)	-21.1	25	T	μ=0	60Lb	
-2.0	3.200(0)	-21.3	30	T	μ=0	60Lb	
-2.1	3.216(0)	-21.5	35	T	μ=0	60Lb	
-2.2	3.239(0)	-21.7	40	T	μ=0	60Lb	

ΔH, kcal /mole	pK	ΔS, cal /mole°K	T, °C	M	Conditions	Ref.	Remarks
			G				
G1 GALACTOSE			$C_6H_{12}O_6$		$CH_2OHCH(CHOH)_3CHOH$ (ring –O–)		
10.10	12.82	-23.0	10	C	μ=0	70Cd	
9.73	12.35	-23.9	25	C	μ=0	70Cd	
9.0	12.48	-26.9	25	C	μ=0	66Ib	
G2 GERMANIUM DIOXIDE			O_2Ge		GeO_2		
11.85	8.77	-0.4	25	-	--	69La	
G3 GLUCOSE			$C_6H_{12}O_6$		$HOCH_2CH(CHOH)_3CHOH$ (ring –O–)		
9.20	12.72	-25.7	10	C	μ=0	70Cd	
8.77	12.28	-26.8	25	C	μ=0	70Cd	
7.7	12.46	-31.3	25	C	μ=0	66Ib	
7.3	12.337	-32	25	T	μ=0.05	52Ka	
G4 GLUCOSE, 2-amino-(D,β)			$C_6H_{13}O_5N$		$HOCH_2CH(CHOH)_2CH(NH_3)CHOH$ (ring –O–)		
0.7	2.20(+1)	-9.83	25	C	μ=0.05M(KCl)	72Gb	
10.2	9.08(0)	-7.33	25	C	μ=0.05M(KCl)	72Gb	
10.000	7.526(0)	-1.375	25	T	μ=0.05	69Nb	
G5 GLUCOSE, 3-amino-			$C_6H_{13}O_5N$		$HOCH_2CHCH(OH)CH(NH_2)CH(OH)CHOH$ (ring –O–)		
10.050	7.703	0.504	25	T	μ=0.05	69Nb	
G6 GLUCOSE, 2-deoxy-			$C_6H_{12}O_5$		$HOCH_2(CHOH)_3CH_2CHO$		
8.51	12.89	-28.9	10	C	μ=0	70Cd	
8.2	12.52	-29.7	25	C	μ=0	70Cd	
8.2	12.52	-29.7	25	C	μ=0	71Hc	
7.4	12.28	-32.7	40	C	μ=0	70Cd	
G7 GLUCOSE, 1-phosphate			$C_6H_{13}O_9P$		$HOCH_2CH(CHOH)_3CHOPO_3H_2$ (ring –O–)		
0.474	6.506	-28.1	5	T	μ=0	55Aa	
0.254	6.500	-28.85	10	T	μ=0	55Aa	
0.029	6.499	-29.6	15	T	μ=0	55Aa	
-0.199	6.500	-30.4	20	T	μ=0	55Aa	
-0.431	6.504	-31.2	25	T	μ=0	55Aa	
-0.668	6.510	-32.0	30	T	μ=0	55Aa	
-0.908	6.519	-32.8	35	T	μ=0	55Aa	
-1.005	6.524	-33.1	37	T	μ=0	55Aa	
-1.152	6.531	-33.6	40	T	μ=0	55Aa	
-1.400	6.545	-34.35	45	T	μ=0	55Aa	
-1.652	6.561	-35.1	50	T	μ=0	55Aa	
G8 GLUCOSE, 6-phosphate			$C_6H_{13}O_9P$		$H_2O_3POCH_2CH(CHOH)_3CHOH$ (ring –O–)		
8.4	11.71	-25.0	25	C	μ=0	66Ib	

ΔH, kcal /mole	pK	ΔS, cal /mole°K	T, °C	M	Conditions	Ref.	Remarks
G9 GLUTAMIC ACID			$C_5H_9O_4N$		$HO_2CCH_2CH_2CH(NH_2)CO_2H$		
0.947	2.186(+1)	-6.6	5	T	μ=0	63La	
-0.064	2.162(+1)	-10.1	25	T	μ=0	63La	
-1.427	2.204(+1)	-14.5	50	T	μ=0	63La	
-2.040	2.22(+1)	-18.3	78	T	μ=0	37Wa	
-3.630	2.32(+1)	-20.6	100	T	μ=0	37Wa	
-5.250	2.44	-24.6	118	T	μ=0	37Wa	
0.950	(+1)	-----	25	C	μ=0.01	59Ka	
0.8	2.12(+1)	-7	25	C	μ=0.2M(KCl)	74Na	e: α-CO_2H
1.083	4.326(0)	-15.9	5	T	μ=0	69La	
0.373	4.272(0)	-18.3	25	T	μ=0	69La	
-0.584	4.282(0)	-21.4	50	T	μ=0	69La	
-3.090	4.46(0)	-32.7	78	T	μ=0	37Wa	
-4.780	4.60(0)	-33.9	100	T	μ=0	37Wa	
-6.500	3.77(0)	-33.9	118	T	μ=0	37Wa	
1.040	4.25(0)	-16.0	25	T	μ=0.02	30Ma	
0.8	4.11(0)	-16	25	C	μ=0.2M(KCl)	74Na	e: γ-CO_2H
9.578	9.358(-1)	-10.7	25	T	μ=0	63La	
9.15	(-1)	-----	20	C	μ≃0.01	71Ma	e: α-NH_3^+
11.20	9.67(-1)	-6.7	25	T	μ=0.02	30Ma	
9.99	(-1)	-----	25	C	μ=0.10M	71Ba	e:NH_3^+
9.7	9.51(-1)	-11	25	C	μ=0.2M(KCl)	74Na	e:NH_3^+
1.6	4.12(+1)	-13	25	C	μ=0.2M(KCl) HA=$[Cu(II)O_2CCH(NH_3^+)$-$CH_2CH_2CO_2]^+$	74Na	
G10 GLUTAMIC ACID RESIDUE 35							
1.9	5.22(25°)	-18	1.6-40	T	μ=0.15M(KCl) HA=Glu-35 residue of lysozyme	72Pe	c,q: ±1.2
2.9	6.01(25°)	-18	1.6-40	T	μ=0.15M(KCl) HA=Glu-35 residue of lysozyme	72Pe	c,q: ±0.6
G11 GLUTAMIC ACID, lysyl-			$C_{11}H_{21}O_5N_3$		$HO_2CCH_2CH_2CH(CO_2H)NHOCCH$-$(NH_2)(CH_2)_4NH_2$		
0.750	2.93(25°,+2)	-10.9	25	T	μ=0.01	33Ga	c,e: CO_2H
0.0	4.47(25°,+1)	-20.40	25	T	μ=0.01	33Ga	c,e: CO_2H
10.50	7.75(25°,0)	-0.2	25	T	μ=0.01	33Ga	c,e: NH_3^+
11.95	10.50(25°,-1)	-8.0	25	T	μ=0.01	33Ga	c,e: NH_3^+
G12 GLUTAMINE			$C_5H_{10}O_3N_2$		$HO_2CCH(NH_2)CH_2CH_2CONH_2$		
0.82	2.15(+1)	-7.1	25	C	μ=0.2(KCl)	75Ga	
1.06	2.72(+1)	-8.94	25	C	μ=3.00M($NaClO_4$)	74Bc	e:CO_2H
9.24	9.911(0)	-12.12	5	T	μ=0.1(KCl)	73Ra	e:NH_3^+
9.67	9.412(0)	-10.61	25	T	μ=0.1(KCl)	73Ra	e:NH_3^+
9.94	8.959(0)	-9.72	45	T	μ=0.1(KCl)	73Ra	e:NH_3^+
9.94	9.00(0)	-7.6	25	C	μ=0.2(KCl)	75Ga	
12.16	9.64(0)	-3.35	25	C	μ=3.00M($NaClO_4$)	74Bc	e:NH_3^+

ΔH, kcal /mole	pK	ΔS, cal /mole°K	T, °C	M	Conditions	Ref.	Remarks

G13 GLYCEROL, 1-phosphoric acid $C_3H_9O_6P$ $HOCH_2CH(OH)CH_2OP(O)(OH)_2$

ΔH, kcal /mole	pK	ΔS, cal /mole°K	T, °C	M	Conditions	Ref.	Remarks
0.185	6.642(-1)	-29.7	5	T	μ=0	58Db	
-0.045	6.641(-1)	-30.5	10	T	μ=0	58Db	
-0.274	6.643(-1)	-31.4	15	T	μ=0	58Db	
-0.510	6.648(-1)	-32.2	20	T	μ=0	58Db	
-0.749	6.656(-1)	-33.0	25	T	μ=0	58Db	
-0.993	6.666(-1)	-33.8	30	T	μ=0	58Db	
-1.241	6.679(-1)	-34.6	35	T	μ=0	58Db	
-1.341	6.685(-1)	-34.9	37	T	μ=0	58Db	
-1.493	6.695(-1)	-35.4	40	T	μ=0	58Db	
-1.749	6.713(-1)	-36.2	45	T	μ=0	58Db	
-2.009	6.733(-1)	-37.0	50	T	μ=0	58Db	

G14 GLYCEROL, 2-phosphoric acid $C_3H_9O_6P$ $HOCH_2CH[OPO(OH)_2]CH_2OH$

ΔH, kcal /mole	pK	ΔS, cal /mole°K	T, °C	M	Conditions	Ref.	Remarks
-1.396	1.223(0)	-10.6	5	T	μ=0	54Ab	
-1.760	1.245(0)	-11.9	10	T	μ=0	54Ab	
-2.132	1.271(0)	-13.2	15	T	μ=0	54Ab	
-2.509	1.301(0)	-14.5	20	T	μ=0	54Ab	
-2.893	1.335(0)	-15.8	25	T	μ=0	54Ab	
-3.284	1.372(0)	-17.1	30	T	μ=0	54Ab	
-3.681	1.413(0)	-18.4	35	T	μ=0	54Ab	
-3.842	1.430(0)	-19.0	37	T	μ=0	54Ab	
-4.085	1.457(0)	-19.7	40	T	μ=0	54Ab	
-4.495	1.504(0)	-21.0	45	T	μ=0	54Ab	
-4.912	1.554(0)	-22.3	50	T	μ=0	54Ab	
0.630	6.657(-1)	-28.2	5	T	μ=0	54Ab	
0.376	6.650(-1)	-29.1	10	T	μ=0	54Ab	
0.118	6.646(-1)	-30.0	15	T	μ=0	54Ab	
-0.145	6.646(-1)	-30.9	20	T	μ=0	54Ab	
-0.412	6.650(-1)	-31.8	25	T	μ=0	54Ab	
-0.684	6.657(-1)	-32.7	30	T	μ=0	54Ab	
-0.961	6.666(-1)	-33.6	35	T	μ=0	54Ab	
-1.073	6.71(-1)	-34.0	37	T	μ=0	54Ab	
-1.242	6.679(-1)	-34.5	40	T	μ=0	54Ab	
-1.528	6.694(-1)	-35.4	45	T	μ=0	54Ab	
-1.818	6.712(-1)	-36.3	50	T	μ=0	54Ab	

G15 GLYCINE $C_2H_5O_2N$ $H_2NCH_2CO_2H$

ΔH, kcal /mole	pK	ΔS, cal /mole°K	T, °C	M	Conditions	Ref.	Remarks
1.41	2.397(+1)	-5.8	10	C	μ=0	68Cd	
1.670	(+1)	-----	10	T	μ=0	34Hc	
1.43	2.35(+1)	-6	25	C	μ=0	64Ic	
0.930	2.35(+1)	-8	25	C	μ=0	41Sa	
0.98	2.351(+1)	-7.5	25	C	μ=0	67Ca	
						68Cd	
1.175	2.351(+1)	-6.8	25	T	μ=0 ΔCp=-32.8	39Hb	1
						34Hc	
1.043	2.347(+1)	-7.25	25	T	μ=0	45Ja	
1.100	2.348(+1)	-7.3	25	T	μ=0	39Ea	
1.16	2.358(+1)	-6.9	25	T	μ=0	62Aa	
0.950	2.349(+1)	-7.6	25	T	μ=0	57Ka	
						51Ka	
0.958	2.350(+1)	-7.54	25	T	μ=0.000	45Ka	
1.156	2.352(+1)	-6.9	25	T	μ=0	50Ha	
						34Oa	
0.47	2.327(+1)	-9.2	40	C	μ=0	68Cd	
2.370	(+1)	-----	40	T	μ=0	34Hc	
0.630	(+1)	-----	40	T	μ=0	34Hc	
1.10	2.39(+1)	-7.2	25	C	μ=0.1M(KNO_3)	71La	
0.977	2.353(+1)	-7.49	25	T	μ=0.100(NaCl)	45Ka	
1.05	2.36(+1)	-7.28	25	C	μ=0.2M(KCl)	73Ga	
1.005	2.359(+1)	-7.43	25	T	μ=0.300(NaCl)	45Ka	

ΔH, kcal /mole	pK	ΔS, cal /mole°K	T,°C	M	Conditions	Ref.	Remarks
G15, cont.,							
1.083	2.385(+1)	-7.28	25	T	μ=0.725(NaCl)	45Ka	
1.201	2.430(+1)	-7.09	25	T	μ=1.250(NaCl)	45Ka	
1.388	2.501(+1)	-6.79	25	T	μ=2.000(NaCl)	45Ka	
1.635	2.605(+1)	-6.44	25	T	μ=3.000(NaCl)	45Ka	
1.18	2.635(+1)	-8.1	25	T	S: 20 w % Dioxane	62Aa	
0.99	3.108(+1)	-10.9	25	T	S: 45 w % Dioxane	62Aa	
0.83	3.974(+1)	-15.4	25	T	S: 70 w % Dioxane	62Aa	
10.78	10.34(0)	-8.6	5	T	μ=0	680b	
10.85	10.193(0)	-8.32	10	C	μ=0	72Ib	
10.68	10.05(0)	-8.9	15	T	μ=0	680b	
10.640	9.779(0)	-8.90	25	C	μ=0	67Ac	
10.6	9.78(0)	-9	25	C	μ=0	64Ic	
10.55	9.780(0)	-9.36	25	C	μ=0	72Ib	
10.53	(0)	-----	25	C	μ=0	71Ha	e:NH_3^+
10.600	9.768(0)	-9.2	25	T	μ=0	39Ea	
10.550	9.777(0)	-9.4	25	T	μ=0	51Ka	
10.58	9.78(0)	-9.3	25	T	μ=0	58Da	
10.806	9.780(0)	-8.5	25	T	μ=0	340a	
						50Ha	
10.46	9.53(0)	-9.7	35	T	μ=0	58Da	
10.38	9.412(0)	-9.91	40	C	μ=0	72Ib	
			10-40	C	μ=0 ΔCp=-16	72Ib	
10.34	9.30(0)	-10.0	45	T	μ=0	58Da	
10.68	9.59(0)	-8.1	25	C	μ=0.010	68Ce	
10.01	(0)	-----	20	C	μ≃0.01	71Ma	e:NH_3^+
10.70	9.60(0)	-8.1	25	T	μ=0.02	30Ma	
						30Ba	
10.3	10.2(0)	-9.2	0	T	μ=0.09(KCl)	57Ma	
10.2	9.44(0)	-9.5	30	T	μ=0.09(KCl)	57Ma	
10.69	10.231(0)	-8.33	5	T	μ=0.1(KCl)	73Ra	e:NH_3^+
10.2	9.56(0)	-9.5	25	C	μ=0.1M($NaClO_4$ or KNO_3)	72Ia	e:NH_3^+
						71La	
10.749	9.603(0)	-7.9	25	T	μ=0.1(NaCl)	51Ka	
11.15	9.655(0)	-6.77	25	T	μ=0.1(KCl)	73Ra	e:NH_3^+
11.39	9.135(0)	-5.98	45	T	μ=0.1(KCl)	73Ra	e:NH_3^+
10.3	9.55(25°,0)	9.2	15-70	T	μ=0.1M(KNO_3)	71Ia	
11.12	9.55(0)	-6.40	25	C	μ=0.2M(KCl)	73Ga	
10.879	9.577(0)	-7.3	25	T	μ=0.3(NaCl)	51Ka	
10.640	9.777(0)	8.90	25	C	--	67Ac	
10.76	9.77(0)	-8.6	25	C	--	67Ac	
10.22	9.46(0)	-10.7	40	C	--	67Ac	

G16 GLYCINE, amide $C_2H_6ON_2$ $H_2NCH_2CONH_2$

ΔH, kcal /mole	pK	ΔS, cal /mole°K	T,°C	M	Conditions	Ref.	Remarks
9.8	8.03(0)	-3.8	25	C	=0.1M(KNO_3)	72Bi	

G17 GLYCINE, ethyl ester $C_4H_9O_2N$ $H_2NCH_2CO_2C_2H_5$

ΔH, kcal /mole	pK	ΔS, cal /mole°K	T,°C	M	Conditions	Ref.	Remarks
7.60	7.75	-9.96	25	T	μ=0	65Cc	
10.53	7.660(25°,+1)	0.3	0-50	T	μ≃0	67Wb	c
10.60	7.69(25°,+1)	0.4	25-50	T	μ=0.1M	67Hb	c,e: NH_3^+

G18 GLYCINE, methyl ester $C_3H_7O_2N$ $H_2NCH_2CO_2CH_3$

ΔH, kcal /mole	pK	ΔS, cal /mole°K	T,°C	M	Conditions	Ref.	Remarks
11.50	7.66(25°,1)	3.5	25-50	T	μ=0.1M	67Hb	c,e: NH_3^+

ΔH, kcal /mole	pK	ΔS, cal /mole°K	T, °C	M	Conditions	Ref.	Remarks
G19 GLYCINE, N-acetyl-			$C_4H_7O_3N$		$CH_3CONHCH_2COOH$		
-0.150	3.6698	-17.3	25	T	μ=0	57Ka	
-0.210	3.68	-17.48	25	T	μ=0	67Ac	
G20 GLYCINE, N-alanyl-			$C_5H_{10}O_3N_2$		$CH_3CH(NH_2)CONHCH_2CO_2H$		
1.0	3.11(+1)	-17.6	25	C	μ=0.1(KNO_3)	72Bi	q:±0.2
10.2	8.11(0)	-2.9	25	C	μ=0.1M(KNO_3)	72Bi	
6.5	3.86(+1)	4.2	25	C	μ=0.1M(KNO_3) HA=$[Cu(II)CH_3CH(NH_2)$-$CONHCH_2CO_2]^+$	72Bi	
G21 GLYCINE, carbamoyl-			$C_3H_6O_3N_2$		$H_2NCONHCH_2CO_2H$		
0.290	3.8758(+1)	-16.76	25	T	μ=0	56Kb	
G22 GLYCINE, N,N-di(2-hydroxyethyl)-			$C_6H_{13}O_4N$		$(OHCH_2CH_2)_2NCH_2CO_2H$		
6.278	8.3335(0)	-17.08	25	T	μ=0	64Da	
G23 GLYCINE, N,N-dimethyl-			$C_4H_9O_2N$		$(CH_3)_2NCH_2CO_2H$		
-0.155	2.146(+1)	-9.3	25	T	μ=0	64Sc	
7.38	10.34(0)	-20.8	5	T	μ=0	58Da	
7.52	10.14(0)	-20.3	15	T	μ=0	58Da	
7.66	9.94(0)	-19.8	25	T	μ=0	58Da	
7.80	9.76(0)	-19.3	35	T	μ=0	58Da	
7.95	9.59(0)	-18.9	45	T	μ=0	58Da	
G24 GLYCINE, N-glycyl-			$C_4H_8O_3N_2$		$NH_2CH_2CONHCH_2CO_2H$		
1.190	3.201(+1)	-10.4	1	T	μ=0	42Sb	
0.862	3.166(+1)	-11.5	12.5	T	μ=0	42Sb	
0.391	3.126(+1)	-12.9	25	T	μ=0	42Sb	
-0.128	3.141(+1)	-14.8	37.5	T	μ=0	42Sb	
-0.736	3.159(+1)	-16.7	50	T	μ=0	42Sb	
0.68	3.12(+1)	-12.0	0&25	T	μ=0.01	33Ga	b,c
0.26	3.20(25°,+1)	-13.8	10-40	T	μ=0.06M	66Vb 65Va	c
0.032	3.08(+1)	-13.9	25	C	μ=0.1	68Bd	
10.370	8.265(0)	-3.0	25	T	μ=0	75Kb	e:NH_3^+
10.30	8.17(0)	-2.84	0&25	T	μ=0.01	33Ga	b,c
9.83	8.130(0)	-4.23	0&25	T	μ=0.02	30Ba	b,c
10.5	8.22(25°,0)	-2.2	10-40	T	μ=0.06M	66Vb 65Va	c
9.0	8.72(0)	-7		T	0.09F(KCl)	57Ma	
10.660	8.594(0)	-2.0	12.5	C	μ=0.1	68Bd	
10.600	8.252(0)	-2.0	25	C	μ=0.1	68Bd	
10.400	7.948(0)	-2.8	37.5	C	μ=0.1	68Bd	
10.060	7.668(0)	-4.0	50	C	μ=0.1	68Bd	
6.9	4.06(+1)	4.5	25	C	μ=0.1M(KNO_3) $[Cu(II)NH_2CH_2CONHCH_2CO_2]^+$	72Bi	
G25 GLYCINE, glycylglycyl-			$C_6H_{11}O_4N_3$		$H_2NCH_2CONHCH_2CONHCH_2CO_2H$		
0.2	3.18	-13.9	25	C	μ=0.1	68Bd	
10.1	7.86	-2.1	25	C	μ=0.1	68Bd	
7.5	5.06(+1)	2.0	25	C	μ=0.1 HA=[Cu(II)-(glycylglycylglycine)]$^+$	68Bd	

ΔH, kcal /mole	pK	ΔS, cal /mole°K	T, °C	M	Conditions	Ref.	Remarks
G25, cont.							
7.4	6.78(0)	-6.2	25	C	μ=0.1 HA=[Cu(II)-(glycylglycylglycine)]°	68Bd	
G26 GLYCINE, N-glycylglycylglycyl-			$C_8H_{14}O_5N_4$		$H_2NCH_2(CONHCH_2)_3CO_2H$		
0.18	3.17	-13.9	25	C	μ=0.10	69Na	
10.40	7.87	-1.2	25	C	μ=0.10	69Na	
7.5	5.40(+1)	0.4	25	C	μ=0.10 HA=[Cu(II)-(N-glycylglycylglycyl-glycine]$^+$	69Na	
6.6	6.80(0)	-9.0	25	C	μ=0.10 HA=[Cu(II)-(N-glycylglycylglycyl-glycine]°	69Na	
8.9	9.15(-1)	-12.0	25	C	μ=0.10 HA=[Cu(II)-(N-glycylglycylglycyl-glycine]$^-$	69Na	
G27 GLYCINE, histidyl-			$C_8H_{12}O_3N_4$		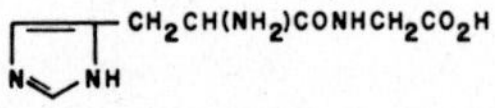		
0.300	2.40	-9.98	0&25	T	μ=0.01	33Ga	b,c,e: CO_2H
7.50	5.80	-1.4	0&25	T	μ=0.01	33Ga	b,c,e: Imidazole
10.8	7.82	0.5	0&25	T	μ=0.01	33Ga	b,c,e: NH_3^+
G28 GLYCINE, N-tris(hydroxymethyl) methane-			$C_6H_{13}O_5N$		$(CH_2OH)_3CNHCH_2CO_2H$		
2.076	2.092(+1)	-2.2	10	T	μ=0 ΔCp=-44	73Rc	e: CO_2H
1.405	2.023(+1)	-4.6	25	T	μ=0 ΔCp=-46	73Rc	
0.701	1.989(+1)	-6.9	40	T	μ=0 ΔCp=-48	73Rc	
2.371	2.970(+1)	-5.2	10	T	μ=0 S: 50 w % Methanol ΔCp=-33	74Bb	
1.870	2.887(+1)	-6.9	25	T	μ=0 S: 50 w % Methanol ΔCp=-34	74Bb	
1.344	2.832(+1)	-8.7	40	T	μ=0 S: 50 w % Methanol ΔCp=-36	74Bb	
7.700	8.430(0)	-11.4	10	T	μ=0 ΔCp=-12	73Rc	e: NH_2^+
7.520	8.135(0)	-12.0	25	T	μ=0 ΔCp=-12	73Rc	
7.330	7.873(0)	-12.7	40	T	μ=0 ΔCp=-13	73Rc	
8.097	8.622(0)	-10.8	10	T	μ=0 S: 50 w % Methanol ΔCp=-26	74Bb	
7.693	8.313(0)	-12.2	25	T	μ=0 S: 50 w % Methanol ΔCp=-28	74Bb	
7.267	8.051(0)	-13.6	40	T	μ=0 S: 50 w % Methanol ΔCp=-29	74Bb	
6.667	7.755(0)	-15.5	60	T	μ=0 S: 50 w % Methanol ΔCp=-31	74Bb	

ΔH, kcal /mole	pK	ΔS, cal /mole°K	T,°C	M	Conditions	Ref.	Remarks
G29 GLYCINE, phenylalanyl-			$C_{11}H_{14}O_3N_2$		$HO_2CCH_2NHCOCH(NH_2)CH_2C_6H_5$		
0.68	3.10(25°)	-11.9	0&25	T	μ=0.01	33Ga	b,c,e: CO_2H
10.0	7.71(25°)	-1.7	0&25	T	μ=0.01	33Ga	b,c,e: NH_3^+
G30 GLYCINE, N-propionyl-			$C_5H_9O_3N$		$CH_3CH_2CONHCH_2CO_2H$		
-0.140	3.717	-17.5	25	T	μ=0	56Kc	
G31 GLYOXAL, 1,2-dimethyl-, dioxime			$C_4H_8O_2N_2$		$HON{:}C(CH_3)C(CH_3){:}NOH$		
2.6	11.238(25°,-1̲)	42.6	20-40	T	μ=1M($NaNO_3$) HA=[Co(III)(DH)$_2$(NO_2)$_2$] DH=1,2-dimethyl dioxime glyoxal	72Zb	q:±0.3
5.6	11.780(45°,-1̲)	36.3	40-55	T	μ=1M($NaNO_3$) HA=[Co(III)(DH)$_2$$NO_2$OH]$^-$ DH=1,2-dimethyl dioxime glyoxal	72Zb	
16.3	12.611(35°,-1̲)	-4.9	30-40	T	μ=1M HA=[Co(III)(DH)$_2$$H_2O$$NO_2$]$^-$ DH=1,2-dimethyl dioxime glyoxal	73Za	
G32 GUANINE			$C_5H_5ON_5$		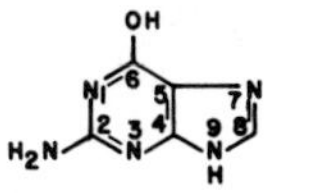		
10.1	9.42	-9.1	25	T	μ=0.1	64Sd	
G33 GUANOSINE			$C_{10}H_{13}O_5N_5$		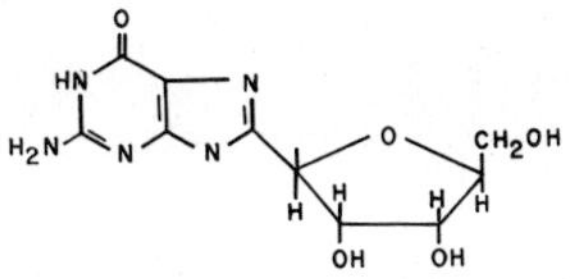		
3.2	1.9(+1̲)	2.1	25	C	μ=0	70Cb	e:N_7
1.0	1.6(+1̲)	-4.0	25	T	μ=0.1	64Sd	
2.22	2.174(+1̲)	-2.5	25	T	μ=0.1	66Bi	
0.99	1.6(+1̲)	-4.0	25	C	μ=0.1(NaCl)	60Ra	d
2.08	2.593(+1̲)	-4.9	25	T	μ=0.1 S: D_2O	66Bi	
7.65	9.25(0̲)	-16.7	25	C	μ=0	70Cb	e: N_1-C_6O
3.6	9.24(0̲)	-13.0	25	T	μ=0.1	64Sd	
11.04	12.83(-1̲)	-19.7	10	C	μ=0	70Cb	e: Ribose OH
10.85	12.33(-1̲)	-20.0	25	C	μ=0	70Cb	e: Ribose OH
10.86	11.60(-1̲)	-18.4	40	C	μ=0	70Cb	e: Ribose OH

ΔH, kcal /mole	pK	ΔS, cal /mole°K	T,°C	M	Conditions	Ref.	Re-marks
G34 GUANOSINE, 2'-deoxy-			$C_{10}H_{13}O_4N_5$		see G33		
1.910	2.5	-5.0	25	C	μ=0.1	60Ra	
G35 GUANOSINE, 5'-diphosphoric acid			$C_{10}H_{15}O_{11}N_5P_2$		see G33		
-1.48	7.19(-2)	-37.7	25	T	μ=0	65Pe	
G36 GUANOSINE, 5'-monophosphoric acid			$C_{10}H_{14}O_8N_5P$		see G33		
-1.45	6.66(-1)	-35.3	25	T	μ=0	65Pe	
G37 GUANOSINE, 5'-monophosphoric acid-2'-deoxy			$C_{10}H_{14}O_7N_5P$		see G33		
0.140	2.9	-12.8	25	C	μ=0.1	60Ra	
G38 GUANOSINE-5'-triphosphoric acid			$C_{10}H_{16}O_{14}N_5P_3$		see G33		
4.3	3.0(25°,-2)	1	25-45	T	μ=0.10M(KNO_3)	69Ka	c
-1.75	7.65(-3)	-40.8	25	T	μ=0	65Pe	
3.1	7.10(25°,-3)	-22	25-45	T	μ=0.10M(KNO_3)	69Ka	c,q:±0.3

ΔH, kcal /mole	pK	ΔS, cal /mole°K	T,°C	M	Conditions	Ref.	Remarks
			H				
H1 HAEMIN-WATER (in metmyglobin)							
3.85	8.55	-27.0	25	T	--	52Ga	
H2 HAEMOGLOBIN A (human)							
3.400	8.81	-28.7	20	T	μ=0	64Bc	
H3 HAEMOGLOBIN C (human)							
4.880	8.66(20°)	-23.0	20	T	μ=0	64Bc	
H4 HAEMOGLOBIN S (human)							
4.090	8.74	-26.0	20	T	μ=0	64Bc	
H5 HEPTANEDIOIC ACID			$C_7H_{12}O_4$		$HO_2C(CH_2)_5CO_2H$		
-0.33	4.484(0)	-21.6	25	C	μ=0	67Ca	
-0.93	5.424(-1)	-27.9	25	C	μ=0	67Ca	
H6 2,4-HEPTANEDIONE			$C_7H_{12}O_2$		$CH_3CH_2CH_2COCH_2COCH_3$		
4.1	8.43(25°,0)	-25	5-45	T	Enol form of HA	68Cb	c
2.5	9.15(25°,0)	-33	5-45	T	Keto form of HA	68Cb	c
H7 3,5-HEPTANEDIONE			$C_7H_{12}O_2$		$CH_3CH_2COCH_2COCH_2CH_3$		
-2.80	11.21(0)	41.8	5-45	T	S: 50 v % Dioxane	73Aa	c
H8 2,4-HEPTANEDIONE, 5-methyl-			$C_7H_{14}O_2$		$CH_3CH_2CH(CH_3)COCH_2COCH_3$		
3.1	8.52(25°)	-29	5-45	T	Enol form of HA	68Cb	c
1.5	9.10(25°)	-36	5-45	T	Keto form of HA	68Cb	c
H9 HEPTANOIC ACID			$C_7H_{14}O_2$		$CH_3(CH_2)_5CO_2H$		
-0.02	4.794(0)	-22.0	10	C	μ=0	70Ce	q:±0.01
-0.61	4.893(0)	-24.4	25	C	μ=0	70Ce	
-1.01	4.88(0)	-25.5	40	C	μ=0	70Ce	
	(0)	-----	10-40	C	μ=0 ΔCp=-33	70Ce	
H10 HEPTANOIC ACID, 7-amino-			$C_7H_{15}O_2N$		$H_2NCH_2(CH_2)_5CO_2H$		
-0.47	4.502(0)	-22.2	25	C	μ=0	68Cd	
H11 HEXANE, 1-amino-			$C_6H_{15}N$		$CH_3(CH_2)_5NH_2$		
13.85	(+1)	-----	0.50	C	μ=0	75Bb	
13.86	11.37(+1)	-2.4	5.00	C	μ=0 ΔH=13.855 + 0.00711t 0<t °C<125	75Bb	
14.09	10.64(+1)	-1.4	25.00	C	μ=0	75Bb	

ΔH, kcal /mole	pK	ΔS, cal /mole°K	T, °C	M	Conditions	Ref.	Remarks
H11, cont.							
14.26	9.84(+1)	-0.92	50.00	C	μ=0	75Bb	
14.28	9.14(+1)	-0.79	75.00	C	μ=0 ΔCp=7.2	75Bb	
14.55	8.53(+1)	-0.02	100.00	C	μ=0	75Bb	
14.79	8.00(+1)	0.55	125.00	C	μ=0	75Bb	
H12 HEXANE, 1,6-diamino-			$C_6H_{16}N_2$		$H_2N(CH_2)_6NH_2$		
11.360	(+2)	-----	0	T	μ=0	52Eb	
13.820	9.830(+2)	1.3	25	T	μ=0	52Eb	
13.71	10.242(+2)	-1.0	25	C	μ=0.5M(KNO_3)	70Ba	
11.410	(+1)	-----	0	T	μ=0	52Eb	
13.910	10.930(+1)	-3.3	25	T	μ=0	52Eb	
13.91	11.024(+1)	-3.8	25	C	μ=0.5M(KNO_3)	70Ba	
H13 HEXANE, 1,6-diamino-N,N,N',N'-tetraacetic acid			$C_{14}H_{24}O_8N_2$		$(HO_2CH_2C)_2N(CH_2)_6N(CH_2CO_2H)_2$		
6.24	9.79	-23.5	20	C	0.1M(KNO_3)	64Ab	
7.91	10.84	-22.5	20	C	0.1M(KNO_3)	64Ab	
H14 HEXANE, 1,1-dinitro-			$C_6H_{12}O_4N_2$		$CH_3(CH_2)_4CH(NO_2)_2$		
4.02	5.386(25°)	-11.08	5-60	T	--	62Sa	c
H15 HEXANEDIOIC ACID			$C_6H_{10}O_4$		$HO_2C(CH_2)_4CO_2H$		
-0.30	4.418(0)	-21.5	25	C	μ=0	67Ca	
-1	4.43(0)	-23	25	T	--	63Ia	
5.76	6.74(0)	-11.57	25	T	μ=0 S: Formamide	75Db	
-0.64	5.412(-1)	-26.9	25	C	μ=0	67Ca	
-1	5.42(-1)	-27	25	T	--	63Ia	
9.21	8.01(-1)	-5.72	25	T	μ=0 S: Formamide	75Db	
H16 2,4-HEXANEDIONE			$C_6H_{10}O_2$		$CH_3CH_2C(O)CH_2C(O)CH_3$		
4.8	8.49(25°)	-22	5-45	T	Enol form of HA	68Cb	c
3.1	9.32(25°)	-31	5-45	T	Keto form of HA	68Cb	c
H17 2,4-HEXANEDIONE, 5,5-dimethyl			$C_8H_{14}O_2$		$(CH_3)_3CC(O)CH_2C(O)CH_3$		
3.5	10.01	-55	25	T	--	61La	
H18 2,4-HEXANEDIONE, 5-methyl			$C_7H_{12}O_2$		$(CH_3)_2CHC(O)CH_2C(O)CH_3$		
2.6	9.44	-50	25	T	--	61La	
3.8	8.66(25°)	-26	5-45	T	Enol form of HA	68Cb	c
2.0	9.31(25°)	-35	5-45	T	Keto form of HA	68Cb	c
H19 HEXANOIC ACID			$C_6H_{12}O_2$		$CH_3(CH_2)_4CO_2H$		
-0.03	4.829(0)	-22.2	10	C	μ=0	70Ce	q:±0.02
-0.197	(0)	-----	10	C	μ=0	48Ca	
-0.493	(0)	-----	20	C	μ=0	48Ca	
-0.644	4.872(0)	-24.5	25	C	μ=0	48Ca	
-0.57	4.86(0)	-24.1	25	C	μ=0	74Md	
-0.64	4.879(0)	-24.5	25	C	μ=0	70Ce	

ΔH, kcal /mole	pK	ΔS, cal /mole°K	T, °C	M	Conditions	Ref.	Remarks
H19, cont.							
-0.700	4.856(0)	-24.6	25	T	μ=0	57Ka	
-0.793	(0)	-----	30	C	μ=0	48Ca	
-1.03	4.890(0)	-25.7	40	C	μ=0	70Ce	
	(0)	-----	10-40	C	μ=0 ΔCp=-33	70Ce	
-0.52	4.95(0)	-24.4	25	C	μ=0 S: 5 w % tert-Butanol	74Md	
-0.48	5.15(0)	-25.1	25	C	μ=0 S: 10 w % tert-Butanol	74Md	
-1.4	5.36(0)	-29.2	25	C	μ=0 S: 15 w % tert-Butanol	74Md	
-1.9	5.56(0)	-31.8	25	C	μ=0 S: 17.5 w % tert-Butanol	74Md	
-1.1	5.67(0)	-29.9	25	C	μ=0 S: 20 w % tert-Butanol	74Md	
0.31	5.98(0)	-26.3	25	C	μ=0 S: 25 w % tert-Butanol	74Md	
0.88	6.12(0)	-25.1	25	C	μ=0 S: 30 w % tert-Butanol	74Md	
1.05	6.30(0)	-25.3	25	C	μ=0 S: 35 w % tert-Butanol	74Md	
1.07	6.46(0)	-26.0	25	C	μ=0 S: 40 w % tert-Butanol	74Md	

H20 HEXANOIC ACID, 2-amino- $C_6H_{13}O_2N$ $CH_3(CH_2)_3CH(NH_2)CO_2H$

ΔH, kcal /mole	pK	ΔS, cal /mole°K	T, °C	M	Conditions	Ref.	Remarks
1.300	2.394(+1)	-6.3	1.0	T	μ=0	37Sa	
0.92	2.363(+1)	-7.6	10	C	μ=0	68Cd	
0.980	2.356(+1)	-7.3	12.5	T	μ=0	37Sa	
0.43	2.335(+1)	-9.2	25	C	μ=0	68Cd	
0.560	2.335(+1)	-8.8	25	T	μ=0 ΔCp=-36.9	37Sa 39IIb	
0.060	2.324(+1)	-10.4	37.5	T	μ=0	37Sa	
-0.7	2.324(+1)	-10.7	40	C	μ=0	68Cd	
-0.540	2.328(+1)	-12.3	50	T	μ=0	37Sa	
10.950	10.546(0)	-8.3	1.0	T	μ=0	37Sa	
11.030	10.190(0)	-8.0	12.5	T	μ=0	37Sa	
11.050	9.834(0)	-7.9	25	T	μ=0	37Sa	
10.840	9.513(0)	-8.6	37.5	T	μ=0	37Sa	
10.540	9.224(0)	-9.6	50	T	μ=0	37Sa	

H21 HEXANOIC ACID, 6-amino- $C_6H_{13}O_2N$ $H_2NCH_2(CH_2)_4CO_2H$

ΔH, kcal /mole	pK	ΔS, cal /mole°K	T, °C	M	Conditions	Ref.	Remarks
0.818	4.420(+1)	-17.2	1	T	μ=0	42Sb	
0.10	4.392(+1)	-19.8	10	C	μ=0	68Cd	
0.459	4.387(+1)	-18.2	12.5	T	μ=0	42Sb	
-0.32	4.373(+1)	-21.1	25	C	μ=0	68Cd	
-0.008	4.373(+1)	-20.0	25	T	μ=0	42Sb	
0	4.37(+1)	-20	25	T	μ=0	63Ea	
-0.561	4.384(+1)	-21.9	37.5	T	μ=0	42Sb	
-0.79	4.388(+1)	-22.6	40	C	μ=0	68Cd	
-1.204	4.410(+1)	-23.9	50	T	μ=0	42Sb	
13.120	11.666(0)	-5.6	1	T	μ=0	42Sb	
13.390	11.244(0)	-4.5	12.5	T	μ=0	42Sb	
13.560	10.804(0)	-4.0	25	T	μ=0	42Sb	
13.620	10.406(0)	-3.8	37.5	T	μ=0	42Sb	
13.550	10.036(0)	-3.9	50	T	μ=0	42Sb	

H22 HISTAMINE $C_5H_9N_3$

H22, cont.

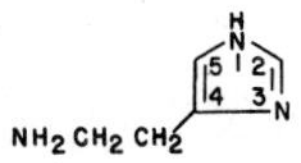

ΔH, kcal /mole	pK	ΔS, cal /mole°K	T,°C	M	Conditions	Ref.	Remarks
6.9	5.784(+2)	-3	25	T	μ=0	59Sb	
7.4	5.595(+2)	-2	37	T	μ=0	59Sb	
8.0	6.04(25°,+2)	-0.9	12-38	T	μ=0	70Pd	c,e: Imidazole group
10.1	5.87(30°,+2)	7	10-40	T	μ=0	61Na 64Sc	c
8.1	5.358(+2)	0	50	T	μ=0	59Sb	
7.57	6.058(+2)	-2.3	25	C	μ=0.1M	70Ec	
7.42	6.25(+2)	3.7	25	C	μ=0.16M(KNO_3)	70Me	e: Imidazole group
8.52	6.16(+2)	0.39	25	C	μ=0.2M(KCl)	73Ga	
9	7.09(25°,+2)	15	0-25	T	μ=0	55Ma	c
11.0	9.756(+1)	-8	25	T	μ=0	59Sb	
11.9	9.386(+1)	-5	37	T	μ=0	59Sb	
11.7	9.75(25°,+1)	-5.4	12-38	T	μ=0	70Pd	c,e: NH_3^+
13.4	9.70(30°,+1)	0	10-40	T	μ=0	61Na 64Sc	c
12.9	9.047(+1)	-1	50	T	μ=0	59Sb	
11.95	9.826(+1)	-4.8	25	C	μ=0.1M	70Ec	
12.55	9.97(+1)	3.5	25	C	μ=0.16M(KNO_3)	70Me	e:NH_3^+
13.01	9.85(+1)	-1.43	25	C	μ=0.2M(KCl)	73Ga	

H23 HISTAMINE, 2,5-diiodo- $C_5H_7N_3I_2$ see H22

ΔH, kcal /mole	pK	ΔS, cal /mole°K	T,°C	M	Conditions	Ref.	Remarks
5.8	2.31(25°,+2)	9.0	12-38	T	μ=0	70Pd	c,e: Imidazole group
9.3	8.20(25°,+1)	-6.3	12-38	T	μ=0	70Pd	c,e: NH_3^+
13.4	10.11(25°,0)	-1.3	12-38	T	μ=0	70Pd	c,e: Imino group

H24 HISTAMINE, 5-iodo- $C_5H_8N_3I$ see H22

ΔH, kcal /mole	pK	ΔS, cal /mole°K	T,°C	M	Conditions	Ref.	Remarks
4.9	4.06(25°,+2)	-2.0	12-38	T	μ=0	70Pd	c,e: Imidazole-group
11.9	9.20(25°,+1)	-2.2	12-38	T	μ=0	70Pd	c,e: Ammonium group
15.8	11.88(25°,0)	-2.3	12-35	T	μ=0	70Pd	c,e: Imino group

ΔH, kcal /mole	pK	ΔS, cal /mole°K	T, °C	M	Conditions	Ref.	Remarks
H25 HISTIDINE			$C_6H_9O_2N_3$		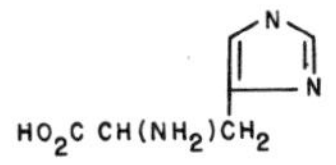		
0.72	2.02(+2)	6.8	25	C	μ=0.16M(KNO_3)	70Me	e:CO_2H
0.25	2.282(+2)	-9.6	25	C	μ=3.00M($NaClO_4$)	70Wb	e:CO_2H q:±0.10
-0.25	2.00(+2)	-9.60	37	C	μ=3.00($NaClO_4$) ΔCp=42.1	71Jb	e:CO_2H
7.14	6.00(+1)	3.1	25	C	μ=0	69Cc	
7.39	(+1)	-----	20	C	μ≃0.01	71Ma	e: Imidazole group
6.9	6.04(+1)	-4.5	0&25	T	μ=0.01	33Ga	b,c
7.20	7.99(+1)	-12.5	25	T	μ=0.02	30Ma	
6.87	(+1)	-----	25	C	μ=0.10M	71Ba	
7.3	(+1)	-----	25	C	μ=0.1M(KNO_3)	69Ta	e: Imidazole group
6.2	(+1)	-----	25	T	μ=0.1M(KCl)	30Sa	
7.5	6.02(25°,+1)	-2.4	25-50	T	μ=0.1M(KCl)	71Hd	
7.01	6.16(+1)	4.7	25	C	μ=0.16M(KNO_3)	70Me	e: group
6.3	5.8(20°,+1)	-5.5	10-32	T	μ=0.2(NaCl) Values valid for histidine in 119 position in ribonuclease.	69Rc	c
4.0	6.3(20°,+1)	-15.3	10-32	T	μ=0.2(NaCl) Values valid for histidine in 12 position in ribonuclease.	69Rc	c
8.2	6.7(20°,+1)	-3.5	10-32	T	μ=0.2(NaCl) Values valid for histidine in 105 position in ribonuclease.	69Rc	c
11.4	5.7(41°,+1)	11.5	32-41	T	μ=0.2(NaCl) Values valid for histidine in 119 position in ribonuclease.	69Rc	c
21.1	6.1(41°,+1)	41.0	32-41	T	μ=0.2(NaCl) Values valid for histidine in 12 position in ribonuclease.	69Rc	c
8.2	6.8(41°,+1)	-3.5	32-41	T	μ=0.2(NaCl) Values valid for histidine in 105 position in ribonuclease.	69Rc	c
8.75	6.970(+1)	-2.5	25	C	μ=3.00M($NaClO_4$)	70Wb	e: Imidazole group
6.95	6.680(+1)	-6.98	37	C	μ=3.00($NaClO_4$) ΔCp=-150	71Jb	e: Imino group q:±2.0
6.9	6.00(25°,+1)	4.12	19-40	T	--	49Ha	c
10.43	9.16(0)	-6.5	25	C	μ=0	69Cc	
10.68	(0)	-----	20	C	μ≃0.01	71Ma	e:α-NH_3^+
9.40	9.16(0)	-10.5	25	T	μ=0.02	30Ma	
11.0	(0)	-----	25	C	μ=0.1M(KNO_3)	69Ta	e:α-NH_3^+

ΔH, kcal /mole	pK	ΔS, cal /mole°K	T,°C	M	Conditions	Ref.	Re-marks
H25, cont.							
10.56	(0)	-----	25	C	μ=0.10M	71Ba	
9.4	(0)	-----	25	T	μ=0.1M(KCl)	30Sa	
10.3	9.31(25°,0)	-8.0	25-50	T	μ=0.1M(KCl)	71Hd	c
10.43	9.21(0)	-7.1	25	C	μ=0.16M(KNO_3)	70Me	e:NH_3^+
9.66	9.630(0)	-11.7	25	C	μ=3.00M($NaClO_4$)	70Wb	e:NH_3^+
8.28	9.365(0)	-14.51	37	C	μ=3.00($NaClO_4$) ΔCp=-115	71Jb	e:NH_3^+
9.4	9.17(25°,0)	1.05	14-45	T	--	49Ha	c
0.47	7.41(+3)	-32	25	C	μ=3.00M(ClO_4) $PrLH^{3+}$ = PrL^{2+} + H^+ HA=[Pr(III)Histidine]$^{3+}$	71Ja 70Ja	
-----	(+3)	-----	31	C	μ=3.00M(ClO_4^-) ΔCp=-33.5 HA=[Pr(III)Histidine]$^{3+}$	70Ja 71Ja	
0.069	7.35(+3)	-33	37	C	μ=3.00M(ClO_4^-) HA=[Pr(III)Histidine]$^{3+}$	70Ja 71Ja	
0.02	7.65(+3)	-35	25	C	μ=3.00M(ClO_4^-) HA=[La(III)Histidine]$^{3+}$	70Ja 71Ja	
			31	C	μ=3.00M(ClO_4^-) ΔCp=-20.1 HA=[La(III)Histidine]$^{3+}$	70Ja 71Ja	
-0.22	7.67(+3)	-36	37	C	μ=3.00M(ClO_4^-) HA=[La(III)Histidine]$^{3+}$	70Ja 71Ja	
0.51	7.37(+3)	-32	25	C	μ=3.00M(ClO_4^-) HA=[Nd(III)Histidine]$^{3+}$	70Ja 71Ja	
-----	(+3)	-----	31	C	μ=3.00M(ClO_4^-) ΔCp=-25.0 HA=[Nd(III)Histidine]$^{3+}$	70Ja 71Ja	
0.21	7.25(+3)	-32	37	C	μ=3.00M(ClO_4^-) HA=[Nd(III)Histidine]$^{3+}$	70Ja 71Ja	
1.96	7.32(+3)	-27	25	C	μ=3.00M(ClO_4^-) HA=[Sm(III)Histidine]$^{3+}$	70Ja 71Ja	
-----	(+3)	-----	31	C	μ=3.00M(ClO_4^-) ΔCp=-68.4 HA=[Sm(III)Histidine]$^{3+}$	70Ja 71Ja	
1.14	6.81(+3)	-27	37	C	μ=3.00M(ClO_4) HA=[Sm(III)Histidine]$^{3+}$	70Ja 71Ja	
2.69	7.08(+3)	-23	25	C	μ=3.00M(ClO_4^-) HA=[Gd(III)Histidine]$^{3+}$	70Ja 71Ja	
-----	(+3)	-----	31	C	μ=3.00M(ClO_4^-) ΔCp=-29.2 HA=[Gd(III)Histidine]$^{3+}$	70Ja 71Ja	
2.34	6.36(+3)	-22	37	C	μ=3.00M(ClO_4^-) HA=[Gd(III)Histidine]$^{3+}$	70Ja 71Ja	
1.22	6.76(+3)	-26.8	25	C	μ=3.00M(ClO_4^-) HA=[Dy(III)Histidine]$^{3+}$	70Ja 71Ja	
-----	(+3)	-----	31	C	μ=3.00M(ClO_4^-) ΔCp=10.0 HA=[Dy(III)Histidine]$^{3+}$	70Ja 71Ja	
1.34	6.47(+3)	-25.3	37	C	μ=3.00M(ClO_4^-) HA=[Dy(III)Histidine]$^{3+}$	70Ja 71Ja	
1.32	6.69(+3)	-26.0	25	C	μ=3.00M(ClO_4^-) HA=[Er(III)Histidine]$^{3+}$	71Ja	
-----	(+3)	-----	31	C	μ=3.00M(ClO_4^-) ΔCp=-39.9 HA=[Er(III)Histidine]$^{3+}$	71Ja	
0.839	6.41(+3)	-26.5	37	C	μ=3.00M(ClO_4^-) HA=[Er(III)Histidine]$^{3+}$	71Ja	
1.29	7.17(+3)	-28.4	25	C	μ=3.00M(ClO_4^-) HA=[Yb(III)Histidine]$^{3+}$	70Ja 71Ja	
			31	C	μ=3.00M(ClO_4^-) ΔCp=-50.9 HA=[Yb(III)Histidine]$^{3+}$	70Ja 71Ja	

ΔH, kcal /mole	pK	ΔS, cal /mole°K	T,°C	M	Conditions	Ref.	Remarks
H25, cont.							
0.679	6.84(+3)	-29.2	37	C	μ=3.00M(ClO_4^-) HA=[Yb(III)Histidine]$^{3+}$	70Ja 71Ja	
5.52	5.52	-6.76	25	C	μ=3.00($NaClO_4$) HA=[Cu(II)Histidine]^{o}	72Wa	q:±1.7
5.66	6.86	-12.4	25	C	μ=3.00($NaClO_4$) HA=Cu(II)(Histidine)$_2$]$^-$	72Wa	q:±1.8
5.28	4.86	-4.52	25	C	μ=3.00($NaClO_4$) HA=[Cu(II)(Histidine)$_2$]^{o}	72Wa	q:±3.0
H26 HISTIDINE, methyl ester (1)			$C_7H_{11}O_2N_3$		see H25		
8.2	5.01(25°,+2)	4.6	25-50	T	μ=0.1M(KCl)	71Hd	
7.57	5.39(+2)	-0.7	25	C	μ=0.16M(KNO_3)	70Me	e: Imid-azole group
9.9	7.23(25°,+1)	0.3	25-50	T	μ=0.1M(KCl)	71Hd	c
9.97	7.37(+1)	0.3	25	C	μ=0.16M(KNO_3)	70Me	e:NH_3^+
H27 HISTIDINE, N(1)-acetyl-			$C_8H_{11}O_3N_3$		see H25		
7.21	-----	-----	20	C	μ≃0.01	71Ma	
H28 HISTIDINE, 1-methyl-			$C_7H_{11}O_2N_3$		see H25		
6.9	6.3(37°)	-6.1	22-37	T	μ=0.3	56Bb	c
H29 HOMOCYSTINE			$C_8H_{16}O_4N_2S_2$		[S-(CH_2)$_2$-CH(NH_2)-COOH]$_2$		
2.917	2.523(+1)	-1.76	25	T	μ=0.1M(KCl)	72Bg	e:CO_2H
10.711	8.676(0)	-3.76	25	T	μ=0.1M(KCl)	72Bg	e:NH_3^+
11.916	9.413(-1)	-3.09	25	T	μ=0.1M(KCl)	72Bg	e:NH_3^+
H30 HUMAN PLASMA ALBUMIN							
-0.835	3.506	-18.75	35	C	μ=0.005	42Sc	
-9.91	3.438	-48.9	55	T	--	54Ga	
H31 HYDRAZINE			H_4N_2		NH_2NH_2		
8.9	-0.67(+2)	33	25	C	μ=0.42	69Cc	
15.1	8.263(+1)	14.6	15	T	μ=0	67Sa	
9.7	7.94(+1)	-3.8	18	T	μ=0	65La	
12.9	8.142(+1)	6.9	20	T	μ=0	67Sa	
9.97	7.956(+1)	-3.0	25	C	μ=0	69Cc	
10.6	7.968(+1)	-0.7	25	T	μ=0	67Sa	

ΔH, kcal /mole	pK	ΔS, cal /mole°K	T,°C	M	Conditions	Ref.	Remarks
H31, cont.							
6.0	7.753(+1)	-16.1	35	T	μ=0	67Sa	

H32 HYDROAZOIC ACID HN_3 HN_3

ΔH, kcal /mole	pK	ΔS, cal /mole°K	T,°C	M	Conditions	Ref.	Remarks
3.60	4.72	-7.8	25	T	μ=0	65La	
3.8	4.64	-8.6	25	T	μ=0	66Bh	

H33 HYDROBROMIC ACID HBr HBr

ΔH, kcal /mole	pK	ΔS, cal /mole°K	T,°C	M	Conditions	Ref.	Remarks
-28.71	-20.68(0)	32.2	25	T	0.0555M(HBr)	66Fa	
-28.67	-19.89(0)	28.7	25	T	0.139M(HBr)	66Fa	
-28.61	-19.34(0)	26.4	25	T	0.277M(HBr)	66Fa	
-28.53	-18.73(0)	23.8	25	T	0.555M(HBr)	66Fa	
-28.52	-18.21(0)	21.6	25	T	0.925M(HBr)	66Fa	
-28.55	-18.03(0)	20.6	25	T	1.117M(HBr)	66Fa	
-28.55(0)	-17.78	19.5	25	T	1.388M(HBr)	66Fa	
-28.22	-17.39(0)	18.8	25	T	1.850M(HBr)	66Fa	
-27.86	-16.77(0)	17.2	25	T	2.775M(HBr)	66Fa	
-27.54	-16.24(0)	15.8	25	T	3.700M(HBr)	66Fa	
-27.03	-15.54(0)	15.3	25	T	4.629M(HBr)	66Fa	
-26.91	-15.0(0)	12.3	25	T	5.551M(HBr)	66Fa	

H34 HYDROCHLORIC ACID HCl HCl

ΔH, kcal /mole	pK	ΔS, cal /mole°K	T,°C	M	Conditions	Ref.	Remarks
13.34	(0)	-----	25	C	μ=4.727 x 10^{-3}M	66Ae	
13.45	(0)	-----	25	C	μ=7 x 10^{-3}M	66Ae	

H35 HYDROCYANIC ACID CHN HCN

ΔH, kcal /mole	pK	ΔS, cal /mole°K	T,°C	M	Conditions	Ref.	Remarks
11.33	9.63(0)	-4.05	10	C	μ=0	70Ca	
10.43	9.21(0)	-7.19	25	C	μ=0 ΔCp=-60	70Ca 62Ib	
10.4	9.19(0)	-7	25	C	μ=0	67Aa	
9.9	9.19(0)	-9.2	26	T	μ=0	66Bh	
9.57	8.88(0)	-10.26	40	C	μ=0	70Ca	
10.9	9.17(0)	-4.8	20	C	μ=0.1(KNO_3)	65Sb	
10.96	(0)	-----	18	C	--	82Ta	
-5.016	8.86(-2)	-57	30	T	μ=0 ΔCp=-85	73Bc	
					HA=$[Mo(IV)(OH)_2(CN)_4]^{2-}$		
-1.8	1.50(-2)	-12.8	25	C	μ=3($LiClO_4$)	70La	
					HA=$[Fe(II)H_2(CN)_6]^{2-}$		
-1	2.3(-2)	-14	25	C	μ≃0	67Ha	q:±1
					HA=$[Fe(II)H_2(CN_6]^{2-}$		
0.0	2.27(-2)	-10.3	25	C	μ≃0	62Ja	
					HA=$[Fe(II)H_2(CN)_6]^{2-}$		
-0.3	2.87(-3)	-14.2	25	C	μ=3($LiClO_4$)	70La	
					HA=$[Fe(II)H(CN)_6]^{3-}$		
-0.5	4.28(-3)	-21	25	C	μ=0	67Ha	q:±0.5
					HA=$[Fe(II)H(CN)_6]^{3-}$		
0.0	4.18(-3)	-19.1	25	C	μ≃0	62Ja	
					HA=$[Fe(II)H(CN)_6]^{3-}$		
0.98	4.70(0)	-18.2	30	T	μ=0 ΔCp=-338.2	74Ka	
					HA=$[Ni(II)H_2(CN)_4]$		
-1.7	6.63(-1)	-35.9	30	T	μ=0 ΔCp=-185.6	74Ka	
					HA=$[Ni(II)H(CN)_4]^-$		

H36 HYDROFLUORIC ACID HF HF

ΔH, kcal /mole	pK	ΔS, cal /mole°K	T,°C	M	Conditions	Ref.	Remarks
-0.700	0.524(0)	-4.73	25	C	μ=1.00M($NaClO_4$)	71Aa	

ΔH, kcal /mole	pK	ΔS, cal /mole°K	T,°C	M	Conditions	Ref.	Remarks
H36, cont.							
-2.76	(0)	-----	5	C	μ=0	73Vc	
-2.96	(0)	-----	15	C	μ=0	73Vc	
-3.18	3.17(0)	-25.2	25	C	μ=0	67Aa	
-3.19	3.13(0)	-25.02	25	C	μ=0 Equations are given for calculating ΔH and ΔS at other temperatures (5-25°C)	73Vc	
-3.180	3.177(0)	-25.2	25	C	μ=0	53Ha	
-3.1	3.16(0)	-24.8	25	C	μ=0	73Kd	
-3.03	3.17(25°,0)	-24.3	5-40	T	μ=0	70Bb	
-2.93	2.90(0)	-23.1	25	T	μ=0.05	67Aa	
-3.22	2.92(0)	-24.2	25	C	μ=0.5	73Kd	
-2.6	3.959(0)	-22	1.5-35	T	μ=0.5($NaClO_4$)	59Kb	
-2.44	(0)	-----	5	C	μ=1($NaClO_4$)	73Vc	
-2.60	(0)	-----	15	C	μ=1($NaClO_4$)	73Vc	
-2.54	2.96(0)	-22.07	25	C	μ=1(NaCl)	73Vc	
-2.86	2.96(0)	-23.14	25	C	μ=1($NaClO_4$)	73Vc	
-3.25	3.00(0)	-24.6	25	C	μ=1.0	73Kd	
-2.91	2.95(0)	-23.3	25	C	μ=1.00M($NaClO_4$)	71Aa	
-3.24	2.944(25°,0)	-24.3	25	C	μ=1.0M($NaClO_4$)	70Wa	
-3.5	2.944(25°,0)	-25.2	25-55	T	μ=1.0M($NaClO_4$)	70Wa	q:±0.5
-2.18	(0)	-----	5	C	μ=2($NaClO_4$)	73Vc	
-2.42	(0)	-----	15	C	μ=2($NaClO_4$)	73Vc	
-2.08	3.12(0)	-21.6	25	C	μ=2(NaCl)	73Vc	
-2.62	3.12(0)	-23.06	25	C	μ=2($NaClO_4$)	73Vc	
-3.50	3.14(0)	-26.0	25	C	μ=2.0	73Kd	
-1.98	(0)	-----	5	C	μ=3($NaClO_4$)	73Vc	
-2.21	(0)	-----	15	C	μ=3($NaClO_4$)	73Vc	
-2.85	3.33(0)	-24.8	25	C	μ=3.0	73Kd	
-2.37	3.32(0)	-23.14	25	C	μ=3($NaClO_4$)	73Vc	
-1.59	3.32(0)	-20.53	25	C	μ=3(NaCl)	73Vc	
-3.18	3.13(0)	-25.2	25	C	--	53Ha	
-2.75	3.96(0)	-22.5	25	T	--	69Ma	
-4.2	3.56	-30.4	25	C	μ=0 R: $HF_2^- = H^+ + 2F^-$	73Vc	
-3.32	3.53	-27.3	25	C	μ=1(NaCl) R: $HF_2^- = H^+ + 2F^-$	73Vc	
-3.74	3.53	-28.7	25	C	μ=1($NaClO_4$) R: $HF_2^- = H^+ + 2F^-$	73Vc	
-2.69	3.83	-26.6	25	C	μ=2(NaCl) R: $HF_2^- = H^+ + 2F^-$	73Vc	
-3.60	3.83	-29.6	25	C	μ=2($NaClO_4$) R: $HF_2^- = H^+ + 2F^-$	73Vc	
-2.08	4.18	-26.1	25	C	μ=3(NaCl) R: $HF_2^- = H^+ + 2F^-$	73Vc	
-3.20	4.18	29.9	25	C	μ=3($NaClO_4$) R: $HF_2^- = H^+ + 2F^-$	73Vc	
-5.2	5.24	-35	25	T	R: $HF_2^- = H^+ + 2F^-$	69Ma	

H37 HYDROGEN PEROXIDE H_2O_2 H_2O_2

ΔH, kcal /mole	pK	ΔS, cal /mole°K	T,°C	M	Conditions	Ref.	Remarks
8.2	11.75	-25.7	20	T	μ=0	49Ea	
7.36	11.58	-28	25	-	--	57Sb	

H38 HYDROGEN SULFIDE H_2S H_2S

ΔH, kcal /mole	pK	ΔS, cal /mole°K	T,°C	M	Conditions	Ref.	Remarks
5.3	6.96(0)	-14.2	25	-	--	69La	
11.9	12.90(-1)	-19	25	-	--	69La	

H39 HYDROXIDE RADICAL HO OH•

ΔH, kcal /mole	pK	ΔS, cal /mole°K	T,°C	M	Conditions	Ref.	Remarks
10	11.9(25°,0)	-20.9	1.6-38	T	HA=OH•	71Bb	c,q:±2

ΔH, kcal /mole	pK	ΔS, cal /mole°K	T,°C	M	Conditions	Ref.	Remarks
H40 HYDROXYLAMINE			NH_3O		NH_2OH		
9.5	6.186	4.6	15	T	μ=0	65La	
9.4	6.063	4.2	20	T	μ=0	65La	
9.3	5.948	3.8	25	T	μ=0	65La	
9.0	5.730	3.0	35	T	μ=0	65La	
H41 HYPOBROMOUS ACID			HOBr		HBrO		
9.300	8.83(0)	-8.2	15	T	μ=0	56Ka	
6.800	8.60(0)	-16.6	25	T	μ=0	56Ka	
5.000	8.47(0)	-22.6	35	T	μ=0	56Ka	
3.800	8.37(0)	-26.2	45	T	μ=0	56Ka	
H42 HYPOCHLOROUS ACID			HOCl		HClO		
3.320	7.424(0)	-22.8	25	T	μ=0	37Pa	
3.90	7.537(0)	-21.4	25	T	μ=0	66Md	
H43 HYPOXANTHINE			$C_5H_4ON_4$		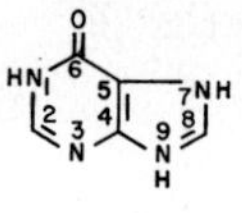		
2.5	1.9(+1)	0.3	25	C	μ=0	70Wd	
2.95	1.79(+1)	1.7	25	C	μ=0	70Cb	e:N_7
8.0	8.8(0)	-13.4	25	C	μ=0	70Wd	
7.88	8.91(0)	-14.4	25	C	μ=0	70Cb	e: N_1-C_6O
7.2	8.8(0)	-16.1	25	T	μ=0.01	64Sd	
9.81	12.64(-1)	-23.2	10	C	μ=0	70Cb	e:N_9
10.0	12.0(-1)	-21.5	25	C	μ=0	70Wd	
9.53	12.07(-1)	-23.3	25	C	μ=0	70Cb	e:N_9
9.00	11.81(-1)	-25.3	40	C	μ=0	70Cb	e:N_9

ΔH, kcal /mole	pK	ΔS, cal /mole$^\circ K$	T,$^\circ$C	M	Conditions	Ref.	Re- marks

I

I1 IMIDAZOLE $C_3H_4N_2$

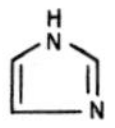

ΔH, kcal /mole	pK	ΔS, cal /mole$^\circ K$	T,$^\circ$C	M	Conditions	Ref.	Re- marks
8.66	7.581(+$\underline{1}$)	-2.99	0	T	μ=0	66Da	
8.70	7.467(+$\underline{1}$)	-2.82	5	T	μ=0	66Da	
8.74	7.334(+$\underline{1}$)	-2.69	10	T	μ=0	66Da	
8.88	7.210(+$\underline{1}$)	-2.17	15	T	μ=0 ΔCp=4.81	69Lb	
8.77	7.216(+$\underline{1}$)	-2.59	15	T	μ=0	66Da	
8.91	7.095(+$\underline{1}$)	-2.08	20	T	μ=0 ΔCp=4.90	69Lb	
8.79	7.103(+$\underline{1}$)	-2.53	20	T	μ=0	66Da	
8.78	6.99(+$\underline{1}$)	-2.5	25	C	μ=0	69Cc	
8.71	6.993(+$\underline{1}$)	-2.8	25	C	μ=0	70Wd	
8.93	6.983(+$\underline{1}$)	-2.00	25	T	μ=0 ΔCp=4.98	69Lb	
7.7	6.95(+$\underline{1}$)	-6.0	25	T	μ=0	57Na	
8.79	6.993(+$\underline{1}$)	-2.51	25	T	μ=0	66Da	
8.79	6.887(+$\underline{1}$)	-2.52	30	T	μ=0	66Da	
8.77	6.784(+$\underline{1}$)	-2.57	35	T	μ=0	66Da	
8.98	6.771(+$\underline{1}$)	-1.83	35	T	μ=0 ΔCp=5.15	69Lb	
7.750	6.98(+$\underline{1}$)	-6	15-35	T	μ=0	64Gb	c
8.75	6.685(+$\underline{1}$)	-2.66	40	T	μ=0	66Da	
8.70	6.589(+$\underline{1}$)	-2.79	45	T	μ=0	66Da	
8.65	6.497(+$\underline{1}$)	-2.95	50	T	μ=0	66Da	
8.53	(+$\underline{1}$)	-----	20	C	μ=0.01	71Ma	
8.78	6.986(+$\underline{1}$)	-2.5	25	C	μ=0.011	68Ce	
7.5	6.37(25°,+$\underline{1}$)	-4	0-25	T	μ=0.135	55Ma	c
7.5	7.12(25°,+$\underline{1}$)	-6.7	15-35	T	μ=0.15	53Ta	c
6.0	7.14(23°,+$\underline{1}$)	-0.51	4.5-22.5	T	μ=0.16	54Ea	c
9.1	7.09(+$\underline{1}$)	-2	25	C	0.16M(KNO$_3$)	64Ba	
8.79	(+$\underline{1}$)	-----	25	C	μ=0,2(NaCl)	62Wa	
9.03	7.06(+$\underline{1}$)	-2.0	25	C	0.3M(NaClO$_4$)	67Hc	
7.75	6.98(+$\underline{1}$)	-6	25	-	--	66Ha	
17.6	14.44($\underline{0}$)	-7	25	T	μ=0	66Ha	
17.6	10.58($\overline{25}^\circ$,$\underline{0}$)	-7	15-35	T	μ=0	64Gb	
11.4	7.58(25°)	-9	15-35	T	pH=9-12 HA=[Fe(III)(My)(Imida-zole)] My=Sperm whale ferrimyoglobin	64Gb	c

I2 IMIDAZOLE, 4(5)-2'-aminoethyl- $C_5H_9N_3$ see I1

ΔH, kcal /mole	pK	ΔS, cal /mole$^\circ K$	T,$^\circ$C	M	Conditions	Ref.	Re- marks
9.25	6.09	3.1	25	C	0.3M(NaClO$_4$)	67Hc	
10.28	9.82	-10.4	25	C	0.3M(NaClO$_4$)	67Hc	

I3 IMIDAZOLE, 4(5)-aminomethyl- $C_4H_7N_3$ see I1

ΔH, kcal /mole	pK	ΔS, cal /mole$^\circ K$	T,$^\circ$C	M	Conditions	Ref.	Re- marks
7.43	4.77(+$\underline{2}$)	3.1	25	C	0.3M(NaClO$_4$)	67Hc	
9.73	9.15(+$\underline{1}$)	-9.3	25	C	0.3M(NaClO$_4$)	67Hc	

I4 IMIDAZOLE, 2,4-dimethyl- $C_5H_8N_2$ see I1

ΔH, kcal /mole	pK	ΔS, cal /mole$^\circ K$	T,$^\circ$C	M	Conditions	Ref.	Re- marks
9.2	8.38(+$\underline{1}$)	-7.4	25	T	μ=0	57Na	

I5 IMIDAZOLE, 2-ethyl-4-methyl- $C_6H_{10}N_2$ see I1

ΔH, kcal /mole	pK	ΔS, cal /mole$^\circ K$	T,$^\circ$C	M	Conditions	Ref.	Re- marks
9.45	(+$\underline{1}$)	-----	20	C	$\mu\approx$0.01	71Ma	

ΔH, kcal /mole	pK	ΔS, cal /mole°K	T, °C	M	Conditions	Ref.	Remarks
I6 IMIDAZOLE, 2-hydroxymethyl-			$C_4H_6ON_2$		see I1		
9.99	7.660(+1)	-1.54	25	C	μ=3.0(ClO_4^-)	69Wa	
I7 IMIDAZOLE, 4-hydroxymethyl-			$C_4H_6ON_2$		see I1		
9.32	7.419(+1)	-2.69	25	C	μ=3.0(ClO_4^-)	69Wa	
I8 IMIDAZOLE, 1-methyl-			$C_4H_6N_2$		see I1		
7.93	(+1)	-----	20	C	μ≈0.01	71Ma	
I9 IMIDAZOLE, 2-methyl-			$C_4H_6N_2$		see I1		
9.24	(+1)	-----	20	C	μ≈0.01	71Ma	
I10 IMIDAZOLE, 4-methyl-			$C_4H_6N_2$		see I1		
8.6	7.55(+1)	-5.7	25	T	μ=0	57Na	
I11 IMIDAZOLE, 2-(2'-pyridyl)			$C_8H_7N_3$		see I1		
9.2	8.98(25°)	-10	0-40	T	μ=0.005	55Ha	c
13.7	9.01	4.73	25	T	μ=0.044	62Hb	
6.88	5.47(+1)	-1.9	25	C	μ=0.1M(NO_3^-)	70Eb	
10.01	8.54(25°)	-5	0-40	T	μ=0.05 S: 50 w % Dioxane	55Ha	
6.9	6.46	-6.42	25	T	μ=0.107 HA=[Fe(II)(2-(2'-pyridyl)imidazole)]	56Fa	
I12 IMIDAZOLE, 4-(2'-pyridyl)-			$C_8H_7N_3$		see I1		
5.76	5.49(+1)	-5.4	25	C	μ=0.1M(NO_3^-)	70Eb	
I13 INOSINE			$C_{10}H_{11}O_5N_4$		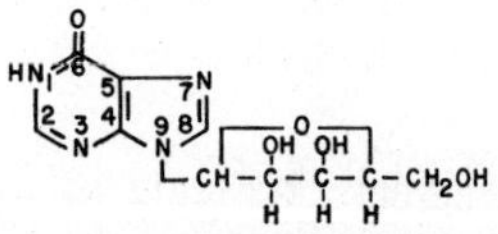		
6.50	8.96(0)	-19.2	25	C	μ=0	70Cb	e: N_1-C_6O
7.2	8.9(0)	-16.4	25	T	μ=0.1	64Sd	
10.4	12.99(-1)	-22.8	10	C	μ=0	70Cb	e: Ribose OH
10.65	12.36(-1)	-20.9	25	C	μ=0	70Cb	e: Ribose OH
10.60	11.84(-1)	-20.3	40	C	μ=0	70Cb	e: Ribose OH
I14 INOSINE, 5'-diphosphoric acid			$C_{10}H_{13}O_{11}N_4P_2$		see I13		
-1.34	7.18(-2)	-37.4	25	T	μ=0	65Pe	

ΔH, kcal /mole	pK	ΔS, cal /mole°K	T, °C	M	Conditions	Ref.	Remarks
I15 INOSINE, 5'-monophosphoric acid IMP			$C_{10}H_{12}O_8N_4P$		see I13		
-1.43	6.66(-1)	-35.3	25	T	μ=0	65Pe	
I16 INOSINE, 5'-triphosphoric acid			$C_{10}H_{14}O_{14}N_4P_3$		see I13		
2.3	2.2(25°,-2)	-2	25-45	T	μ=0.10M(KNO_3)	69Ka	c
1.4	6.92(25°,-3)	-27	25-45	T	μ=0.10M(KNO_3)	69Ka	c,q: ±0.1
-1.61	7.68(-4)	-40.5	25	T	μ=0	65Pe	
I17 INSULIN, iodinated beef							
13.4	11.12(25°)	-5.95	0-35	T	μ=2	59Gc	c,e:ε-amino group
14.3	12.50(25°)	-9.21	0-35	T	μ=2	59Gc	c,e: Quanidinium group
I18 IODIC ACID			HO_3I		HIO_3		
2.48	0.815(30°,0)	4.33	25-35	T	μ=0	75Da	g:41La
-2.4	0.81(0)	-11.6	25	-	--	65Wa	
5.06	0.920(30°,0)	12.51	25-35	T	μ=0 S: Formamide	75Da	c
I19 IODIC ACID, per			H_5O_6I		H_5IO_6		
2.55	(0)	-----	25	C	μ=0.13	68Ma	
I20 ISOHISTAMINE			$C_5H_9N_3$		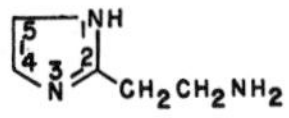		
8.61	6.036(+2)	1.3	25	C	μ=0.1M	70Ec	
11.60	9.274(+1)	-3.4	25	C	μ=0.1M	70Ec	
I21 ISOQUINOLINE			C_9H_7N		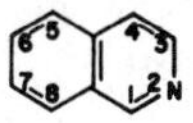		
5.925	5.07	-3.30	25	C	--	60Sa	
I22 ISOQUINOLINE, 1-amino-			$C_9H_8N_2$		see I21		
10.5	7.62(20°)	0.1	5-35	T	C=0.01M	59Eb	c
I23 ISOQUINOLINE, 3-amino-			$C_9H_8N_2$		see I21		
5.0	5.05(20°)	-6.0	5-35	T	C=0.005M	59Eb	c

ΔH, kcal /mole	pK	ΔS, cal /mole°K	T,°C	M	Conditions	Ref.	Remarks
I24 ISOSERINE			$C_3H_7O_3N$		$H_2NCH_2CH(OH)CO_2H$		
10.09	9.25(0)	8.5	25	C	μ=0.16	70Lb	e:NH_3^+

L

L1 LEUCINE $C_6H_{13}O_2N$ $(CH_3)_2CHCH_2CH(NH_2)CO_2H$

ΔH, kcal /mole	pK	ΔS, cal /mole°K	T, °C	M	Conditions	Ref.	Remarks
1.180	2.383(+1)	-6.5	1.0	T	μ=0	37Sa	
0.860	2.348(+1)	-7.7	12.5	T	μ=0	37Sa	
0.420	2.328(+1)	-9.2	25	T	μ=0 ΔCp=-37.8	37Sa	
						39Hb	
-0.090	2.327(+1)	-10.9	37.5	T	μ=0	37Sa	
-0.700	2.333(+1)	-12.8	50	T	μ=0	37Sa	
10.870	10.454(0)	-8.2	1.0	T	μ=0	37Sa	
10.940	10.095(0)	-7.9	12.5	T	μ=0	37Sa	
10.900	9.744(0)	-8.0	25	T	μ=0	37Sa	
10.730	9.434(0)	-8.6	37.5	T	μ=0	37Sa	
10.410	9.142(0)	-9.6	50	T	μ=0	37Sa	
10.8	(0)	-----	20	C	μ≃0.01	71Ma	e:α-NH_3^+
10.73	9.54(0)	-7.6	25	C	μ=0.16(KNO_3)	70Md	

L2 LEUCINE, ethyl ester $C_8H_{17}O_2N$ $(CH_3)_2CHCH_2CH(NH_2)CO_2CH_2CH_3$

ΔH, kcal /mole	pK	ΔS, cal /mole°K	T, °C	M	Conditions	Ref.	Remarks
10.46	7.66(25°,+1)	0	25-50	T	μ=0.1M	67Hb	c,e: NH_3^+

L3 LEUCINE, N-glycyl- $C_8H_{16}O_3N_2$ $(CH_3)_2CHCH_2CH(NHCOCH_2NH_2)CO_2H$

ΔH, kcal /mole	pK	ΔS, cal /mole°K	T, °C	M	Conditions	Ref.	Remarks
-0.752	3.1800(+1)	-17.07	25	T	μ=0	57Ka	
10.600	8.327(0)	-2.5	25	T	μ=0	75Kb	e:NH_3^+

L4 LYSINE $C_6H_{14}O_2N_2$ $H_2N(CH_2)_4CH(NH_2)CO_2H$

ΔH, kcal /mole	pK	ΔS, cal /mole°K	T, °C	M	Conditions	Ref.	Remarks
0.30	2.18(25°,+2)	-9.0	0&25	T	μ=0.01	33Ga	b,c
12.88	(+1)	-----	20	C	μ≃0.01	71Ma	e:ε-NH_3^+
12.80	8.95(25°,+1)	2.0	0&25	T	μ=0.01	33Ga	b,c
12.3	8.95(+1)	0	25	T	0.1M(KCl)	30Sa	
10.68	9.116(25°,+1)	-5.9	25-50	T	μ=0.1M	72Ha	c
11.0	9.53(25°,+1)	-6.70	0&25	T	μ=2	59Gc	c
10.50	(0)	-----	20	C	μ≃0.01	71Ma	e:α-NH_3^+
11.60	10.53(25°,0)	-9.30	0&25	T	μ=0.01	33Ga	b,c
11.60	10.55(0)	-9.3	25	T	μ=0.02	30Ma	
11.6	10.53(0)	-9	25	T	0.1M(KCl)	30Sa	
13.05	10.902(25°,0)	-6.1	25-50	T	μ=0.1M	72Ha	c,e: α-NH_3^+
13.3	10.94(25°,0)	-5.43	0-35	T	μ=2	59Gc	c

L5 LYSINE, methyl ester $C_7H_{16}O_2N_2$ $H_2N(CH_2)_4CH(NH_2)CO_2CH_3$

ΔH, kcal /mole	pK	ΔS, cal /mole°K	T, °C	M	Conditions	Ref.	Remarks
11.64	6.964(25°,+1)	7.2	25-50	T	μ=0.1M	72Ha	c,e: α-NH_3^+, q:+0.85
12.89	10.251(25°,0)	-3.7	25-50	T	μ=0.1M	72Ha	c,e: ε-NH_3^+

L6 LYSINE, N^{α}-acetyl- $C_8H_{16}O_3N_2$ $H_2N(CH_2)_4CH(NHCOCH_3)CO_2H$

ΔH, kcal /mole	pK	ΔS, cal /mole°K	T, °C	M	Conditions	Ref.	Remarks
12.52	-----	-----	20	C	μ≃0.01	71Ma	e:ε-NH_3^+

L7 LYSINE, lysyl- $C_{12}H_{26}O_3N_4$ $H_2N(CH_2)_4CH(NH_2)CONH(CH_2)_4$-$CH(NH_2)CO_2H$

ΔH, kcal /mole	pK	ΔS, cal /mole°K	T, °C	M	Conditions	Ref.	Remarks
L7, cont.							
2.00	1.95(25°,+2)	-2.2	0&25	T	μ=0.01	33Ga	b,c
12.7	8.17(25°,+1)	5.4	0&25	T	μ=0.01	33Ga	b,c
11.35	9.45(25°,0)	-5.3	0&25	T	μ=0.01	33Ga	b,c
13.30	10.63(25°,-1)	-4.0	0&25	T	μ=0.01	33Ga	b,c
L8 LYSOZYME							
3.1	4.42(25°)	-14	1.6-40	T	μ=0.15M(KCl) Ionization from Asp-52 group	72Pe	c
1.6	5.21(25°)	-22	1.6-40	T	μ=0.15M(KCl) Ionization from Glu-35 group	72Pe	c
0.358	-----	-----	25	C	μ=0.15 pH 2.0-4.6 Amino acid residue - 1 α-carboxyl	72Bc	
1.004	-----	-----	25	C	μ=0.15 pH 2.0-4.6 Amino acid residue - 4 Asp	72Bc	
0.120	-----	-----	25	C	μ=0.15 pH 2.0-4.6 Amino acid residue - 1 Glu	72Bc	
1.076	-----	-----	25	C	pH 4.6-6.5 Amino acid residue - 1 Asp	72Bc	
0.120	-----	-----	25	C	pH 4.6-6.5 Amino acid residue - 1 Glu	72Bc	
0.836	-----	-----	25	C	pH 4.6-6.5 Amino acid residue - 2 Unidentified	72Bc	
5.975	-----	-----	25	C	pH 6.5-7.5 Amino acid residue - 1 His	72Bc	
10.516	-----	-----	25	C	pH 7.5-8.25 Amino acid residue - 1 α-amino	72Bc	
5.975	-----	-----	25	C	pH 8.25-10.25 Amino acid residue - 1 Tyr	72Bc	
12.428	-----	-----	25	C	pH 8.25-10.25 Amino acid residue - 4 Lys	72Bc	
5.975	-----	-----	25	C	pH 10.25-12.0 Amino acid residue - 1 Tyr	72Bc	
12.667	-----	-----	25	C	pH 10.25-12.0 Amino acid residue - 2 Lys	72Bc	
L9 LYXOSE			$C_5H_{10}O_5$		$HOCH_2(CHOH)_3CHO$		
8.42	12.48	-27.4	10	C	μ=0	70Cd	
8.0	12.11	-28.6	25	C	μ=0	70Cd 66Ib	

ΔH, kcal /mole	pK	ΔS, cal /mole°K	T, °C	M	Conditions	Ref.	Remarks

M

ΔH, kcal /mole	pK	ΔS, cal /mole°K	T, °C	M	Conditions	Ref.	Remarks
M1 MALEIC ACID			$C_4H_4O_4$		$HO_2CCH:CHCO_2H$		
0.08	1.910(0)	-8.5	25	C	μ=0	67Ca	
-0.26	1.91(0)	-10	25	C	μ=0	63Ea	
-0.15	1.599(0)	-7.36	25	C	μ=1M($NaClO_4$)	73Dc	q:±0.04
-0.83	6.332(-1)	-31.8	25	C	μ=0	67Ca	
-0.21	6.33(-1)	-29	25	C	μ=0	63Ea	
-0.18	5.618(-1)	-26.4	25	C	μ=1M($NaClO_4$)	73Dc	q:±0.03
5.3	2.73(24°,+2)	5.3	0-30	T	μ=1.0M($LiClO_4$) HA=[Cr(III)$(NH_3)_5$-(Maleic acid)]$^{2+}$	71Da	c
-1.23	2.462(0)	-15.4	25	C	μ=0 HA=[Co(III)$(NH_3)_5$-(Maleic acid)]^{o}	74La	
M2 MALEIC ACID, ethylene copolymer			$[C_6H_8O_4]_n$		$[-CH_2-CH_2-CH(CO_2H)CH(CO_2H)-]_n$		
0.24	4.86	-21.4	25	C	$C_L \simeq 10^{-2}$M	71Fa	
					For following reactions C_L=1.7 x 10^{-2} monomoles per liter α=1 corresponds to half-neutralization 1:1 Ethylene-maleic acid		
0.13	3.80	-17.0	25	C	α=0.0	73Cb	
0.10	4.21	-19.0	25	C	α=0.1	73Cb	
0.07	4.34	-19.6	25	C	α=0.2	73Cb	
0.04	4.44	-20.2	25	C	α=0.3	73Cb	
0.00	4.50	-20.6	25	C	α=0.4	73Cb	
-0.02	4.55	-20.9	25	C	α=0.5	73Cb	
-0.06	4.58	-21.2	25	C	α=0.6	73Cb	
-0.09	4.61	-21.4	25	C	α=0.7	73Cb	
					For following reactions S: 0.05M[$(CH_3)_4NClO_4$] α=degree of neutralization with $(CH_3)_4NOH$		
0.0	4.26	-19.5	25	C	α=0.70	74Cc	
-0.21	4.31	-20.4	25	C	α=0.80	74Cc	
					For following reactions S: 0.05M[$(CH_3)_4NClO_4$] α=degree of neutralization with $(CH_3)_4NOH$		
-1.65	7.47	-39.7	25	C	α=1.20	74Cc	
-1.65	7.66	-40.6	25	C	α=1.30	74Cc	
M3 MALEIC ACID, ethyl vinyl ether copolymer			$[C_{10}H_{14}O_5]_n$		$[-C(C_2H_5)(OCH:CH_2)CH_2CH(CO_2H)-CH(CO_2H)-]_n$		
					For following reactions C_L=1.7 x 10^{-2} monomoles per liter α=Degree of neutralization α=1 corresponds to half-neutralization		
0.05	4.773	-15.8	25	C	α=0.0	73Cb	
0.09	5.577	-18.4	25	C	α=0.1	73Cb	
0.14	6.136	-20.1	25	C	α=0.2	73Cb	
0.52	7.082	-22.0	25	C	α=0.7	73Cb	
M4 MALEIC ACID, isobutene copolymer			$[-C_8H_{12}O_4-]_n$		see M2		
					For following reactions S: 0.05M[$(CH_3)_4NClO_4$] α=degree of neutralization		
-0.98	3.29	-18.3	25	C	α=0.70	74Cc	
-0.76	3.33	-17.8	25	C	α=0.80	74Cc	

ΔH, kcal /mole	pK	ΔS, cal /mole°K	T,°C	M	Conditions	Ref.	Remarks
M4, cont.							
1.66	9.91	-39.9	25	C	α=1.20	74Cc	
1.66	10.10	-40.6	25	C	α=1.30	74Cc	
M5 MALEIC ACID, propylene copolymer $[C_7H_{10}O_4]_n$					$[-CH(CH_3)CH_2CH(CO_2H)CH-(CO_2H)-]_n$		
					For following reactions C_L=1.3 x 10^{-2} monomoles per liter α=Degree of neutralization α=1 corresponds to half-neutralization		
0.0	3.50	-16.0	25	C	α=0.0	73Cb	
-0.22	3.92	-18.7	25	C	α=0.1	73Cb	
-0.40	4.09	-20.1	25	C	α=0.2	73Cb	
-0.51	4.20	-21.0	25	C	α=0.3	73Cb	
-0.56	4.29	-21.5	25	C	α=0.4	73Cb	
-0.56	4.36	-21.8	25	C	α=0.5	73Cb	
-0.53	4.41	-22.0	25	C	α=0.6	73Cb	
-0.49	4.44	-22.0	25	C	α=0.7	73Cb	
					For following reactions S: 0.05M[$(CH_3)_4NClO_4$] α=Degree of neutralization with $(CH_3)_4NOH$		
-0.09	4.02	-18.7	25	C	α=0.70	74Cc	
-0.09	4.06	-18.9	25	C	α=0.80	74Cc	
-0.09	8.60	-39.7	25	C	α=1.20	74Cc	
-0.09	8.83	-40.7	25	C	α=1.30	74Cc	
0.09	4.04	18.8	25	C	α=0.70-0.80	74Cc	
0.09	8.71	40.2	25	C	α=1.20-1.30	74Cc	
M6 MALONIC ACID			$C_3H_4O_4$		$CH_2(CO_2H)_2$		
0.29	2.826(0)	-12.0	25	C	μ=0	67Ca	
0.37	2.59(0)	-10.6	25	C	μ=1.00M($NaClO_4$)	73Da	
4.61	4.38(0)	-4.59	25	T	μ=0 S: Formamide	75Db	
-0.48	5.06(-1)	-24.8	25	C	μ=1.00M($NaClO_4$)	73Da	
-0.92	(-1)	-----	--	-	μ=0.10M($NaClO_4$)	72Db	
4.19	8.34(-1)	-24.12	25	T	μ=0 S: Formamide	75Db	
-0.275	3.116(0)	-15.2	25	C	μ=0 HA=[Co(III)$(NH_3)_5$-(Malonic acid)]	74La	
M7 MALONIC ACID, dinitrile-			$C_3H_2N_2$		$CH_2(CN)_2$		
13.4	11.20(0)	-6.4	25	C	μ=0	65Bb	
M8 MALONIC ACID, mononitrile-			$C_3H_3O_2N$		$HOOCCH_2CN$		
-0.032	2.4447(0)	-11.3	5	T	μ=0	56Fa	
-0.283	2.4467(0)	-12.2	10	T	μ=0	56Fa	
-0.502	2.4523(0)	-13.0	15	T	μ=0	56Fa	
-0.705	2.4597(0)	-13.7	20	T	μ=0	56Fa	
-0.888	2.4600(0)	-14.3	25	T	μ=0	56Fa	
-0.840	2.468(0)	-14.1	25	T	μ=0	57Ka	
-1.074	2.4818(0)	-14.9	30	T	μ=0	56Fa	
-1.255	2.4955(0)	-15.5	35	T	μ=0	56Fa	
-1.465	2.5107(0)	-16.2	40	T	μ=0	56Fa	
-1.705	2.5281(0)	-16.9	45	T	μ=0	56Fa	

ΔH, kcal /mole	pK	ΔS, cal /mole°K	T, °C	M	Conditions	Ref.	Remarks
M8, cont.							
-0.024	2.44(0)	-11.3	5	T	--	66Cc	
-0.890	2.47(0)	-14.3	25	T	--	66Cc	
-0.508	2.458(35°,0)	-13	0&35	T	--	39Wa	b,c
-1.695	2.53(0)	-16.9	45	T	--	66Cc	
M9 MALONIC ACID, 2-amino-2(hydroxy-methyl)-	$C_4H_7O_5N$				$HO_2CC(CH_2OH)(NH_2)CO_2H$		
11.35	8.07(0)	-1.12	25	C	μ=0	68Ob	
M10 MALONIC ACID, 2-amino-2-methyl-	$C_4H_7O_4N$				$HO_2CC(CH_3)(NH_2)CO_2H$		
11.93	8.80	-0.26	25	C	μ=0	68Ob	
M11 MALONIC ACID, 1-cyano-2(p-dicyanomethylphenyl)-, dinitrile	$C_{14}H_5N_5$				$(NC)_2C{:}C(CN)C_6H_4CH(CN)_2$		
-0.4	0.75(0)	-4.7	25	T	μ=0	65Bb	
M12 MALONIC ACID, diethyl-	$C_7H_{12}O_4$				$(C_2H_5)_2C(CO_2H)_2$		
-0.506	2.1290(0)	-11.5	10	T	μ=0	70Ib	
-0.570	2.1360(0)	-11.8	15	T	μ=0	70Ib	
-0.585	2.1439(0)	-11.8	20	T	μ=0	70Ib	
-1.25	2.211(0)	-14.3	25	C	μ=0	67Ca	
-0.657	2.1513(0)	-12.0	25	T	μ=0	70Ib	
-0.849	2.1602(0)	-12.7	30	T	μ=0	70Ib	
-1.170	2.1721(0)	-13.7	35	T	μ=0	70Ib	
-1.564	2.1874(0)	-15.0	40	T	μ=0	70Ib	
-0.033	7.3998(-1)	-34.0	10	T	μ=0	70Ib	
-0.326	7.4012(-1)	-35.0	15	T	μ=0	70Ib	
-0.598	7.4079(-1)	-35.9	20	T	μ=0	70Ib	
-0.82	7.292(-1)	-36.1	25	C	μ=0	67Ca	
-0.844	7.4166(-1)	-36.8	25	T	μ=0	70Ib	
-1.063	7.4281(-1)	-37.5	30	T	μ=0	70Ib	
-1.248	7.4414(-1)	-38.1	35	T	μ=0	70Ib	
-1.397	7.4575(-1)	-38.6	40	T	μ=0	70Ib	
M13 MALONIC ACID, diisopropyl-	$C_9H_{16}O_4$				$[(CH_3)_2CH]_2C(CO_2H)_2$		
0.618	2.1410(0)	-7.6	10	T	μ=0	70Ib	
0.543	2.1341(0)	-7.9	15	T	μ=0	70Ib	
0.341	2.1283(0)	-8.6	20	T	μ=0	70Ib	
0.063	2.1240(0)	-9.5	25	T	μ=0	70Ib	
-0.230	2.1267(0)	-10.5	30	T	μ=0	70Ib	
-0.468	2.1307(0)	-11.3	35	T	μ=0	70Ib	
-0.573	2.1361(0)	-11.6	40	T	μ=0	70Ib	
-0.046	8.8274(-1)	-40.5	10	T	μ=0	70Ib	
-0.397	8.8300(-1)	-41.8	15	T	μ=0	70Ib	
-0.709	8.8375(-1)	-42.8	20	T	μ=0	70Ib	
-0.975	8.8480(-1)	-43.8	25	T	μ=0	70Ib	
M14 MALONIC ACID, diisopropyl-, mononitrile	$C_9H_{15}O_2N$				$CH_3CH(CH_3)C[CH(CH_3)_2](CN)CO_2H$		
-2.783	2.3931(0)	-20.95	5	T	μ=0 ΔCp=-31.9	69La 65Ia	

ΔH, kcal /mole	pK	ΔS, cal /mole°K	T, °C	M	Conditions	Ref.	Remarks
M14, cont.							
-2.941	2.4326(0)	-21.51	10	T	μ=0 ΔCp=-31.5	69La 65Ia	
-3.098	2.4734(0)	-22.06	15	T	μ=0 ΔCp=-31.0	69La 65Ia	
-3.252	2.5143(0)	-22.59	20	T	μ=0 ΔCp=-30.4	69La 65Ia	
-3.402	2.5557(0)	-23.10	25	T	μ=0 ΔCp=-29.6	69La 65Ia	
-3.548	2.5980(0)	-23.58	30	T	μ=0 ΔCp=-28.7	69La 65Ia	
-3.689	2.6403(0)	-24.05	35	T	μ=0 ΔCp=-27.6	69La 65Ia	
-3.824	2.6828(0)	-24.48	40	T	μ=0 ΔCp=-26.3	69La 65Ia	
-3.952	2.7255(0)	-24.89	45	T	μ=0 ΔCp=-24.9	69La 65Ia	
M15 MALORIC ACID, ethyl (isoamyl)-			$C_{10}H_{18}O_4$		$CH_3CH_2C(CO_2H)_2(CH_2)_2CH(CH_3)$-$CH_3$		
-1.31	2.50(0)	-14.2	25	C	μ=0	67Ca	
-0.36	7.31(-1)	-34.7	25	C	μ=0	67Ca	
M16 MALONIC ACID, isopropyl-, mononitrile			$C_6H_9O_2N$		$CH_3CH(CH_3)CH(CN)CO_2H$		
-1.609	2.299(0)	-16.3	5	T	μ=0 ΔCp=-26.1	70Ia	
-1.742	2.320(0)	-16.8	10	T	μ=0 ΔCp=-27.3	70Ia	
-1.882	2.343(0)	-17.3	15	T	μ=0 ΔCp=-28.6	70Ia	
-2.029	2.365(0)	-17.8	20	T	μ=0 ΔCp=-29.9	70Ia	
-2.181	2.401(0)	-18.3	25	T	μ=0 ΔCp=-31.3	70Ia	
-2.341	2.427(0)	-18.8	30	T	μ=0 ΔCp=-32.6	70Ia	
-2.508	2.452(0)	-19.4	35	T	μ=0 ΔCp=-34.0	70Ia	
-2.682	2.481(0)	-19.9	40	T	μ=0 ΔCp=-35.5	70Ia	
-2.863	2.511(0)	-20.5	45	T	μ=0 ΔCp=-36.9	70Ia	
M17 MANNOSE			$C_6H_{12}O_6$		$HOCH_2CH(OH)CH(OH)CH(OH)CH(OH)$-CHO		
8.26	12.45(0)	-27.8	10	C	μ=0	70Cd	
7.9	12.08(0)	-28.9	25	C	μ=0	70Cd 66Ib	
7.4	11.81(0)	-30.5	40	C	μ=0	70Cd	
M18 METAMYGLOBIN (Horse)							
5.75	9.038(20°)	-21.7	20-37	T	μ=0.0 pH=7.8-10.2 Borate Buffer	52Gb	c
M19 METHANE, amino-			CH_5N		H_2NCH_3		
12.91	11.31(+1)	-5.3	5	T	μ=0	58Da	
13.00	10.96(+1)	-5.0	15	T	μ=0	58Da	
13.29	10.641(+1)	-4.1	25	C	μ=0	69Cc	
13.238	10.62(+1)	-4.2	25	C	μ=0	46Aa	
13.5	10.6(+1)	-3	25	C	μ=0	64Wa	
13.25	10.59(+1)	-4.0	25	C	μ=0	66Pe	
13.09	10.62(+1)	-4.7	25	T	μ=0	65La	
13.092	10.62(+1)	-4.7	25	T	μ=0	41Ea	

ΔH, kcal /mole	pK	ΔS, cal /mole°K	T, °C	M	Conditions	Ref.	Remarks

M19, cont.

13.09	10.63(+1)	-4.7	25	T	μ=0	58Da	
13.18	10.32(+1)	-4.4	35	T	μ=0	58Da	
13.27	10.02(+1)	-4.1	45	T	μ=0	58Da	
16.45	12.27(+1)	-1.0	25	T	μ=0 S: Ethylene glycol ΔCp=-6	70Kc	
12.790	9.752(+1)	-1.73	25	T	μ=0 S: 60 w % Methanol	52Ec	

M20 METHANE, dimorpholine $C_9H_{18}O_2N_2$ $(C_4H_8ON)CH_2(C_4H_8ON)$

4.4	7.38(25°,+1)	-18.6	6-44	T	μ=1.0M(KCl)	72Ka	c

M21 METHANE, dinitro- $CH_2O_4N_2$ $O_2NCH_2NO_2$

2.220	3.60(0)	-9.22	20	T	--	61Sc	

M22 METHANE, nitro- CH_3O_2N CH_3NO_2

5.93	10.2(0)	-26.9	25	C	μ=0	73Mc	
6.220	10.212(25°,0)	-25.8	10-25	T	μ<0.007	43Ta	c

M23 METHANE, trinitro- CHO_6N_3 $O_2NCH(NO_2)_2$

1.592	0.17(20°,0)	4.64	5-60	T	μ=0	61Sb	c

M24 METHANEDIPHOSPHONIC ACID $CH_6O_6P_2$ $CH_2[PO(OH)_2]_2$

-3.9	10.69(25°,-3)	-62.0	25-50	T	μ=0	62Ia	c,q: ±2.5

M25 METHANEDITHIOL CH_4S_2 $CH_2(SH)_2$

-7.35	10.99(-1)	-74.9	25	T	μ=0 HA=[(Se)(Methane-diethiol)]⁻ ΔCp=-47.6 pK=-53.316 + 1495.8/T + 23.960 log T ΔH=6.841.8 - 0.047.59 T ΔS=0.19628 - 0.10959 log T 273<T °K<298	71Gc	

M26 METHANOL CH_4O CH_3OH

12.5	12.22(0)	-14.0	18-37	T	--	62Bb	c
12.0	16.66(0)	-36.0	25	T	0.1M(MeOH)	25Bb	
14.369	22.61(0)	-22.8	-95	C	μ=0 S: Methanol ΔCp=-24.2	70Le	
14.127	21.74(0)	-24.4	-85	C	μ=0 S: Methanol ΔCp=-24.5	70Le	
13.882	20.97(0)	-25.9	-75	C	μ=0 S: Methanol ΔCp=-25.6	70Le	
13.626	20.28(0)	-27.3	-65	C	μ=0 S: Methanol ΔCp=-27.7	70Le	
13.349	19.67(0)	-30.1	-55	C	μ=0 S: Methanol ΔCp=-29.0	70Le	
13.059	19.12(0)	-30.2	-45	C	μ=0 S: Methanol ΔCp=-31.6	70Le	

ΔH, kcal /mole	pK	ΔS, cal /mole°K	T, °C	M	Conditions	Ref.	Remarks
M26, cont.							
12.743	18.63(0)	-31.7	-35	C	μ=0 S: Methanol ΔCp=-34.9	70Le	
12.394	18.19(0)	-33.3	-25	C	μ=0 S: Methanol ΔCp=-35.5	70Le	
12.039	17.79(0)	-34.8	-15	C	μ=0 S: Methanol ΔCp=-40.1	70Le	
11.638	17.44(0)	-36.4	-5	C	μ=0 S: Methanol ΔCp=-43.0	70Le	
11.208	17.12(0)	-38.4	5	C	μ=0 S: Methanol ΔCp=-46.0	70Le	
10.748	16.84(0)	-39.7	15	C	μ=0 S: Methanol ΔCp=-52.1	70Le	
11.4	17.01(0)	-38	15	T	μ=0 S: Methanol	57Kb	
11.1	16.88(0)	-39	20	T	μ=0 S: Methanol	57Kb	
10.227	16.71(0)	-42.2	25	C	μ=0 S: Methanol ΔCp=-56.8	70Le	
10.70	16.708(0)	-40.56	25	C	S: Methanol	72Pd	
10.8	16.73(0)	-40	25	T	μ=0 S: Methanol	57Kb	
9.932	16.65(0)	-43.4	30	C	μ=0 S: Methanol ΔCp=-59.0	70Le	
10.5	16.61(0)	-41	30	T	μ=0 S: Methanol	57Kb	
9.611	16.53(0)	-44.4	35	C	μ=0 S: Methanol ΔCp=-64.2	70Le	
10.2	16.49(0)	-42	35	T	μ=0 S: Methanol	57Kb	
9.9	16.37(0)	-43	40	T	μ=0 S: Methanol	57Kb	
M27 METHEMOGLOBINS							
7.14	8.000	-12.2	20	T	μ=0.051 Mouse	64Bb	
3.34	8.036	-25.3	20	T	μ=0.051 Rat	64Bb	
4.66	8.112	-21.2	20	T	μ=0.051 Baboon	64Bb	
5.18	8.130	-19.5	20	T	μ=0.051 Patas monkey	64Bb	
4.93	8.163	-20.5	20	T	μ=0.051 Mona monkey	64Bb	
7.87	8.211	-10.7	20	T	μ=0.051 Pig	64Bb	
6.09	8.266	-17.0	20	T	μ=0.051 Dog	64Bb	
5.50	8.295	-19.2	20	T	μ=0.051 Hyena	64Bb	
5.65	8.300	-18.7	20	T	μ=0.051 Tantalus monkey	64Bb	
4.35	8.301	-23.1	20	T	μ=0.051 Cat	64Bb	
9.16	8.290	-6.68	20	T	μ=0.051 Pigeon	64Bb	
3.66	8.405	-26.0	20	T	μ=0.051 Horse	64Bb	
3.40	8.136	-25.6	20	T	μ=0.051 Human	64Bb	
4.09	8.189	-23.5	20	T	μ=0.051 Human	64Bb	
4.88	8.146	-20.6	20	T	μ=0.051 Human	64Bb	
5.860	8.213	-17.6	20	T	μ=0.051 Guinea pig	65Ba	
4.120	8.260	-23.7	20	T	μ=0.051 Lizard	65Ba	
8.340	8.288	-9.45	20	T	μ=0.051 Bat	65Ba	
10.220	8.280	-3.00	20	T	μ=0.051 Shrew	65Ba	
6.550	8.260	-15.45	20	T	μ=0.051 Duck	65Ba	
8.110	8.204	-9.86	20	T	μ=0.051 Turkey	65Ba	
6.650	8.247	-15.0	20	T	μ=0.051 Chicken	65Ba	
6.830	8.244	-14.4	20	T	μ=0.051 Guinea fowl	65Ba	
1.570	8.230	-32.3	20	T	μ=0.051 Cow	65Ba	
4.120	8.090	-23.0	20	T	μ=0.051 Mangabey monkey	65Ba	
1.760	8.020	-30.7	20	T	μ=0.051 Rabbit	65Ba	
M28 METHIONINE			$C_5H_{11}O_2NS$		$CH_3SCH_2CH_2CH(NH_2)CO_2H$		
0.140	2.125(25°,+1)	-9.3	10-40	T	μ=0.05	57Pa 60Pb 60Pc	c
3.05	3.88(+1)	-7.5	25	T	--	60Pc	
3.5	3.09(+1)	-2.3	25	T	S: 44.6 w % Dioxane	60Pc	

M28, cont.

ΔH, kcal /mole	pK	ΔS, cal /mole°K	T, °C	M	Conditions	Ref.	Remarks
0.65	3.51(+1)	-13.9	25	T	S: 59.7 w % Dioxane	60Pc	
3.05	3.88(+1)	-7.5	25	T	S: 69 w % Dioxane	60Pc	
10.23	(0)	-----	20	C	μ≃0.01	71Ma	e:α-NH_3^+
10.4	9.28(25°,0)	-7.5	10-40	T	μ=0.05	57Pa 60Pb 60Pc	c
9.3	9.75(25°,0)	-13.2	10-40	T	S: 44.6 w % Dioxane	60Pc	c
10.10	10.19(25°,0)	-12.8	10-40	T	S: 59.7 w % Dioxane	60Pc	c
9.00	10.83	(25°,0)	10-40	T	S: 69 w % Dioxane	60Pc	c

M29 METHYL ORANGE — $C_{14}H_{14}O_3N_3SNa$ — $(CH_3)_2NC_6H_4N{:}NC_6H_4SO_3H$

ΔH, kcal /mole	pK	ΔS, cal /mole°K	T, °C	M	Conditions	Ref.	Remarks
4.28	3.438(0)	-1.37	25	T	μ=0 ΔCp=8	73Bf	
3.35	3.43(25°,0)	-4.4	20-50	T	In Tartrate Buffer	65Pd	c

M30 o-METHYL ORANGE — $C_{14}H_{14}O_3N_3SNa$ — $(CH_3)_2NC_6H_4N{:}NC_6H_4SO_3H$

ΔH, kcal /mole	pK	ΔS, cal /mole°K	T, °C	M	Conditions	Ref.	Remarks
1.123	3.482(0)	-12.16	25	T	μ=0 ΔCp=-11	73Bf	

M31 MOLYBDIC ACID — $H_6O_{24}Mo_7$ — $H_6Mo_7O_{24}$

ΔH, kcal /mole	pK	ΔS, cal /mole°K	T, °C	M	Conditions	Ref.	Remarks
0.6	2.53(-3)	-10	25	C	μ=3M($NaClO_4$)	68Ab 70Aa	q:±1.3
-0.8	3.54(-4)	-19	25	C	μ=3M($NaClO_4$)	68Ab 70Aa	q:±0.5
-2.6	4.40(-5)	-29	25	C	μ=3M($NaClO_4$)	68Ab 70Aa	q:±0.3
-4.0	-----	-----	25	C	μ=3M($NaClO_4$) R: $H_2Mo_7O_{24}^{4-}$ = $Mo_7O_{24}^{6-}$ + $2H^+$	74Kc	
56.0	-57.75	-76	25	C	3M($NaClO_4$) R: $Mo_6O_{26}^{6-}$ + $4H_2O$ = $8H^+$ + $7MoO_4^{2-}$	68Ab	
-14	3.89	-65	25	C	μ=3M($NaClO_4$) R: $HMoO_4^-$ = H^+ + MoO_4^{2-}	70Aa 68Ab	q:±7

M33 MORPHOLINE — C_4H_9ON — $\underline{OCH_2CH_2NHCH_2CH_2}$

ΔH, kcal /mole	pK	ΔS, cal /mole°K	T, °C	M	Conditions	Ref.	Remarks
9.33	8.492(+1)	7.58	25	T	μ=0	66Hc	
9.83	8.75(+1)	-7.08	25	C	μ=0.5M(KNO_3)	75Bc	
8.9	8.88(25°,+1)	-10.6	6-44	T	μ=1.0M(KCl)	72Ka	c

M34 MOROPHOLINE, N-amino- — $C_4H_{10}ON_2$ — $C_4H_8ON(NH_2)$

ΔH, kcal /mole	pK	ΔS, cal /mole°K	T, °C	M	Conditions	Ref.	Remarks
5.67	4.19(+1)	-7.20	25	C	μ=0.5M(KNO_3)	75Bc	

M35 MORPHOLINE, N-(2-aminoethyl)- — $C_6H_{14}ON_2$

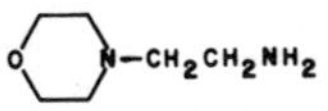

ΔH, kcal /mole	pK	ΔS, cal /mole°K	T, °C	M	Conditions	Ref.	Remarks
-2.9	4.06(30°,+2)	-9	10-40	T	μ=0	63Ba	c
-9.0	9.15(30°,+1)	-12	10-40	T	μ=0	63Ba	c

ΔH, kcal /mole	pK	ΔS, cal /mole°K	T,°C	M	Conditions	Ref.	Remarks
M36 MORPHOLINE, 2,6-dimethyl-			$C_6H_{13}ON$		$C_4H_7ON(CH_3)_2$		
9.81	8.75(+1)	-7.16	25	C	μ=0.5M(KNO_3)	75Bc	
M37 MORPHOLINE, N-ethyl-			$C_6H_{13}ON$		$C_4H_8ON(C_2H_5)$		
6.56	7.77(+1)	-13.50	25	C	μ=0.5M(KNO_3)	75Bc	
M38 MORPHOLINE, N-hydroxymethyl-			$C_5H_{11}O_2N$		$OCH_2CH_2N(CH_2OH)CH_2CH_2$ (ring)		
3.5	6.09(25°,+1)	-16.0	6-44	T	μ=1.0M(KCl)	72Ka	c
M39 MORPHOLINE, N-methyl-			$C_5H_{11}ON$		$C_4H_8ON(CH_3)$		
6.50	7.73(+1)	-13.60	25	C	μ=0.5M(KNO_3)	75Bc	
M40 MYOGLOBIN, sperm whale							
1.6	4.40	-14.8	30	C	μ=0.15(KCl)	65Hb	e:CO_2H
7.1	6.62	-6.9	30	C	μ=0.15(KCl)	65Hb	e:Histidine
11	7.8	0.6	30	C	μ=0.15(KCl)	65Hb	e:α-NH_3^+
6	8.9	-20.9	30	C	μ=0.15(KCl)	65Hb	e:Hemic acid
6.1	10.0	-25.6	30	C	μ=0.15(KCl)	65Hb	e:Tyrosine
12.7	10.0	-3.9	30	C	μ=0.15(KCl)	65Hb	e: Lysine

ΔH, kcal /mole	pK	ΔS, cal /mole°K	T,°C	M	Conditions	Ref.	Remarks
			N				
N1 NAPHTHALENE, 1-amino-			$C_{10}H_9N$		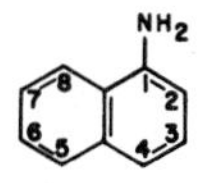		
5.9	3.40(20°,+1)	4.6	0-20	T	C=0.01M S: 50 w % Ethanol	59Eb	c
N2 NAPHTHALENE, 2-amino-			$C_{10}H_9N$		see N1		
4.8	3.77(20°,+1)	-0.9	0-20	T	C=0.01M S: 50 w % Ethanol	59Eb	c
N3 NAPHTHALENE, 1,4-dihydroxy-			$C_{10}H_8O_2$		see N1		
6.2	9.37(26.4°,0)	-22	13	T	μ=0.65	53Bb	
5.5	10.93(26.6°, -1)	-32	25	T	μ=0.65	53Bb	
N4 NAPHTHALENE, 2-hydroxy-1(2'-hydroxy-5'-sulphophenylazo)-			$C_{16}H_{12}O_5N_2S$		see N1		
3.4	6.94(30°,-1)	-22	25	T	μ=0.1	61Cb	
2.2	6.83(-1)	-26	50	T	μ=0.1	61Cb	
10.8	12.78(30°,-2)	-25	25	T	μ=0.1	61Cb	
8.2	12.36(-2)	-33	50	T	μ=0.1	61Cb	
7.5	6.89(25°,0)	-7	25-40	T	μ=0 HA=[Cr(III)(H_2O)$_3$-(2-hydroxy-1(2'-hydroxy-5'-sulphophenylazo)]	62Cd	c
9.7	9.82(25°,0)	-13	25-40	T	μ=0 HA=[Cr(III)(H_2O)$_2$-(OH)(2-hydroxy-1(2'-hydroxy-5'-sulphophenylazo)]-	62Cd	c
N5 1,3-NAPHTHALENEDISULPHONIC ACID, 8-amino-N(2',4'-dimethylphenylazo)-			$C_{18}H_{17}O_6N_3S_2$		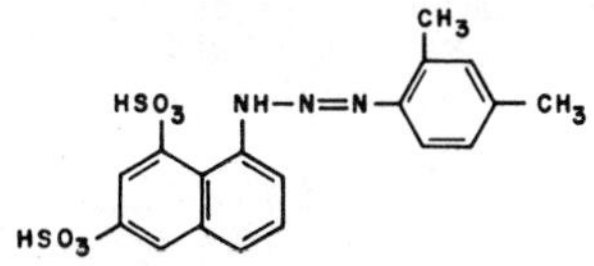		
6.4	3.59(40°)	2.0	40-80	T	μ=0	43Bc	c
N6 1,3-NAPHTHALENEDISULFONIC ACID, 8-amino-N(2',5'-dimethylphenylazo)-			$C_{18}H_{17}O_6N_3S_2$		see N5		
5.40	3.64(49°)	0.4	40-80	T	μ=0	43Bc	c
N7 1,3-NAPHTHALENEDISULFONIC ACID, 8-amino-N(2',6'-dimethylphenylazo)-			$C_{18}H_{17}O_6N_3S_2$		see N5		
-1.2	3.56(40°)	-10.0	40-80	T	μ=0	43Bc	c

ΔH, kcal /mole	pK	ΔS, cal /mole°K	T,°C	M	Conditions	Ref.	Remarks
N8 1,3-NAPHTHALENEDISULFONIC ACID, 8-amino-N(3'-carboxaldehyde-phenylazo)-			$C_{17}H_{13}O_7N_3S_2$		see N5		
8.3	3.96(40°)	4.2	40-80	T	μ=0	43Bc	c
N9 1,3-NAPHTHALENEDISULFONIC ACID, 8-amino-N(2'-carboxyphenylazo)-			$C_{17}H_{13}O_8N_3S_2$		see N5		
3.8	6.46(40°)	-20.9	40-80	T	μ=0	43Bc	c
N10 1,3-NAPHTHALENEDISULFONIC ACID, 8-amino-N(3'-carboxyphenylazo)-			$C_{17}H_{13}O_8N_3S_2$		see N5		
5.8	3.70(40°)	0.7	40-80	T	μ=0	43Bc	c
N11 1,3-NAPHTHALENEDISULFONIC ACID, 8-amino-N(4'-carboxyphenylazo)-			$C_{17}H_{13}O_8N_3S_2$		see N5		
3.2	3.69(40°)	-3.3	40-80	T	μ=0	43Bc	c
N12 1,3-NAPHTHALENEDISULFONIC ACID, 8-amino-N(2'-chlorophenylazo)-			$C_{16}H_{12}O_6N_3ClS_2$		see N5		
5.8	3.17(40°)	2.1	40-80	T	μ=0	43Bc	c
N13 1,3-NAPHTHALENEDISULFONIC ACID, 8-amino-N(3'-chlorophenylazo)-			$C_{16}H_{12}O_6N_3ClS_2$		see N5		
8.7	3.84(40°)	5.2	40-80	T	μ=0	43Bc	c
N14 1,3-NAPHTHALENEDISULFONIC ACID, 8-amino-N(4'-chlorophenylazo)-			$C_{16}H_{12}O_6N_3ClS_2$		see N5		
8.1	3.75(40°)	4.5	40-80	T	μ=0	43Bc	c
N15 1,3-NAPHTHALENEDISULFONIC ACID, 8-amino-N(4'-hydroxyphenylazo)-			$C_{16}H_{13}O_7N_3S_2$		see N5		
6.7	3.42(40°)	2.9	40-80	T	μ=0	43Bc	c
N16 1,3-NAPHTHALENEDISULFONIC ACID, 8-amino-N(2'-methoxyphenylazo)-			$C_{17}H_{15}O_7N_3S_2$		see N5		
6.3	3.59(40°)	1.7	40-80	T	μ=0	43Bc	c
N17 1,3-NAPHTHALENEDISULFONIC ACID, 8-amino-N(3'-methoxyphenylazo)-			$C_{17}H_{15}O_7N_3S_2$		see N5		
9.60	3.88(40°)	4.5	40-80	T	μ=0	43Bc	c

ΔH, kcal /mole	pK	ΔS, cal /mole°K	T, °C	M	Conditions	Ref.	Remarks
N18 1,3-NAPHTHALENEDISULFONIC ACID, 8-amino-N(4'-methoxyphenylazo)-			$C_{17}H_{15}O_7N_3S_2$		see N5		
8.5	3.57(40°)	5.4	40-80	T	μ=0	43Bc	c
N19 1,3-NAPHTHALENEDISULFONIC ACID, 8-amino-N(2'-methylphenylazo)-			$C_{17}H_{15}O_6N_3S_2$		see N5		
1.7	3.44(40°)	-5.1	40-80	T	μ=0	43Bc	c
N20 1,3-NAPHTHALENEDISULFONIC ACID, 8-amino-N(3'-methylphenylazo)-			$C_{17}H_{15}O_6N_3S_2$		see N5		
9.4	3.71(40°)	6.4	40-80	T	μ=0	43Bc	c
N21 1,3-NAPHTHALENEDISULFONIC ACID, 8-amino-N(4'-methylphenylazo)-			$C_{17}H_{15}O_6N_3S_2$		see N5		
8.8	3.58(40°)	5.9	40-80	T	μ=0	43Bc	c
N22 1,3-NAPHTHALENEDISULFONIC ACID, 8-amino-(3'-nitrophenylazo)-			$C_{16}H_{12}O_8N_4S_2$		see N5		
4.2	3.60(40°)	-1.5	40-80	T	μ=0	43Bc	c
N23 1,3-NAPHTHALENEDISULPHONIC ACID, 8-amino-(4'-nitrophenylazo)-			$C_{16}H_{12}O_8N_4S_2$		see N5		
4.2	3.91(40°)	-2.5	40-80	T	μ=0	43Bc	c
N24 1,3-NAPHTHALENEDISULFONIC ACID, 8-amino-N-phenylazo-			$C_{16}H_{13}O_6N_3S_2$		see N5		
7.1	3.56(40°)	3.2	40-80	T	μ=0	43Bc	c
N25 1,3-NAPHTHALENEDISULFONIC ACID, 8-amino-N(2'-sulphophenylazo)-			$C_{16}H_{13}O_9N_3S_3$		see N5		
-2.6	3.37(40°)	-12.1	40-80	T	μ=0	43Bc	c
N26 1,3-NAPHTHALENEDISULFONIC ACID, 8-amino-N(3'-sulphophenylazo)-			$C_{16}H_{13}O_9N_3S_3$		see N5		
3.4	3.29(40°)	-2.3	40-80	T	μ=0	43Bc	c
N27 1,3-NAPHTHALENEDISULFONIC ACID, 8-amino-N(4'-sulphophenylazo)-			$C_{16}H_{13}O_9N_3S_3$		see N5		
3.5	3.31(40°)	-2.0	40-80	T	μ=0	43Bc	c

ΔH, kcal /mole	pK	ΔS, cal /mole°K	T, °C	M	Conditions	Ref.	Remarks

N28 3,6-NAPHTHALENEDISULFONIC ACID, 1,9-di-hydroxy- $C_{10}H_8O_8S_2$

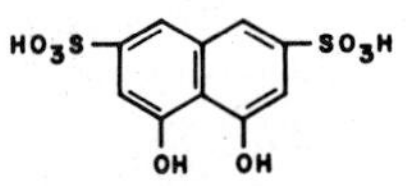

ΔH, kcal /mole	pK	ΔS, cal /mole°K	T, °C	M	Conditions	Ref.	Remarks
3.2	5.32	-13.6	25	C	μ=0.1(KNO_3)	70Kb	

N29 1-NAPHTHALENESULFONIC ACID, 4-amino- $C_{10}H_9O_3NS$

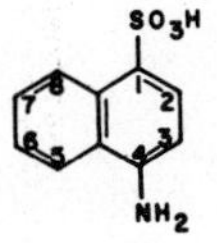

ΔH, kcal /mole	pK	ΔS, cal /mole°K	T, °C	M	Conditions	Ref.	Remarks
4.101	2.559(35°)	2	0&35	-	--	39Wa	b,c

N30 1-NAPHTHALENESULFONIC ACID, 3-hydroxy-2-nitroso- $C_{10}H_7O_5NS$ see N29

ΔH, kcal /mole	pK	ΔS, cal /mole°K	T, °C	M	Conditions	Ref.	Remarks
4.82	6.39(30°,-$\underline{1}$)	-13.70	20-35	T	μ=0.1	72Lb	c
2.66	6.65(30°,-$\underline{1}$)	-21.71	20-35	T	μ=0.1 S: 25.32 w % Methanol	72Lb	c
3.38	6.56(30°,-$\underline{1}$)	-18.81	20-35	T	μ=0.1 S: 25.30 w % Ethanol	72Lb	c
2.58	6.77(30°,-$\underline{1}$)	-22.44	20-35	T	μ=0.1 S: 25.32 w % Acetone	72Lb	c

N31 1-NAPHTHOIC ACID, 2-hydroxy- $C_{11}H_8O_3$

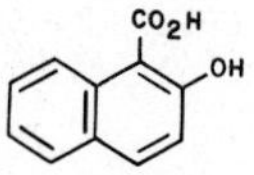

ΔH, kcal /mole	pK	ΔS, cal /mole°K	T, °C	M	Conditions	Ref.	Remarks
2.013	3.29($\underline{0}$)	-8.58	20	T	μ=0.01M	72Pa	
2.148	3.24($\underline{0}$)	-8.12	30	T	μ=0.01M	72Pa	
2.288	3.19($\underline{0}$)	-7.66	40	T	μ=0.01M	72Pa	
2.433	3.26($\underline{0}$)	-7.21	50	T	μ=0.01M	72Pa	
2.581	3.47($\underline{0}$)	-6.75	60	T	μ=0.01M	72Pa	
1.291	9.68(-$\underline{1}$)	-40.72	20	T	μ=0.01M	72Pa	
1.480	9.65(-$\underline{1}$)	-40.08	30	T	μ=0.01M	72Pa	
1.676	9.61(-$\underline{1}$)	-39.44	40	T	μ=0.01M	72Pa	
1.881	9.58(-$\underline{1}$)	-38.80	50	T	μ=0.01M	72Pa	
2.09	9.56(-$\underline{1}$)	-38.16	60	T	μ=0.01M	72Pa	

N32 NITRIC ACID HO_3N HNO_3

ΔH, kcal /mole	pK	ΔS, cal /mole°K	T, °C	M	Conditions	Ref.	Remarks
-3.2	-1.65($\underline{0}$)	-4.0	0	T	μ=0 ΔCp=-34	75Ma	g:55Ka, k
-4.1	-1.38($\underline{0}$)	-7.1	25	T	μ=0 ΔCp=-37	75Ma	g:55Ka, k
-3.30	-1.44($\underline{0}$)	-4.46	25	T	μ=0	65La	
-5.0	-1.20($\underline{0}$)	-10.2	50	T	μ=0 ΔCp=-40	75Ma	g:55Ka, k
-7.2	-0.55($\underline{0}$)	-16.4	100	T	μ=0 ΔCp=-46	75Ma	g:55Ka, k
-8.4	-0.26($\underline{0}$)	-19.5	125	T	μ=0 ΔCp=-49	75Ma	g:55Ka, k
-9.7	-0.08($\underline{0}$)	-22.6	150	T	μ=0 ΔCp=-53	75Ma	g:55Ka, k
-12.4	0.51($\underline{0}$)	-28.8	200	T	μ=0 ΔCp=-59	75Ma	g:55Ka, k

ΔH, kcal /mole	pK	ΔS, cal /mole°K	T, °C	M	Conditions	Ref.	Remarks
N32, cont.							
-13.5	0.85(0)	-31.1	218	T	μ=0 ΔCp=-61	75Ma	g:07Na, k
-15.5	1.11(0)	-35.1	250	T	μ=0 ΔCp=-65	75Ma	g:55Ka, k
-16.0	0.99(0)	-35.2	250	T	μ=0 ΔCp=-190	75Mb	g:55Ka, k
-21.8	1.35(0)	-45.9	275	T	μ=0 ΔCp=-270	75Mb	g:55Ka, k
-19.0	1.97(0)	-41.3	300	T	μ=0 ΔCp=-71	75Ma	g:55Ka, k
-29.9	1.79(0)	-60.4	300	T	μ=0 ΔCp=-380	75Mb	g:55Ka, k
-19.4	1.84(0)	-42.2	306	T	μ=0 ΔCp=-72	75Ma	g:07Na, k
-40.9	2.35(0)	-79.1	325	T	μ=0 ΔCp=-500	75Mb	g:55Ka, k
-55.3	3.05(0)	-103	350	T	μ=0 ΔCp=-650	75Mb	g:55Ka, k
-22.7	3.58(0)	-47.5	350	T	μ=0 ΔCp=-77	75Ma	g:55Ka, k
-62.2	3.37(0)	-114	360	T	μ=0 ΔCp=-720	75Mb	g:55Ka, k
-69.7	3.73(0)	-125	370	T	μ=0 ΔCp=-790	75Mb	g:55Ka, k

N33 NITROUS ACID NHO_2 HO_2N

3.8	3.224(0)	-1.6	15	T	μ=0	68Ta
3.6	3.230(0)	-2.2	15	T	μ=0	65La
3.0	3.177(0)	-4.4	20	T	μ=0	68Ta
3.1	3.203(0)	-4.2	20	T	μ=0	65La
2.2	3.138(0)	-7.1	25	T	μ=0	68Ta
2.5	3.148(0)	-6.1	25	T	μ=0	65La
0.47	3.100(0)	-12.7	35	T	μ=0	68Ta
1.3	3.113(0)	-10.0	35	T	μ=0	65La

N34 NITROUS ACID, hypo- $N_2H_2O_2$ $H_2O_2N_2$

3.0	6.92(0)	-22	25	T	μ=0.1	63Hf
8	11.54(-1)	-26	25	T	μ=0	65La
8.0	9.90(-1)	-19	25	T	μ=0.1	63Hf
11.10	18.05	-9.9	25	C	μ=0 R: $H_2O_2N_2 = O_2N_2^{2-} + 2H^+$	39La

ΔH, kcal /mole	pK	ΔS, cal /mole°K	T,°C	M	Conditions	Ref.	Remarks

O

01 OCTANE, 1,8-diamino-3,6-dioxa-N,N,N',N'-tetraacetic acid — $C_{14}H_{24}O_{10}N_2$ — $(HO_2CCH_2)_2N(CH_2CH_2O)_2CH_2CH_2$-$N(CH_2CO_2H)_2$

ΔH, kcal /mole	pK	ΔS, cal /mole°K	T,°C	M	Conditions	Ref.	Remarks
5.75	8.84(-2)	-20.8	20	C	0.1M(KNO_3)	64Ab	
5.40	8.95(-2)	-22.5	20	T	μ=0.1	65Bc	
4.88	8.88(-2)	-24.2	25	T	μ=0.1	65Bc	
5.4	9.09(25°,-2)	-24	20-40	T	μ=0.1M(KNO_3)	69Sb	c
5.84	9.44(-3)	-23.3	20	C	0.1M(KNO_3)	64Ab	
6.10	9.60(-3)	-23.1	20	T	μ=0.1	65Bc	
6.33	9.53(-3)	-22.4	25	T	μ=0.1	65Bc	
6.0	9.17(25°,-3)	-22	20-40	T	μ=0.1M(KNO_3)	69Sb	c

02 OCTANE, 1,8-diamino-3,6-dithio- — $C_6H_{16}N_2S_2$ — $H_2NCH_2CH_2SCH_2CH_2SCH_2CH_2NH_2$

ΔH, kcal /mole	pK	ΔS, cal /mole°K	T,°C	M	Conditions	Ref.	Remarks
12.2	8.43(30°,+2)	-2	10-40	T	μ=0	59Ma	c
12.2	9.31(30°,+1)	2	10-40	T	μ=0	59Ma	c

03 OCTANE, 1,8-diamino-3-oxa-6-thio- — $C_6H_{16}ON_2S$ — $H_2NCH_2CH_2SCH_2CH_2OCH_2CH_2NH_2$

ΔH, kcal /mole	pK	ΔS, cal /mole°K	T,°C	M	Conditions	Ref.	Remarks
12.0	8.54(30°,+2)	1	10-40	T	μ=0	59La	c
12.0	9.46(30°,+1)	-4	10-40	T	μ=0	59La	c

04 OCTANE, 1,8-diamino, N,N,N',N'-tetraacetic acid — $C_{16}H_{28}O_8N_2$ — $(HO_2CH_2C)_2N(CH_2)_8N(CH_2CO_2H)_2$

ΔH, kcal /mole	pK	ΔS, cal /mole°K	T,°C	M	Conditions	Ref.	Remarks
5.79	9.89(-2)	-25.5	20	C	0.1	64Ab	
8.09	10.84(-3)	-22.0	20	C	0.1	64Ab	

05 OCTANE, 1,4-diazabicyclo[2,2,2]- — $C_6H_{12}N_2$

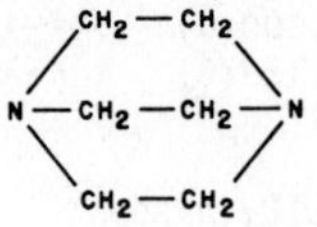

ΔH, kcal /mole	pK	ΔS, cal /mole°K	T,°C	M	Conditions	Ref.	Remarks
2.97	2.90(+2)	-3.6	25	C	μ=0	66La	
3.01	2.97(+2)	-3.5	25	C	0.1M(KCl)	65Pb	
7.29	8.60(+1)	-15.2	25	C	μ=0	66La	
7.30	8.82(+1)	-15.9	25	C	0.1M(KCl)	65Pb	

06 OCTANEDIOIC ACID — $C_8H_{14}O_4$ — $HO_2C(CH_2)_6CO_2H$

ΔH, kcal /mole	pK	ΔS, cal /mole°K	T,°C	M	Conditions	Ref.	Remarks
-0.39	4.512(0)	-21.9	25	C	μ=0	67Ca	
-0.64	5.404(-1)	-26.9	25	C	μ=0	67Ca	

07 OCTANEDIOIC ACID, 3,6-oxa- — $C_6H_{10}O_6$ — $HO_2CCH_2OCH_2CH_2OCH_2CO_2H$

ΔH, kcal /mole	pK	ΔS, cal /mole°K	T,°C	M	Conditions	Ref.	Remarks
0.24	3.055(0)	-13.2	25	C	μ=1.00($NaClO_4$)	74Gb	
0.26	3.676(-1)	-15.9	25	C	μ=1.00($NaClO_4$)	74Gb	

08 3,6-OCTANEDIONE-, 1,8-diamino- — $C_8H_{16}O_2N_2$ — $H_2N(CH_2)_2CO(CH_2)_2CO(CH_2)_2NH_2$

ΔH, kcal /mole	pK	ΔS, cal /mole°K	T,°C	M	Conditions	Ref.	Remarks
11.0	8.60(30°,+2)	-3	10-40	T	μ=0	59La	c
11.4	9.57(30°,+1)	-6	10-40	T	μ=0	59La	c

ΔH, kcal /mole	pK	ΔS, cal /mole°K	T,°C	M	Conditions	Ref.	Re-marks
09 OCTANOIC ACID			$C_8H_{16}O_2$		$CH_3(CH_2)_6COOH$		
-0.62	4.895(0)	-24.5	25	C	μ=0	70Ce	
-0.077	4.910(35°,0)	-23	0&35	T	--	39Wa	b,c
010 OXALIC ACID			$C_2H_2O_4$		HO_2CCO_2H		
-0.750	1.360(0)	-8.72	25	C	μ=0	71Vb	q:±0.05
-1.02	1.271(0)	-9.2	25	C	μ=0	67Ca	
-0.8	1.30(0)	-9	25	T	μ=0	69Kb	
-0.719	1.45(25°,0)	-9.1	25-90	T	--	67Ma	c
0	1.52(0)	-6.98	90	T	--	67Ma	
0.571	0.79(0)	-5.45	90-160	T	--	67Ma	c
3.84	2.85(0)	-0.16	25	T	μ=0 S: Formamide	75Db	
-0.346	4.201(-1)	-20.5	0	T	μ=0	48Pa	
-0.278	4.228(-1)	-20.4	0	T	μ=0	39Ha	
-0.550	4.210(-1)	-21.3	0	T	μ=0 ΔCp=-42	74Sa	g:39Ha, 48Pa
-0.600	4.207(-1)	-21.4	5	T	μ=0	48Pa	
-0.501	4.235(-1)	-21.2	5	T	μ=0	39Ha	
-0.761	4.216(-1)	-22.0	5	T	μ=0 ΔCp=-43	74Sa	g:39Ha, 48Pa
-0.857	4.218(-1)	-22.3	10	T	μ=0	48Pa	
-0.739	4.244(+1)	-22.0	10	T	μ=0	39Ha	
-0.977	4.227(-1)	-22.8	10	T	μ=0 ΔCp=-44	74Sa	g:39Ha, 48Pa
-1.120	4.231(-1)	-23.2	15	T	μ=0	48Pa	
-0.993	4.255(-1)	-22.9	15	T	μ=0	39Ha	
-1.198	4.240(-1)	-23.6	15	T	μ=0 ΔCp=-45	74Sa	g:39Ha, 48Pa
-1.387	4.247(-1)	-24.2	20	T	μ=0	48Pa	
-1.264	4.268(-1)	-23.8	20	T	μ=0	39Ha	
-1.425	4.254(-1)	-24.3	20	T	μ=0 ΔCp=-46	74Sa	g:39Ha, 48Pa
-1.55	4.27(-1)	-24.7	25	C	μ=0	73Vd	
-1.50	4.266(-1)	-24.6	25	C	μ=0	67Ca	
-1.659	4.266(-1)	-25.1	25	T	μ=0	48Pa	
-1.551	4.286(-1)	-24.8	25	T	μ=0	39Ha	
-1.658	4.272(-1)	-25.1	25	T	μ=0 ΔCp=-47	74Sa	g:39Ha, 48Pa
-1.936	4.287(-1)	-26.0	30	T	μ=0	48Pa	
-1.856	4.308(-1)	-25.8	30	T	μ=0	39Ha	
-1.896	4.295(-1)	-25.9	30	T	μ=0 ΔCp=-48	74Sa	g:39Ha, 48Pa
-2.217	4.312(-1)	-26.9	35	T	μ=0	48Pa	
-2.139	4.318(-1)	-26.7	35	T	μ=0 ΔCp=-49	74Sa	g:39Ha, 48Pa
-2.178	4.33(-1)	-26.9	35	T	μ=0	39Ha	
-2.502	4.338(-1)	-27.8	40	T	μ=0	48Pa	
-2.519	4.356(-1)	-28.0	40	T	μ=0	39Ha	
-2.389	4.349(-1)	-27.5	40	T	μ=0 ΔCp=-51	74Sa	g:39Ha, 48Pa
-2.793	4.369(-1)	-28.8	45	T	μ=0	48Pa	
-2.877	4.388(-1)	-29.1	45	T	μ=0	39Ha	
-2.644	4.379(-1)	-28.3	45	T	μ=0 ΔCp=-52	74Sa	g:39Ha, 48Pa
-3.088	4.399(-1)	-29.7	50	T	μ=0	48Pa	
-3.256	4.417(-1)	-30.3	50	T	μ=0	39Ha	
-2.906	4.409(-1)	-29.2	50	T	μ=0 ΔCp=-53	74Sa	g:39Ha, 48Pa
-1.107	(-1)	-----	25	C	μ=0.5($NaNO_3$)	73Vd	
-0.858	(-1)	-----	25	C	μ=1.0($NaNO_3$)	73Vd	
-0.394	(-1)	-----	25	C	μ=2.0($NaNO_3$)	73Vd	
-2.328	2.61(25°,-1)	-20.578	25-70	T	--	67Ma	c
0.0	3.00(-1)	-13.72	70	T	--	67Ma	

ΔH, kcal /mole	pK	ΔS, cal /mole°K	T, °C	M	Conditions	Ref.	Remarks
010, cont.							
2.830	3.50(-1)	-6.5	70-110	T	--	67Ma	c
2.71	6.39(-1)	-20.14	25	T	μ=0 S: Formamide	75Db	

ΔH, kcal /mole	pK	ΔS, cal /mole°K	T,°C	M	Conditions	Ref.	Remarks

P

P1 PAPAIN (enzyme)

No	∿4	-----	38	T	--	58Sb	e: Carboxyl
44	∿8	-----	38	T	--	58Sb	e: Sulfhydryl

P3 PENTANE, 1-amino- $C_5H_{13}N$ $CH_3(CH_2)_3CH_2NH_2$

13.98	10.64(+1)	-1.8	25	C	μ=0	69Cc	

P4 PENTANE, 3-aza-1,5-diamino-N,N,N',N",N"-pentaacetic acid $C_{14}H_{23}O_{10}N_3$ $(HO_2CCH_2)_2N(CH_2)_2N(CH_2CO_2H)-(CH_2)_2N(CH_2CO_2H)_2$

1.73	4.26	-13.6	20	C	0.1M(KNO_3)	65Ab	
1.49	4.31	-14.7	25	T	μ=0.1(KNO_3)	62Mc	
4.32	7.09	-17.7	20	C	0.1M(KNO_3)	65Ab	
4.86	8.53	-22.7	25	T	μ=0.1(KNO_3)	62Mc	
2.4	8.50	-30.9	27	C	μ=0.1(KNO_3)	68Cc	
7.96	10.57	-21.2	20	C	0.1M(KNO_3)	65Ab	
5.44	10.43	-29.5	25	T	μ=0.1(KNO_3)	62Mc	
8.4	10.41	-19.7	27	C	μ=0.1(KNO_3)	68Cc	

P5 PENTANE, 1,5-diamino- $C_5H_{14}N_2$ $H_2N(CH_2)_5NH_2$

13.41	10.050(+2)	-0.9	25	C	μ=0.5M(KNO_3)	70Ba	
13.86	10.916(+1)	-3.5	25	C	μ=0.5M(KNO_3)	70Ba	

P6 PENTANE, 1,1-dinitro- $C_5H_{10}O_4N_2$ $CH_3(CH_2)_3CH(NO_2)_2$

3.86	5.337(25°,0)	-11.44	5-60	T	--	62Sa	c

P7 PENTANE, 1,5-diamino-N,N,N',N'-tetraacetic acid $C_{13}H_{22}O_8N_2$ $(HO_2CH_2C)_2N(CH_2)_5N(CH_2CO_2H)_2$

6.3	9.52(-2)	-22.1	20	C	0.1M(KNO_3)	64Ab	
7.5	10.68(-3)	-23.3	20	C	0.1M(KNO_3)	64Ab	

P8 PENTANEDIOIC ACID $C_5H_8O_4$ $HO_2C(CH_2)_3CO_2H$

-0.12	4.344(0)	-20.3	25	C	μ=0	67Ca	
0	4.43(0)	-20	25	T	μ=0	63Ea	
7.37	6.22(0)	-3.70	25	T	μ=0 S: Formamide	75Db	
-0.58	5.420(-1)	-26.7	25	C	μ=0	67Ca	
-0.3	5.41(-1)	-25	25	T	μ=0	63Ea	
7.36	8.21(-1)	-12.84	25	T	μ=0 S: Formamide	75Db	

P9 PENTANEDIOIC ACID, 3,3-dimethyl- $C_7H_{12}O_4$ $HO_2CCH_2C(CH_3)_2CH_2CO_2H$

-3	3.70(0)	-27	25	T	μ=0	63Ea	
-2.5	6.34(-1)	-37	25	T	μ=0	63Ea	

ΔH, kcal /mole	pK	ΔS, cal /mole°K	T, °C	M	Conditions	Ref.	Re-marks
P10 PENTANEDIOIC ACID, 3-isopropyl-			$C_8H_{14}O_4$		$(CH_3)_2CHCH(CH_2CO_2H)_2$		
-1	4.30(0)	-23	25	T	μ=0	63Ea	
-1.5	5.51(-1)	-30	25	T	μ=0	63Ea	
P11 PENTANEDIOIC ACID, 3-methyl-			$C_6H_{10}O_4$		$HO_2CCH_2CH(CH_3)CH_2CO_2H$		
-0.3	4.25(0)	-20	25	T	μ=0	63Ea	
-1.0	5.41(-1)	-28	25	T	μ=0	63Ea	
P12 PENTANEDIOIC ACID, 3,3-penta-methylene-			$C_{10}H_{16}O_4$		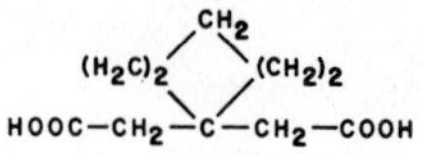		
-2.5	3.49(0)	-24	25	T	μ=0	63Ea	
-1.5	6.96(-1)	-36	25	T	μ=0	63Ea	
P13 2,4-PENTANEDIONE			$C_5H_8O_2$		$CH_3COCH_2COCH_3$		
1.1	9.07(0)	-37.3	0	T	μ=0	63Gc	q:±0.1
1.5	(0)	-----	19.1	T	μ=0	63Gc	q:±0.4
3.549	8.93(0)	-28.9	25	C	μ≈0	69Hb	
1.7	8.97(0)	-35.4	25	T	μ=0	63Gc	
2.0	8.90(0)	-34.6	40	T	μ=0	63Gc	
2.8	9.02(20°,0)	-32	10-40	T	μ=0	55Ib	c
2.5	9.12(0)	-32.7	0	T	--	68Ga	p
2.4	9.02(0)	-33.1	25	T	--	68Ga	p
2.35	8.93(0)	-33.1	40	T	--	68Ga	p
2.6	9.03(25°,0)	-33	10-38	T	--	68La	c
5.3	8.24(25°,0)	-20	5-45	T	Enol form of HA	68Cb	c
2.2	8.93(25°,0)	-35	5-45	T	Keto form of HA	68Cb	c
1.9	9.41(0)	-36.3	0	T	μ=0 S: 21.36 w % Dioxane	63Gc	
2.1	(0)	-----	19.1	T	μ=0 S: 21.36 w % Dioxane	63Gc	q:±0.4
2.3	9.25(0)	-34.9	25	T	μ=0 S: 21.36 w % Dioxane	63Gc	
2.4	9.75(0)	-34.5	40	T	μ=0 S: 21.36 w % Dioxane	63Gc	
2.95	10.37(0)	-38	25	C	μ≈0 S: 50 v % Dioxane	69Hb	
3.21	8.82(0)	-29.6	25	C	μ=0.1($NaClO_4$) S: 50 w % Dioxane	68Cd	
2.5	9.28(0)	-33.3	0	T	S: 9.8 w % Methanol	68Ga	
2.6	9.16(0)	-33.3	25	T	S: 9.9 w % Methanol	68Ga	
2.5	9.00(0)	-33.2	40	T	S: 9.9 w % Methanol	68Ga	
2.4	9.44(0)	-33.4	0	T	S: 22 w % Methanol	68Ga	
2.4	9.31(0)	-34.6	25	T	S: 22.2 w % Methanol	68Ga	
2.4	9.21(0)	-34.4	40	T	S: 22.4 w % Methanol	68Ga	
2.4	9.52(0)	-34.7	0	T	S: 28.9 w % Methanol	68Ga	
2.2	9.46(0)	-36.1	25	T	S: 29.5 w % Methanol	68Ga	
2.2	9.35(0)	-35.8	40	T	S: 29.5 w % Methanol	68Ga	
1.9	9.40(0)	-36.1	0	T	μ=0 S: 35.98 w % Methanol	63Gc	
2.1	(0)	-----	19.1	T	μ=0 S: 35.98 w % Methanol	63Gc	q:±0.4
2.3	9.22(0)	-34.8	25	T	μ=0 S: 35.98 w % Methanol	63Gc	
2.4	9.15(0)	-34.3	40	T	μ=0 S: 35.98 w % Methanol	63Gc	

ΔH, kcal /mole	pK	ΔS, cal /mole°K	T,°C	M	Conditions	Ref.	Re-marks
P13, cont.							
2.3	9.68(0)	-35.8	0	T	S: 37.4 w % Methanol	68Ga	
2.0	9.60(0)	-37.0	25	T	S: 38.3 w % Methanol	68Ga	
2.0	9.49(0)	-37.0	40	T	S: 38.7 w % Methanol	68Ga	
2.1	9.51(0)	-36.0	0	T	μ=0 S: 45.66 w % Methanol	63Gc	
2.3	(0)	-----	19.1	T	μ=0 S: 45.66 w % Methanol	68Gc	q:±0.4
2.5	9.38(0)	-34.6	25	T	μ=0 S: 45.66 w % Methanol	68Gc	
2.6	9.27(0)	-34.1	40	T	μ=0 S: 45.66 w % Methanol	68Gc	
2.3	9.84(0)	-36.6	0	T	S: 47.4 w % Methanol	68Ga	
1.8	9.75(0)	-38.6	25	T	S: 48.5 w % Methanol	68Ga	
1.8	9.70(0)	-38.3	40	T	S: 48.9 w % Methanol	68Ga	
2.2	9.62(0)	-36.3	0	T	μ=0 S: 54.15 w % Methanol	63Gc	
2.4	(0)	-----	19.1	T	μ=0 S: 54.15 w % Methanol	63Gc	q:±0.4
2.6	9.47(0)	-34.7	25	T	μ=0 S: 54.15 w % Methanol	63Gc	
2.7	9.40(0)	-34.3	40	T	μ=0 S: 54.15 w % Methanol	63Gc	
2.1	10.08(0)	-38.8	0	T	S: 59.0 w % Methanol	68Ga	
1.7	10.04(0)	-40.1	25	T	S: 61.0 w % Methanol	68Ga	
1.7	9.98(0)	-40.2	40	T	S: 62.1 w % Methanol	68Ga	
2.4	9.82(0)	-36.1	0	T	μ=0 S: 65.82 w % Methanol	63Gc	
2.7	(0)	-----	19.1	T	μ=0 S: 65.82 w % Methanol	63Gc	q:±0.4
3.0	9.63(0)	-34.5	25	T	μ=0 S: 65.82 w % Methanol	63Gc	
3.1	9.57(0)	-34.1	40	T	μ=0 S: 65.82 w % Methanol	63Gc	
2.4	9.28(0)	-34.8	0	T	S: 5.6 w % 1-Propanol	68Ga	
2.2	9.16(0)	-34.5	25	T	S: 5.6 w % 1-Propanol	68Ga	
2.2	9.07(0)	-34.3	40	T	S: 5.6 w % 1-Propanol	68Ga	
2.1	9.52(0)	-36.0	0	T	S: 13.2 w % 1-Propanol	68Ga	
2.0	9.38(0)	-36.4	25	T	S: 13.4 w % 1-Propanol	68Ga	
2.0	9.35(0)	-36.3	40	T	S: 13.4 w % 1-Propanol	68Ga	
2.1	9.76(0)	-36.9	0	T	S: 18.3 w % 1-Propanol	68Ga	
1.9	9.60(0)	-37.8	25	T	S: 18.8 w % 1-Propanol	68Ga	
1.8	9.56(0)	-38.1	40	T	S: 18.8 w % 1-Propanol	68Ga	
1.9	10.00(0)	-38.7	0	T	S: 25.1 w % 1-Propanol	68Ga	
1.8	9.90(0)	-39.4	25	T	S: 25.6 w % 1-Propanol	68Ga	
1.6	9.77(0)	-40.0	40	T	S: 25.6 w % 1-Propanol	68Ga	
1.8	10.32(0)	-40.6	0	T	S: 33.4 w % 1-Propanol	68Ga	
1.3	10.26(0)	-42.4	25	T	S: 34.6 w % 1-Propanol	68Ga	
1.2	10.12(0)	-42.3	40	T	S: 34.6 w % 1-Propanol	68Ga	
1.5	10.72(0)	-43.7	0	T	S: 45.0 w % 1-Propanol	68Ga	
0.8	10.70(0)	-46.0	25	T	S: 47.3 w % 1-Propanol	68Ga	
0.7	10.61(0)	-46.4	40	T	S: 47.3 w % 1-Propanol	68Ga	

P14 2,4-PENTANEDIONE, 3-methyl- $C_6H_{10}O_2$ $CH_3COCH(CH_3)COCH_3$

5	10.87(25°,0)	-63	10-38	T	--	61La	c

P15 PENTANOIC ACID $C_5H_{10}O_2$ $CH_3(CH_2)_3CO_2H$

-0.11	4.763(0)	-23.0	10	C	μ=0	70Ce	q:±0.01
-0.66	4.842(0)	-24.4	25	C	μ=0	70Ce	
-0.720	4.842(0)	-24.57	25	T	μ=0	52Ea	
-1.07	4.861(0)	-25.7	40	C	μ=0	70Ce	
-----	(0)	-----	10-40	C	μ=0 ΔCp=-32	70Ce	

ΔH, kcal /mole	pK	ΔS, cal /mole°K	T,°C	M	Conditions	Ref.	Remarks
P16	**PENTANOIC ACID, 2-amino-**		$C_5H_{11}O_2N$		$CH_3CH_2CH_2CH(NH_2)CO_2H$		
1.290	2.376(+1)	-6.1	1	T	μ=0	37Sa	
0.83	2.347(+1)	-8.0	10	C	μ=0	68Cd	
0.970	2.340(+1)	-7.3	12.5	T	μ=0	37Sa	
0.553	2.316(+1)	-8.7	25	T	μ=0 ΔCp=-37.0	39Hb	g:37Sa
0.550	2.318(+1)	-8.8	25	T	μ=0	37Sa	
0.39	2.318(+1)	-9.3	25	C	μ=0	68Cd	
0.050	2.309(+1)	-10.4	37.5	T	μ=0	37Sa	
0.06	2.309(+1)	-10.7	40	C	μ=0	68Cd	
-0.540	2.313(+1)	-12.3	50	T	μ=0	37Sa	
0.40	2.34(+1)	-9.36	25	C	μ=0.2M(KCl)	73Ga	
10.840	10.508(0)	-8.5	1	T	μ=0	37Sa	
10.910	10.154(0)	-8.2	12.5	T	μ=0	37Sa	
10.850	9.808(0)	-8.32	25	C	μ=0	67Ac	
10.880	9.808(0)	-8.4	25	T	μ=0	37Sa	
10.700	9.490(0)	-9.0	37.5	T	μ=0	37Sa	
10.380	9.198(0)	-9.9	50	T	μ=0	37Sa	
11.48	9.64(0)	-5.60	25	C	μ=0.2M(KCl)	73Ga	
P17	**PENTANOIC ACID, 5-amino-**		$C_5H_{11}O_2N$		$H_2N(CH_2)_4CO_2H$		
0.28	4.26(+1)	-18.5	10	C	μ=0	68Cd	
-0.22	4.20(+1)	-19.9	25	C	μ=0	68Cd	
0.160	4.229(+1)	-18.8	25	T	μ=0	57Ka	
-0.61	4.25(+1)	-21.4	40	C	μ=0	68Cd	
12.970	10.766(0)	-5.57	25	C	--	67Ac	
P18	**PENTANOIC ACID, 2-amino-3-methyl-**		$C_6H_{13}O_2N$		$CH_3CH_2CH(CH_3)CH(NH_2)CO_2H$		
1.080	2.365(+1)	-6.9	1.0	T	μ=0	37Sa	
0.740	2.338(+1)	-8.1	12.5	T	μ=0	37Sa	
0.300	2.318(+1)	-9.6	25	T	μ=0	37Sa	
0.301	2.320(+1)	-9.6	25	T	μ=0 ΔCp=-38.6	39Hb	g:37Sa
-0.220	2.317(+1)	-11.3	37.5	T	μ=0	37Sa	
-0.840	2.332(+1)	-13.3	50	T	μ=0	37Sa	
10.800	10.460(0)	-8.4	1.0	T	μ=0	37Sa	
10.870	10.100(0)	-8.1	12.5	T	μ=0	37Sa	
10.830	9.758(0)	-8.3	25	T	μ=0	37Sa	
10.650	9.439(0)	-8.9	37.5	T	μ=0	37Sa	
10.330	9.157(0)	-9.9	50	T	μ=0	37Sa	
10.60	(0)	-----	20	C	μ≃0.01	71Ma	e:α-NH_3^+
P19	**PENTANOIC ACID, 2,2-dimethyl-**		$C_7H_{14}O_2$		$CH_3(CH_2)_2C(CH_3)_2CO_2H$		
-0.54	5.021(0)	-24.0	10	C	μ=0	70Ce	q:±0.07
-0.99	4.969(0)	-26.1	25	C	μ=0	70Ce	
-1.46	5.088(0)	-28.0	40	C	μ=0	70Ce	
-----	(0)	-----	10-40	C	μ=0 ΔCp=-31	70Ce	
P20	**PENTANOIC ACID, 2-methyl-**		$C_6H_{12}O_2$		$CH_3(CH_2)_2CH(CH_3)CO_2H$		
-0.96	4.742(0)	-28.4	10	C	μ=0	70Ce	
-1.28	4.782(0)	-26.2	25	C	μ=0	70Ce	
-1.77	4.877(0)	-27.9	40	C	μ=0	70Ce	
-----	(0)	-----	10-40	C	μ=0 ΔCp=-27	70Ce	

ΔH, kcal /mole	pK	ΔS, cal /mole°K	T,°C	M	Conditions	Ref.	Remarks
P21 PENTANOIC ACID, 3-methyl-			$C_6H_{12}O_2$		$CH_3CH_2CH(CH_3)CH_2CO_2H$		
-0.77	4.752(0)	-24.5	10	C	μ=0	70Ce	
-1.12	4.766(0)	-25.9	25	C	μ=0	70Ce	
-1.56	4.821(0)	-27.0	40	C	μ=0	70Ce	
-----	(0)	-----	10-40	C	μ=0 ΔCp=-26	70Ce	
P22 PENTANOIC ACID, 4-methyl-			$C_6H_{12}O_2$		$(CH_3)_2CH(CH_2)_2CO_2H$		
-0.19	4.887(0)	-23.0	10	C	μ=0	70Ce	q:±0.03
-0.61	4.845(0)	-24.2	25	C	μ=0	70Ce 68Cd	
-1.04	4.879(0)	-25.7	40	C	μ=0	70Ce	
-----	(0)	-----	10-40	C	μ=0 ΔCp=-28	70Ce	
P23 PENTANOIC ACID, 4-oxo-			$C_5H_8O_3$		$CH_3COCH_2CH_2CO_2H$		
0.534	4.609(35°,0)	-19	0&35	T	--	39Wa	b,c
P24 1-PENTANOL, 5-amino-3-thio			$C_4H_{11}ONS$		$H_2NCH_2CH_2SCH_2CH_2OH$		
11.7	9.12(30°,0)	-3	10-40	T	μ=0	59La	c
P25 PEPSINOGEN							
6.3	10.97	-24	25	T	μ=0.15	64Pb	
P25a PEPTIDES							
0.6	3.49	-14.1	25	T	μ=0.15M(KCl) HA=[IIe[5]]angiotensin II peptide	74Ja	e: α-CO_2H q:±0.5
0.5	2.95	-11.7	25	T	μ=0.15M(KCl) HA=[IIe[5]]angiotensin II peptide	74Ja	e: β-CO_2H q:±0.1
7.1	6.47	-5.7	25	T	μ=0.15M(KCl) HA=[IIe[5]]angiotensin II peptide	74Ja	e:Imidazole group q:±0.3
8.8	7.60	-5.2	25	T	μ=0.15M(KCl) HA=[IIe[5]]angiotensin II peptide	74Ja	e:NH_3^+ q:±0.5
6.6	10.09	-23.8	25	T	μ=0.15M(KCl) HA=[IIe[5]]angiotensin II peptide	74Ja	e:Phenoxyl group q:±1.0
-0.01	3.34	-15.3	25	T	μ=0.15M(KCl) HA=[IIe[5]]angiotensin II-(2-8)-heptapeptide	74Ja	e: α-CO_2H q:±0.8
7.7	6.38	-3.3	25	T	μ=0.15M(KCl) HA=[IIe[5]]angiotension II-(2-8)-heptapentide	74Ja	e:Imidazole group
11.6	7.33	5.4	25	T	μ=0.15M(KCl) HA=[IIe[5]]angiotension II-(2-8)-heptapentide	74Ja	e:NH_3^+
9.3	10.07	-14.5	25	T	μ=0.15M(KCl) HA=[IIe[5]]angiotension II-(2-8)-heptapentide	74Ja	e:Phenoxyl group q:±0.7

ΔH, kcal /mole	pK	ΔS, cal /mole°K	T, °C	M	Conditions	Ref.	Remarks
P25a, cont.							
-0.6	3.30	-17.2	25	T	μ=0.15M(KCl) HA=[IIe^5]angiotensin II-(3-8)-hexapeptide	74Ja	e: α-CO_2H q:±0.4
6.7	6.38	-6.7	25	T	μ=0.15M(KCl) HA=[IIe^5]angiotensin II-(3-8)-hexapeptide	74Ja	e:Imidazole group
10.8	7.55	0.5	25	T	μ=0.15M(KCl) HA=[IIe^5]angiotensin II-(3-8)-hexapeptide	74Ja	e:NH_3^+
7.2	10.11	-22.3	25	T	μ=0.15M(KCl) HA=[IIe^5]angiotensin II-(3-8)-hexapeptide	74Ja	e:Phenoxyl q:±0.6
0.05	3.32	-15.0	25	T	μ=0.15M(KCl) HA=[IIe^5]angiotensin II-(4-8)-pentapeptide	74Ja	e: α-CO_2H q:±0.1
7.9	6.36	-2.5	25	T	μ=0.15M(KCl) HA=[IIe^5]angiotensin II-(4-8)-pentapeptide	74Ja	e:Imidazole group
9.8	7.40	-1.1	25	T	μ=0.15M(KCl) HA=[IIe^5]angiotensin II-(4-8)-pentapeptide	74Ja	e:NH_3^+
6.5	10.21	-24.4	25	T	μ=0.15M(KCl) HA=[IIe^5]angiotensin II-(4-8)-pentapeptide	74Ja	e:Phenoxyl group q:±0.6
-0.5	3.24	-16.5	25	T	μ=0.15M(KCl) HA=[IIe^5]angiotensin II-(5-8)-tetrapeptide	74Ja	e: α-CO_2H q:±0.4
6.9	6.30	-5.8	25	T	μ=0.15M(KCl) HA=[IIe^5]angiotensin II-(5-8)-tetrapeptide	74Ja	e:Imidazole group q:±0.6
11.9	7.81	4.1	25	T	μ=0.15M(KCl) HA=[IIe^5]angiotensin II-(5-8)-tetrapeptide	74Ja	e:NH_3^+
0.1	3.33	-15.0	25	T	μ=0.15M(KCl) HA=[Arg^1][IIe^5] angiotensin II	74Ja	e: α-CO_2H q:±0.8
7.9	6.38	-2.6	25	T	μ=0.15M(KCl) HA=[Arg^1][IIe^5] angiotensin II	74Ja	e:Imidazole group
10.6	7.22	2.4	25	T	μ=0.15M(KCl) HA=[Arg^1][IIe^5] angiotensin II	74Ja	e:NH_3^+
8.3	9.93	-17.6	25	T	μ=0.15M(KCl) HA=[Arg^1][IIe^5] angiotensin II	74Ja	e:Phenoxyl group q:±2.7
0.04	3.43	-15.6	25	T	μ=0.15M(KCl) HA=[Arg^6][IIe^5] angiotensin II	74Ja	e: α-CO_2H q:±0.04
0.02	2.95	-13.5	25	T	μ=0.15M(KCl) HA=[Arg^6][IIe^5] angiotensin II	74Ja	e: β-CO_2H q:±0.2
-----	-----	------	25	T	μ=0.15M(KCl) HA=[Arg^6][IIe^5] angiotensin II	74Ja	e:Imidazole group
9.3	7.50	-3.0	25	T	μ=0.15M(KCl) HA=[Arg^6][IIe^5] angiotensin II	74Ja	e:NH_3^+
10.1	10.03	-11.4	25	T	μ=0.15M(KCl) HA=[Arg^6][IIe^5] angiotensin II	74Ja	e:Phenoxyl group q:±3.8

ΔH, kcal /mole	pK	ΔS, cal /mole°K	T, °C	M	Conditions	Ref.	Remarks
P25a, cont.							
-1.5	3.36	-20.2	25	T	μ=0.15M(KCl) HA=[Asn^1][IIe^5] angiotensin II	74Ja	e: α-CO_2H q:$\pm$1.1
7.0	6.30	-5.4	25	T	μ=0.15M(KCl) HA=[Asn^1][IIe^5] angiotensin II	74Ja	e:Imidazole group q:$\pm$0.5
9.7	6.82	1.6	25	T	μ=0.15M(KCl) HA=[Asn^1][IIe^5] angiotensin II	74Ja	e:NH_3^+
8.9	10.07	-15.7	25	T	μ=0.15M(KCl) HA=[Asn^1][IIe^5] angiotensin II	74Ja	e:Phenoxyl group q:$\pm$0.5
8.6	6.14	0.8	25	T	μ=0.15M(KCl) HA=[Asn^1][IIe^5] angiotensin II - α-amide	74Ja	e:Imidazole group
7.1	6.77	-5.7	25	T	μ=0.15M(KCl) HA=[Asn^1][IIe^5] angiotensin II - α-amide	74Ja	e:NH_3^+ q:$\pm$0.3
5.5	10.00	-27.4	25	T	μ=0.15M(KCl) HA=[Asn^1][IIe^5] angiotensin II - α-amide	74Ja	e:Phenoxyl group q:$\pm$0.8
0.01	3.26	-14.9	25	T	μ=0.15M(KCl) HA=[Gly^1][IIe^5] angiotensin II	74Ja	e: α-CO_2H q:$\pm$0.4
7.5	6.45	-4.5	25	T	μ=0.15M(KCl) HA=[Gly^1][IIe^5] angiotensin II	74Ja	e:Imidazole group
10.9	7.90	0.3	25	T	μ=0.15M(KCl) HA=[Gly^1][IIe^5] angiotensin II	74Ja	e:NH_3^+
9.1	10.03	-14.5	25	T	μ=0.15M(KCl) HA=[Gly^1][IIe^5] angiotensin II	74Ja	e:Phenoxyl group q:$\pm$1.2
-3.3	3.40	-26.5	25	T	μ=0.15M(KCl) HA=[Gly^1,Gly^2][IIe^5] angiotensin II	74Ja	e: α-CO_2H q:+0.8
7.4	6.46	-4.7	25	T	μ=0.15M(KCl) HA=[Gly^1,Gly^2][IIe^5] angiotensin II	74Ja	e:Imidazole group q:$\pm$0.5
10.1	7.85	-2.3	25	T	μ=0.15M(KCl) HA=[Gly^1,Gly^2][IIe^5] angiotensin II	74Ja	e:NH_3^+
7.7	10.10	-19.7	25	T	μ=0.15M(KCl) HA=[Gly^1,Gly^2][IIe^5] angiotensin II	74Ja	e:Phenoxyl group q:$\pm$0.8
-0.2	3.78	-18.0	25	T	μ=0.15M(KCl) HA=[Leu^8][IIe^5] angiotensin II	74Ja	e: α-CO_2H q:$\pm$0.5
-0.02	2.95	-13.6	25	T	μ=0.15M(KCl) HA=[Leu^8][IIe^5] angiotensin II	74Ja	e: β-CO_2H q:$\pm$0.6
7.5	6.45	-4.3	25	T	μ=0.15M(KCl) HA=[Leu^8][IIe^5] angiotensin II	74Ja	e:Imidazole group
8.3	7.58	-6.8	25	T	μ=0.15M(KCl) HA=[Leu^8][IIe^5] angiotensin II	74Ja	e:NH_3^+

ΔH, kcal /mole	pK	ΔS, cal /mole°K	T, °C	M	Conditions	Ref.	Remarks
P25a, cont.							
6.5	10.08	-23.4	25	T	μ=0.15M(KCl) HA=[Leu[8]][IIe[5]] angiotensin II	74Ja	e:Phenoxyl group q:±0.4
-0.8	3.34	-18.0	25	T	μ=0.15M(KCl) HA=[Pro[3],Pro[5]][IIe[5]] angiotensin II	74Ja	e: α-CO_2H q:±0.5
-1.0	3.02	-17.2	25	T	μ=0.15M(KCl) HA=[Pro[3],Pro[5]][IIe[5]] angiotensin II	74Ja	e: β-CO_2H q:±0.4
6.4	6.54	-8.5	25	T	μ=0.15M(KCl) HA=[Pro[3],Pro[5]][IIe[5]] angiotensin II	74Ja	e:Imidazole group
8.1	7.64	-7.7	25	T	μ=0.15M(KCl) HA=[Pro[3],Pro[5]][IIe[5]] angiotensin II	74Ja	e:NH_3^+
5.2	9.92	-27.6	25	T	μ=0.15M(KCl) HA=[Pro[3],Pro[5]][IIe[5]] angiotensin II	74Ja	e:Phenoxyl group q:±0.3
-1.5	3.26	-20.0	25	T	μ=0.15M(KCl) HA=[Suc[1]][IIe[5]] angiotensin II	74Ja	e: α-CO_2H q:±0.3
-0.8	4.45	-23.2	25	T	μ=0.15M(KCl) HA=[Suc[1]][IIe[5]] angiotensin II	74Ja	e: β-CO_2H q:±0.3
6.2	6.58	-9.4	25	T	μ=0.15M(KCl) HA=[Suc[1]][IIe[5]] angiotensin II	74Ja	e:Imidazole group
-----	-----	-----	25	T	μ=0.15M(KCl) HA=[Suc[1]][IIe[5]] angiotensin II	74Ja	e:NH_3^+
7.6	10.12	-20.5	25	T	μ=0.15M(KCl) HA=[Suc[1]][IIe[5]] angiotensin II	74Ja	e:Phenoxyl group q:±0.9
P26 PEROXIDE RADICAL			HO_2		$HO_2\cdot$		
0	4.9(0̲)	-----	3-75	T	μ=5 x 10^{-3}M	71Bb	c
P27 PHENANTHRENE, 1-amino-			$C_{14}H_{11}N$		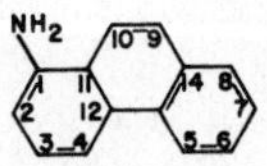		
5.6	3.23(20°,±1)	4.2	0-20	T	C=0.0025 S: 50 w % Ethanol	59Eb	c
P28 PHENANTHRENE, 2-amino-			$C_{14}H_{11}N$		see P27		
4.4	3.60(20°,±1)	-1.3	0-20	T	C=0.005 S: 50 w % Ethanol	59Eb	c

ΔH, kcal /mole	pK	ΔS, cal /mole°K	T,°C	M	Conditions	Ref.	Remarks
P29 PHENANTHRENE, 3-amino-			$C_{14}H_{11}N$		see P27		
4.4	3.59(20°,+1)	-1.3	0-20	T	C=0.005 S: 50 w % Ethanol	59Eb	c
P30 PHENANTHRENE, 9-amino-			$C_{14}H_{11}N$		see P27		
5.6	3.19(20°,+1)	4.3	0-20	T	C=0.0025 S: 50 w % Ethanol	59Eb	c
P31 1,7-PHENANTHROLINE			$C_{12}H_8N_2$		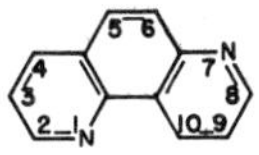		
2.17	4.30(+1)	-11.4	20	T	μ=0	60Pd	
P32 1,10-PHENANTHROLINE			$C_{12}H_8N_2$		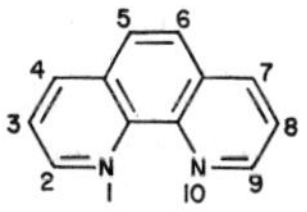		
3.60	4.857(+1)	-10.1	25	C	μ=0	70Ea	
3.45	4.83(+1)	-10.4	25	C	μ=0	71Pb	
3.0	3.95(+1)	-8	25	C	μ=1.0	62Ka	
3.47	4.85(+1)	-9.41	20	T	--	60Pd	
4.07	5.06(25°,+1)	-9.5	25-45	T	--	64La	c
4.84	4.71(+1)	-5.2	25	C	μ=0.5M(KCl) S: 20 v % Dioxane	71Pd	
P33 6,7-PHENANTHROLINE			$C_{12}H_8N_2$		see P31		
3.5	4.857(25°,+1)	-10.2	0-50	T	μ=0	56Na	c
3.95	4.96(+1)	-9.2	20	C	0.1M($NaNO_3$)	55Aa	
P34 4,7-PHENANTHROLINE, 1,10-dimethoxy-3,8-dimethyl-			$C_{16}H_{16}O_2N_2$		see P31		
4.97	7.21	-15.1	20	T	--	60Pd	

ΔH, kcal /mole	pK	ΔS, cal /mole°K	T,°C	M	Conditions	Ref.	Remarks
P35 4,7-PHENANTHROLINE, 3,8-dimethyl			$C_{14}H_{12}N_2$		see P31		
6.50	2.56	11.4	20	T	μ=0	60Pd	
5.28	5.32	-5.4	20	T	μ=0	60Pd	
P36 4,7-PHENANTHROLINE, 1,2,3,8, 9,10-hexamethyl-			$C_{18}H_{20}N_2$		see P31		
4.97	7.26	-15.3	20	T	μ=0	60Pd	
P37 1,10-PHENANTHROLINE, 6-methyl-			$C_{13}H_{10}N_2$		see P31		
4.40	5.11	-8.6	25	T	μ=0	67La	
P38 1,10-PHENANTHROLINE, 5-nitro-			$C_{12}H_7O_2N_3$		see P31		
2.26	3.230(25°,$\underline{+}$1)	-7.0	25-45	T	μ=0	64Lb	c
P39 1,10-PHENANTHROLINE, 6-nitro-			$C_{12}H_7O_2N_3$		see P31		
2.26	3.23	-7.2	25	T	μ=0	64Lc	
P40 6,7-PHENANTHROLINE, 1-nitro-			$C_{12}H_7O_2N_3$		see P31		
2.26	3.23	-7.0	25	T	0.2M Sodium Acetate Buffer	64Lb	

ΔH, kcal /mole	pK	ΔS, cal /mole°K	T,°C	M	Conditions	Ref.	Remarks
P41 4,7-PHENANTHROLINE, 1,3,8,10-tetramethyl-			$C_{16}H_{16}N_2$		see P31		
6.53	3.38	7.74	20	T	μ=0	60Pd	
4.33	5.35	-8.87	20	T	μ=0	60Pd	
P42 PHENAZINE, 3-hydroxy			$C_{12}H_8ON_2$		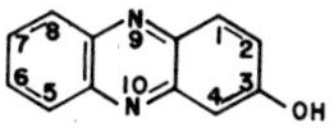		
3.35	2.67	0.09	15	T	--	58Pa	
P43 PHENAZINE, N-oxide			$C_{12}H_8ON_2$		see P42		
6.083	4.882(25°)	1.817	25-90	T	Aqueous H_2SO_4 Solution	74Cb	c
P44 PHENOL			C_6H_6O		C_6H_5OH		
5.55	9.990(0)	-27.1	25	C	μ=0	75Pa	
5.650	9.979(0)	-26.7	25	C	μ=0	59Fa	
						67Mb	
						47Ma	
5.60	9.99(0)	-27.0	25	C	μ=0	59Pa	
5.66	(0)	-----	25	C	μ=0	72Na	
5.65	10.00(0)	-26.8	25	C	μ=0	64Ka	
5.58	(0)	-----	25	C	μ=0	71Ha	
5.66	10.02(0)	-26.9	25	T	μ=0 ΔCp=-27.1	62Cb	
5.8	9.75(0)	-25.1	25	T	μ=0	63Hd	
4.8	9.97(0)	-29.6	25	T	μ=0	67Za	
5.53	10.00(0)	-27.2	25	T	μ=0	61Bc	
5.72	(0)	-----	20	C	μ≃0.01	71Ma	
5.47	9.95(0)	-27.2	5-60	T	μ=0.04	66Bf	
6.1	9.8(0)	-24	25	T	μ=0.10	69Da	
5.36	(0)	-----	25	T	--	61Bc	g: E.H. Binns, Trans. Faraday Soc., 55, 1900 (1959)
5.49	10.083(0)	-27.7	25	C	μ=0 S: 0.059 mole fraction Methanol	75Pa	
5.40	10.286(0)	-28.9	25	C	μ=0 S: 0.123 mole fraction Methanol	75Pa	
5.39	10.530(0)	-30.1	25	C	μ=0 S: 0.194 mole fraction Methanol	75Pa	
5.25	10.747(0)	-31.5	25	C	μ=0 S: 0.273 mole fraction Methanol	75Pa	
4.93	10.956(0)	-33.6	25	C	μ=0 S: 0.360 mole fraction Methanol	75Pa	
4.77	11.078(0)	-34.7	25	C	μ=0 S: 0.458 mole fraction Methanol	75Pa	

ΔH, kcal /mole	pK	ΔS, cal /mole°K	T,°C	M	Conditions	Ref.	Remarks
P44, cont.							
4.65	11.343(0)	-36.3	25	C	μ=0 S: 0.567 mole fraction Methanol	75Pa	
4.39	11.634(0)	-38.6	25	C	μ=0 S: 0.692 mole fraction Methanol	75Pa	
4.33	12.136(0)	-41.0	25	C	μ=0 S: 0.835 mole fraction Methanol	75Pa	
8.50	14.356(0)	-37.2	25	C	μ=0 S: Methanol	75Pa	
8.5	14.46(0)	-38	25	T	S: Methanol	69Rd	
8.87	14.356(0)	-35.8	25	T	μ=0 S: Methanol ΔCp=-70	70Bf	

P45 PHENOL, 2-azo-4-ethyl-, N(2-thiazolyl)- $C_{11}H_{11}ON_3S$

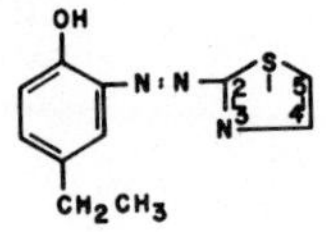

ΔH	pK	ΔS	T,°C	M	Conditions	Ref.	Remarks
2.52	8.12	-28.60	15-35	T	μ=0.1(KNO_3)	75Ka	c

P46 PHENOL, 2-bromo- C_6H_5OBr $C_6H_5O(Br)$

ΔH	pK	ΔS	T,°C	M	Conditions	Ref.	Remarks
4.40	8.452(0)	-23.9	25	T	μ=0 ΔCp=-39	66Bd	
4.40	8.452(25°,0)	-23.9	5-50	T	μ=0	66Be	c

P47 PHENOL, 3-bromo- C_6H_5OBr $C_6H_5O(Br)$

ΔH	pK	ΔS	T,°C	M	Conditions	Ref.	Remarks
5.35	9.031(25°,0)	-23.4	10-60	T	μ=0	66Bg	c

P48 PHENOL, 4-bromo- C_6H_5OBr $C_6H_5O(Br)$

ΔH	pK	ΔS	T,°C	M	Conditions	Ref.	Remarks
5.74	9.340(0)	-23.5	25	C	μ=0	75Pa	
5.74	9.34(25°,0)	-23.5	5-60	T	μ=0	66Bf	
5.48	9.480(0)	-25.0	25	C	μ=0 S: 0.059 mole fraction Methanol	75Pa	
5.49	9.629(0)	-25.6	25	C	μ=0 S: 0.123 mole fraction Methanol	75Pa	
5.71	9.801(0)	-25.7	25	C	μ=0 S: 0.194 mole fraction Methanol	75Pa	
5.81	10.016(0)	-26.3	25	C	μ=0 S: 0.273 mole fraction Methanol	75Pa	
5.06	10.190(0)	-29.6	25	C	μ=0 S: 0.360 mole fraction Methanol	75Pa	
4.65	10.307(0)	-31.5	25	C	μ=0 S: 0.458 mole fraction Methanol	75Pa	
4.25	10.537(0)	-34.0	25	C	μ=0 S: 0.567 mole fraction Methanol	75Pa	
3.91	10.858(0)	-36.6	25	C	μ=0 S: 0.692 mole fraction Methanol	75Pa	
3.92	11.369(0)	-38.9	25	C	μ=0 S: 0.835 mole fraction Methanol	75Pa	
8.26	13.501(0)	-34.0	25	C	μ=0 S: Methanol	75Pa	
8.58	13.501(0)	-33.0	25	T	μ=0 S: Methanol ΔCp=-34	70Bf	
8.3	13.61(0)	-34	25	T	S: Methanol	69Rd	

P49 PHENOL, 4-bromo-2,6-di(t-butyl)- $C_{14}H_{21}OBr$ $C_6H_3O[C(CH_3)_3]_2(Br)$

ΔH, kcal /mole	pK	ΔS, cal /mole°K	T,°C	M	Conditions	Ref.	Remarks
P49, cont.							
10.13	15.804(0)	-38.2	25	T	μ=0 S: Methanol ΔCp=-58	70Bf	
9.9	15.91(0)	-39	25	T	S: Methanol	69Rd	
P50 PHENOL, 2-chloro-			C_6H_5OCl		$C_6H_5O(Cl)$		
5.101	(0)	-----	10	C	μ=0	48Ca	
4.795	(0)	-----	20	C	μ=0	48Ca	
4.630	8.476(0)	-23.5	25	C	μ=0	59Fa 61Oa	
4.26	8.555(0)	-24.9	25	T	μ=0 ΔCp=-89	66Bd	
4.18	8.53(0)	-25.0	25	T	μ=0 ΔCp=-46.9	62Cb	
4.590	8.487(0)	-23.5	25	C	μ=0	48Ca	
4.387	(0)	-----	30	C	μ=0	48Ca	
4.26	8.555(25°,0)	-24.9	5-50	T	μ=0	66Be	c
4.7	6.3(0)	-22	25	T	μ=0.10	69Da	
P51 PHENOL, 3-chloro-			C_6H_5OCl		$C_6H_5O(Cl)$		
5.286	9.08(0)	-23.8	25	C	μ=0	61Oa	
5.35	9.119(25°,0)	-23.8	10-60	T	μ=0	66Bg	c
6.1	8.7(0)	-19	25	T	μ=0.10	69Da	
P52 PHENOL, 4-chloro-			C_6H_5OCl		$C_6H_5O(Cl)$		
5.800	9.379(0)	-23.5	25	C	μ=0	59Fa 61Oa	
5.630	9.39(0)	-24.0	25	C	μ=0	70Bc	g:59Fa
5.727	9.440(0)	-23.9	25	T	μ=0	70Bc	g:66Bf
5.73	9.43(25°,0)	-23.9	5-60	T	μ=0	66Bf	c
6.1	9.1(0)	-21	25	T	μ=0.10	69Da	
P53 PHENOL, 4-chloro-2-(2'-thiazolyazo)-			$C_9H_6ON_3ClS$		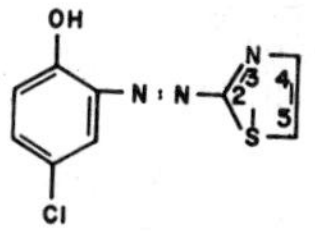		
4.05	7.09(25°)	-18.87	15-35	T	μ=0	74Kb	c
P54 PHENOL, 3,5-dibromo-			$C_6H_4OBr_2$		$C_6H_4O(Br)_2$		
5.251	8.056(0)	-19.25	25	T	μ=0 ΔCp=-79.5	68Bc	
P55 PHENOL, 3,4-dichloro-			$C_6H_4OCl_2$		$C_6H_4O(Cl)_2$		
5.528	8.630(0)	-20.95	25	T	μ=0 ΔCp=-14	70Bc	
P56 PHENOL, 3,5-dichloro-			$C_6H_4OCl_2$		$C_6H_4O(Cl)_2$		
4.880	8.179(0)	-21.05	25	T	μ=0 ΔCp=-24.5	68Bc	
P57 PHENOL, 3,5-diethoxy-			$C_{10}H_{14}O_3$		$C_6H_4O(OCH_2CH_3)_2$		
4.647	9.370(0)	-27.30	25	T	μ=0 ΔCp=-46.7	68Bc	

ΔH, kcal /mole	pK	ΔS, cal /mole°K	T,°C	M	Conditions	Ref.	Remarks
P58 PHENOL, 3,5-diiodo-			$C_6H_4OI_2$		$C_6H_4O(I)_2$		
5.451	8.103(0)	-18.79	25	T	μ=0 ΔCp=-79.4	68Bc	
P59 PHENOL, 3,5-dimethoxy-			$C_8H_{10}O_3$		$C_6H_4O(OCH_3)_2$		
4.479	9.345(0)	-27.74	25	T	μ=0 ΔCp=-48	72Be	
P60 PHENOL, 2,4-dinitro-			$C_6H_4O_5N_2$		$C_6H_4O(NO_2)_2$		
2.250	4.08(0)	-11.2	25	T	μ=0	60Rb	
3.821	6.48(0)	-16.3	15	T	S: Dimethylformamide	68Pb	
3.955	6.41(0)	-15.9	20	T	S: Dimethylformamide	68Pb	
4.091	6.36(0)	-15.4	25	T	S: Dimethylformamide	68Pb	
4.230	6.32(0)	-15.0	30	T	S: Dimethylformamide	68Pb	
4.370	6.27(0)	-14.5	35	T	S: Dimethylformamide	68Pb	
4.513	6.22(0)	-14.0	40	T	S: Dimethylformamide	68Pb	
P61 PHENOL, 2,5-dinitro-			$C_6H_4O_5N_2$		$C_6H_4O(NO_2)_2$		
1.230	3.73(0)	-12.9	25	T	μ=0	60Rb	
6.208	8.96(0)	-19.4	15	T	S: Dimethylformamide	68Pb	
6.425	8.86(0)	-18.7	20	T	S: Dimethylformamide	68Pb	
6.646	8.78(0)	-17.9	25	T	S: Dimethylformamide	68Pb	
6.871	8.71(0)	-17.2	30	T	S: Dimethylformamide	68Pb	
7.099	8.63(0)	-16.4	35	T	S: Dimethylformamide	68Pb	
7.332	8.54(0)	-15.7	40	T	S: Dimethylformamide	68Pb	
P62 PHENOL, 2,6-dinitro-			$C_6H_4O_5N_2$		$C_6H_4O(NO_2)_2$		
3.000	5.23(0)	-13.9	25	T	μ=0	60Rb	
1.740	6.12(0)	-22.0	15	T	S: Dimethylformamide	68Pb	
1.801	6.09(0)	-21.8	20	T	S: Dimethylformamide	68Pb	
1.863	6.07(0)	-21.5	25	T	S: Dimethylformamide	68Pb	
1.926	6.06(0)	-21.3	30	T	S: Dimethylformamide	68Pb	
1.990	6.03(0)	-21.1	35	T	S: Dimethylformamide	68Pb	
2.055	6.01(0)	-20.9	40	T	S: Dimethylformamide	68Pb	
P63 PHENOL, 3,5-dinitro-			$C_6H_4O_5N_2$		$C_6H_4O(NO_2)_2$		
3.757	6.732(0)	-18.21	25	T	μ=0 ΔCp=-16.8	68Bc	
P64 PHENOL, 2-ethoxy-			$C_8H_{10}O_2$		$C_6H_5O(OC_2H_5)$		
6.18	10.109(0)	-25.5	25	T	μ=0 ΔCp=-23	66Bd	
P65 PHENOL, 3-ethoxy-			$C_8H_{10}O_2$		$C_6H_5O(OCH_2CH_3)$		
5.082	9.655(0)	-27.13	25	T	μ=0 ΔCp=-56.5	68Bc	
P66 PHENOL, 2-fluoro-			C_6H_5OF		$C_6H_5O(F)$		
4.66	8.73(0)	-24.3	25	C	μ=0	66Cf	
P67 PHENOL, 3-fluoro-			C_6H_5OF		$C_6H_5O(F)$		
5.52	9.29(0)	-24.0	25	C	μ=0	66Cf	

ΔH, kcal /mole	pK	ΔS, cal /mole°K	T, °C	M	Conditions	Ref.	Remarks
P68 PHENOL, 4-fluoro-			C_6H_5OF		$C_6H_5O(F)$		
5.93	9.89(0)	-25.4	25	C	μ=0	66Cf	
P69 PHENOL, 2-iodo-			C_6H_5OI		$C_6H_5O(I)$		
4.27	8.513(0)	-24.6	25	T	μ=0 ΔCp=-49	66Bd	
P70 PHENOL, 3-iodo-			C_6H_5OI		$C_6H_5O(I)$		
5.53	9.023(25°,0)	-22.7	10-60	T	μ=0.4	66Bg	c
P71 PHENOL, 4-iodo-			C_6H_5OI		$C_6H_5O(I)$		
5.38	9.32(25°,0)	-24.6	5-60	T	μ=0.04	66Bf	c
P72 PHENOL, 2-methoxy-			$C_7H_8O_2$		$C_6H_5O(OCH_3)$		
5.74	9.99(0)	-26.5	25	C	μ=0	64Mb	
P73 PHENOL, 3-methoxy-			$C_7H_8O_2$		$C_6H_5O(OCH_3)$		
5.26	9.65(0)	-26.5	25	C	μ=0	64Mb	
5.020	9.652(0)	-27.33	25	T	μ=0 ΔCp=-35.1	67Bb	
P74 PHENOL, 4-methoxy-			$C_7H_8O_2$		$C_6H_5O(OCH_3)$		
5.70	10.20(0)	-27.6	25	C	μ=0	64Mb	
P75 PHENOL, 4-methoxy-2-(2'-thiazolyazo)-			$C_{10}H_9O_2N_3S$		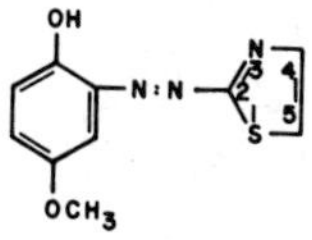		
2.60	7.83(25°)	-27.10	15-35	T	μ=0	74Kb	c
P76 PHENOL, 2-nitro-			$C_6H_5O_3N$		$C_6H_5O(NO_2)$		
4.655	7.222(0)	-17.4	25	C	μ=0	59Fa	
4.59	7.230(25°,0)	-17.67	5-60	T	μ=0.1	63Ra	c
8.576	12.38(0)	-26.8	15	T	S: Dimethylformamide	68Pb	
8.876	12.25(0)	-25.8	20	T	S: Dimethylformamide	68Pb	
9.181	12.14(0)	-24.8	25	T	S: Dimethylformamide	68Pb	
9.492	12.03(0)	-23.7	30	T	S: Dimethylformamide	68Pb	
9.807	11.92(0)	-22.7	35	T	S: Dimethylformamide	68Pb	
10.128	11.81(0)	-21.7	40	T	S: Dimethylformamide	68Pb	
P77 PHENOL, 3-nitro-			$C_6H_5O_3N$		$C_6H_5O(NO_2)$		
5.145	(0)	-----	10	C	μ=0	48Ca	
4.833	(0)	-----	20	C	μ=0	48Ca	
4.705	8.267(0)	-22.1	25	C	μ=0	48Ca 59Fa	
4.791	8.360(0)	-22.18	25	T	μ=0 ΔCp=-20.9	67Bb	

ΔH, kcal /mole	pK	ΔS, cal /mole°K	T,°C	M	Conditions	Ref.	Remarks

P77, cont.

ΔH, kcal /mole	pK	ΔS, cal /mole°K	T,°C	M	Conditions	Ref.	Remarks
4.575	(0)	-----	30	C	μ=0	48Ca	
4.8	8.1(0)	-21	25	T	μ=0.10	69Da	
5.143	8.355(0)	-21.1	25	T	μ=0.1	63Rb	
10.549	14.15(0)	-28.1	15	T	S: Dimethylformamide	68Pb	
10.918	13.99(0)	-26.8	20	T	S: Dimethylformamide	68Pb	
11.294	13.85(0)	-25.5	25	T	S: Dimethylformamide	68Pb	
11.676	13.72(0)	-24.3	30	T	S: Dimethylformamide	68Pb	
12.064	13.58(0)	-23.0	35	T	S: Dimethylformamide	68Pb	
12.459	13.45(0)	-21.7	40	T	S: Dimethylformamide	68Pb	
8.15	12.331(0)	-29.2	25	T	μ=0 S: Methanol ΔCp=-30	70Bf	
7.9	12.44(0)	-30	25	T	S: Methanol	69Rd	

P78 PHENOL, 4-nitro- $C_6H_5O_3N$ $C_6H_5O(NO_2)$

ΔH, kcal /mole	pK	ΔS, cal /mole°K	T,°C	M	Conditions	Ref.	Remarks
4.895	(0)	-----	10	C	μ=0	48Ca	
4.669	(0)	-----	20	C	μ=0	48Ca	
4.570	7.150(0)	-17.4	25	C	μ=0	48Ca	
4.53	7.155(0)	-17.5	25	C	μ=0	75Pa	
4.700	7.143(0)	-16.9	25	C	μ=0	59Fa	
4.466	(0)	-----	30	C	μ=0	48Ca	
4.8	6.8(0)	-15	25	T	μ=0.10	69Da	
4.71	7.156(25°,0)	-16.96	5-60	T	μ=0.1	63Ra 62Ab	c
8.916	12.07(0)	-24.3	15	T	S: Dimethylformamide	68Pb	
9.228	11.95(0)	-23.2	20	T	S: Dimethylformamide	68Pb	
9.545	11.84(0)	-22.1	25	T	S: Dimethylformamide	68Pb	
9.868	11.73(0)	-21.1	30	T	S: Dimethylformamide	68Pb	
10.196	11.60(0)	-20.0	35	T	S: Dimethylformamide	68Pb	
10.530	11.48(0)	-18.9	40	T	S: Dimethylformamide	68Pb	
4.72	7.180(0)	-17.0	25	C	μ=0 S: 0.059 mole fraction Methanol	75Pa	
4.52	7.311(0)	-18.3	25	C	μ=0 S: 0.123 mole fraction Methanol	75Pa	
4.42	7.441(0)	-19.0	25	C	μ=0 S: 0.194 mole fraction Methanol	75Pa	
4.10	7.572(0)	-20.9	25	C	μ=0 S: 0.273 mole fraction Methanol	75Pa	
3.71	7.721(0)	-22.9	25	C	μ=0 S: 0.360 mole fraction Methanol	75Pa	
3.34	7.909(0)	-25.0	25	C	μ=0 S: 0.458 mole fraction Methanol	75Pa	
3.05	8.140(0)	-27.0	25	C	μ=0 S: 0.567 mole fraction Methanol	75Pa	
2.87	8.473(0)	-29.1	25	C	μ=0 S: 0.692 mole fraction Methanol	75Pa	
3.10	9.060(0)	-31.0	25	C	μ=0 S: 0.835 mole fraction Methanol	75Pa	
7.76	11.401(0)	-26.1	25	C	μ=0 S: Methanol	75Pa	
8.25	11.400(0)	-24.4	25	T	μ=0 S: Methanol ΔCp=-22	70Bf	
8.0	11.50(0)	-26	25	T	S: Methanol	69Rd	

P79 PHENOL, 3(trimethylamino)- $C_9H_{14}ON$ $C_6H_5O[N(CH_3)_3]$

ΔH, kcal /mole	pK	ΔS, cal /mole°K	T,°C	M	Conditions	Ref.	Remarks
6.09	8.06(0)	-16.4	25	C	μ=0	64Ka	

P80 PHENOL, 4(trimethylamino)- $C_9H_{14}ON$ $C_6H_5O[N(CH_3)_3]$

ΔH, kcal /mole	pK	ΔS, cal /mole°K	T,°C	M	Conditions	Ref.	Remarks
5.46	8.35(0)	-19.9	25	C	μ=0	64Ka	

ΔH, kcal /mole	pK	ΔS, cal /mole°K	T, °C	M	Conditions	Ref.	Remarks
P81 PHENOL, 2,4,5-trichloro-			$C_6H_3OCl_3$		$C_6H_3O(Cl)_3$		
10.094	12.73(0)	-23.2	15	T	S: Dimethylformamide	68Pb	
10.447	12.59(0)	-22.0	20	T	S: Dimethylformamide	68Pb	
10.807	12.46(0)	-20.8	25	T	S: Dimethylformamide	68Pb	
11.172	12.33(0)	-19.6	30	T	S: Dimethylformamide	68Pb	
11.544	12.20(0)	-18.3	35	T	S: Dimethylformamide	68Pb	
11.921	12.06(0)	-17.1	40	T	S: Dimethylformamide	68Pb	
P82 PHENOL, 3,4,5-trichloro-			$C_6H_3OCl_3$		$C_6H_3O(Cl)_3$		
4.804	7.839(0)	-19.75	25	T	μ=0 ΔCp=-13	70Bc	
P83 PHENOL, 2,4,6-trinitro-			$C_6H_3O_7N_3$		$C_6H_3O(NO_2)_3$		
4.570	0.810	11.4	30	T	μ=0	21Na	
-1.483	0.19(20°)	-6.03	5-60	T	μ=0	61Sb	c
5.330	2.097	7.8	30	T	μ=0 S: 87.76 v % Acetone	21Na	
8.680	2.770	15.6	30	T	μ=0 S: 93.80 v % Acetone	21Na	
8.960	5.947	2.3	30	T	μ=0 S: Acetone	21Na	
P84 PHENYLALANINE			$C_9H_{11}O_2N$		$C_6H_5CH_2CH(NH_2)CO_2H$		
0.6	2.20(20°,+1)	8.3	0-40	T	μ=0	61Ib	c,e: CO_2H
2.33	2.754(+1)	-4.8	25	C	μ=3.0M($NaClO_4$)	72Ga	
11.40	9.75(0)	-4.3	10	C	μ=0	66Ac	
10.67	9.31(0)	-6.7	25	C	μ=0	66Ac	
10.53	8.96(0)	-7.4	40	C	μ=0	66Ac	
10.3	9.38(20°,0)	-7.8	0-40	T	μ=0	61Ib	c,e:NH_3^+
10.10	(0)	-----	20	C	μ≃0.01	71Ma	e:NH_3^+
9.46	9.759(0)	-10.63	5	T	μ=0.1(KCl)	73Ra	e:NH_3^+
11.06	9.220(0)	-5.07	25	T	μ=0.1(KCl)	73Ra	e:NH_3^+
12.81	8.670(0)	-0.62	45	T	μ=0.1(KCl)	73Ra	e:NH_3^+
10.32	9.02(0)	-6.6	25	C	μ=0.16(KNO_3)	70Md	
12.04	9.610(0)	-3.6	25	C	μ=3.0M($NaClO_4$)	72Ga	e:NH_3^+
P85 β-PHENYLSERINE			$C_9H_{11}O_3N$		$C_6H_5CH(OH)CH(NH_2)CO_2H$		
9.33	8.79(0)	-8.9	25	C	μ=0.16	70Lc	e:NH_3^+
P86 PHOSPHINE, dichlorophenyl-, oxide			$C_6H_5OCl_2P$		$(C_6H_5)P(Cl)_2O$		
10.7	(+1)	-----	25	C	C∿10^{-3}M S: Fluorosulfuric acid	74Ab	
P87 PHOSPHINE, trimethyl-, oxide			C_3H_9OP		$(CH_3)_3PO$		
32.2	-----	-----	25	C	C∿10^{-3}M S: Fluorosulfuric acid	74Ab	
P88 PHOSPHINE, triphenyl-, oxide			$C_{18}H_{15}OP$		$(C_6H_5)_3PO$		
20.9	-----	-----	25	C	C∿10^{-3}M S: Fluorosulfuric acid	74Ab	

ΔH, kcal /mole	pK	ΔS, cal /mole°K	T,°C	M	Conditions	Ref.	Re-marks
P89 PHOSPHORAMIDIC ACID			H_4O_3NP		$HOPO(NH_2)OH$		
∿0	3.08(25°)	-14	18-39	T	μ=0	68La	c
5.2	8.63(25°)	-22	18-39	T	μ=0	68La	c
P90 PHOSPHORIC ACID, monofluoro-			H_2O_4FP		H_2FPO_4		
-1.90	4.79(-1)	-28.2	25	-	--	69La	
P91 PHOSPHORIC ACID, ortho-			H_3O_4P		H_3PO_4		
-0.154	2.048(0)	-9.94	0.3	T	μ=0	34Na	
-0.428	2.076(0)	-11.00	12.5	T	μ=0	34Na	
-1.466	2.127(0)	-14.7	15	T	μ=0	51Ba	
-1.646	2.127(0)	-15.3	20	T	μ=0	51Ba	
-1.880	2.124(0)	-16.0	25	C	μ=0	37Pa	
-1.748	2.123(0)	-15.6	25	T	μ=0 ΔCp=-52.4	39Hb	
-0.849	2.124(0)	-12.57	25.0	T	μ=0	34Na	
-1.828	2.148(0)	-16.0	25	T	μ=0	67Ma	
-2.014	2.171(0)	-16.6	30	T	μ=0	67Ma	
-2.203	2.196(0)	-17.2	35	T	μ=0	67Ma	
-1.413	2.185(0)	-14.54	37.5	T	μ=0	34Na	
-2.394	2.224(0)	-17.8	40	T	μ=0	51Ba	
-2.589	2.251(0)	-18.4	45	T	μ=0	51Ba	
-2.121	2.260(0)	-16.90	50	T	μ=0	34Na	
-1.9	2.1(0)	-16	25	C	μ=0.5	66Ia	
-1.529	1.89(25°,0)	-13.8	25-70	T	--	67Ma	
0.0	2.06(0)	-9.4	80-95	T	--	67Ma	
-0.654	2.81(25°,0)	-15.1	25-160	T	--	67Ma	
0.0	2.95(0)	-13.5	160	T	--	67Ma	
1.233	2.26(0)	-6.21	110-200	T	--	67Ma	
4.758	4.05(0)	-2.6	160-200	T	--	67Ma	
0.0	1.92(0)	-8.8	200	T	--	67Ma	
0.0	2.79(0)	-12.7	200-220	T	--	67Ma	
-2.899	1.14(0)	-14.9	200-300	T	--	67Ma	
-2.739	2.04(0)	-18.5	220-300	T	--	67Ma	
2.284	7.3139(-1)	-25.1	0	T	μ=0	43Ba	
2.034	7.2806(-1)	-26.0	5	T	μ=0	43Ba	
1.927	6.79(-1)	-26.4	5	T	μ=0	58Ga	
1.779	7.2533(-1)	-26.9	10	T	μ=0	43Ba	
1.230	(-1)	-----	10	T	μ=0	34Hc	
1.520	7.2305(-1)	-27.8	15	T	μ=0	43Ba	
1.256	7.2129(-1)	-28.7	20	T	μ=0	43Ba	
0.90	7.20(-1)	-30	25	C	μ=0	62Cc	
0.6	7.20(-1)	-31	25	C	μ=0	66Ia	
0.800	7.205(-1)	-30.3	25	C	μ=0	37Pa	
0.987	7.180(-1)	-29.6	25	T	μ=0 ΔCp=-54	65Pa	
0.736	7.205(-1)	-30.5	25	T	μ=0 ΔCp=-35.7	39Hb	
0.76	7.18(-1)	-30.4	25	T	μ=0	63Pc	
						65Pe	
0.740	(-1)	-----	25	T	μ=0	34Hc	
1.003	7.20(-1)	-29.6	25	T	μ=0	58Ga	
0.987	7.1972(-1)	-29.6	25	T	μ=0	43Ba	
0.714	7.1909(-1)	-30.5	30	T	μ=0	43Ba	
0.437	7.1856(-1)	-31.4	35	T	μ=0	43Ba	
0.140	(-1)	-----	40	T	μ=0	34Hc	
0.155	7.1822(-1)	-32.4	40	T	μ=0	43Ba	
-0.132	7.1818(-1)	-33.3	45	T	μ=0	43Ba	
-0.013	7.78(-1)	-33.6	50	T	μ=0	58Ga	
-0.432	7.1851(-1)	-34.2	50	T	μ=0	43Ba	
-0.718	7.1893(-1)	-35.1	55	T	μ=0	43Ba	
-1.019	7.1981(-1)	-36.0	60	T	μ=0	43Ba	
6.000	6.76(-1)	-9.4	5	T	μ=0.7	56Bb	
2.900	6.68(-1)	-20.5	15	T	μ=0.7	56Bb	
-0.200	6.64(-1)	-31.0	25	T	μ=0.7	56Bb	

ΔH, kcal /mole	pK	ΔS, cal /mole°K	T, °C	M	Conditions	Ref.	Remarks
P91, cont.							
-4.200	6.68(-1)	-44.0	37	T	μ=0.7	56Bb	
0.496	8.440(-1)	-37.0	25	T	μ=0 S: 50 w % Methanol ΔCp=-61	65Pa	
3.500	11.946(-2)	-43	25	C	μ=0	37Pa	
2.6	12.36(-2)	-48	25	C	μ=0	66Ia	
4.20	12.40(-2)	-43	25	C	μ=0	66Ca	
3.5	12.0(-2)	-43.4	25	T	μ=0	65Pe	
4.3	11.93(-2)	69.00	25	C	μ=0.22	65Pc	
P92 PHOSPHORIC ACID, ortho-, deuterated			D_3O_4P		D_3PO_4		
-1.815	2.4202(0)	-17.2	25	T	μ=0 S: Deuterium Oxide ΔCp=-38	70Pc	
2.034	7.2810(5°,-1)	-26.0	0	T	μ=0	64Ga	
0.987	7.2005(-1)	-29.6	25	T	μ=0	64Ga	
-0.423	7.1828(-1)	-34.2	30	T	μ=0	64Ga	
2.504	7.885(5°,-o)	-27.1	0	T	μ=0 S: Deuterium Oxide	64Ga	
1.376	7.780(-1)	-31.0	25	T	μ=0 S: Deuterium Oxide	64Ga	
-0.144	7.744(-1)	-35.9	50	T	μ=0 S: Deuterium Oxide	64Ga	
P93 PHOSPHORIC ACID, pyro-			$H_4O_7P_2$		$H_4P_2O_7$		
-6.70	0.91(0)	-26.65	25	T	μ=0	66Mb	
-3.0	-0.44(0)	-8	25	C	μ>1	66Ia	
-8.31	2.10(-1)	-37.56	25	T	μ=0	66Mb	
3.0	2.64(-1)	-22	25	C	μ>1	66Ia	
-0.3	6.7(-2)	-32	25	C	μ=0	62Cc	
0.130	6.700(-2)	-30.2	25	C	μ=0	73Va	
0.30	6.54(-2)	-29.0	25	T	μ=0	65Pe	
-10.82	6.70(-2)	-66.87	25	T	μ=0	66Mb	
0.1	6.76(-2)	-31	25	C	μ=0.13	66Ia	
0.680	(-2)	-----	25	C	μ=0.5($LiNO_3$ + HNO_3)	73Va	
0.630	(-2)	-----	25	C	μ=0.5($LiClO_4$ + $HClO_4$)	73Va	
0.530	(-2)	-----	25	C	μ=0.5(NaCl + HCl)	73Va	
0.470	(-2)	-----	25	C	μ=0.5($NaNO_3$ + HNO_3)	73Va	
0.340	(-2)	-----	25	C	μ=0.5($NaClO_4$ + $HClO_4$)	73Va	
0.650	(-2)	-----	25	C	μ=0.5(KCl + HCl)	73Va	
0.580	(-2)	-----	25	C	μ=0.5(KNO_3 + HNO_3)	73Va	
0.670	(-2)	-----	25	C	μ=1.0(NaCl + HCl)	73Va	
0.490	(-2)	-----	25	C	μ=1.0($NaNO_3$ + HNO_3)	73Va	
0.280	(-2)	-----	25	C	μ=1.0($NaClO_4$ + $HClO_4$)	73Va	
0.790	(-2)	-----	25	C	μ=1.0(KCl + HCl)	73Va	
0.710	(-2)	-----	25	C	μ=1.0(KNO_3 + HNO_3)	73Va	
0.750	(-2)	-----	25	C	μ=1.5(NaCl + HCl)	73Va	
0.520	(-2)	-----	25	C	μ=1.5($NaNO_3$ + HNO_3)	73Va	
0.150	(-2)	-----	25	C	μ=1.5($NaClO_4$ + $HClO_4$)	73Va	
0.930	(-2)	-----	25	C	μ=1.5(KCl + HCl)	73Va	
0.840	(-2)	-----	25	C	μ=1.5(KNO_3 + HNO_3)	73Va	
-1.7	9.4(-3)	-49	25	C	μ=0	62Cc	
0.390	9.38(-3)	-41.6	25	C	μ=0	73Va	
2.8	9.42(-3)	-33.9	25	T	μ=0	65Pe	
-10.00	9.32(-3)	-76.06	25	T	μ=0	66Mb	
0.4	9.41(-3)	-42	25	C	μ=0.22	66Ia	
1.050	(-3)	-----	25	C	μ=0.5($LiNO_3$ + HNO_3)	73Va	
0.980	(-3)	-----	25	C	μ=0.5(NaCl + HCl)	73Va	
0.960	(-3)	-----	25	C	μ=0.5($NaNO_3$ + HNO_3)	73Va	
0.840	(-3)	-----	25	C	μ=0.5($NaClO_4$ + $HClO_4$)	73Va	

ΔH, kcal /mole	pK	ΔS, cal /mole°K	T, °C	M	Conditions	Ref.	Remarks
P93, cont.							
0.860	(-3)	-----	25	C	μ=0.5(KNO_3 + HNO_3)	73Va	
1.120	(-3)	-----	25	C	μ=0.75($LiNO_3$ + HNO_3)	73Va	
1.250	(-3)	-----	25	C	μ=1.0($LiNO_3$ + HNO_3)	73Va	
1.150	(-3)	-----	25	C	μ=1.0(NaCl + HCl)	73Va	
1.080	(-3)	-----	25	C	μ=1.0($NaNO_3$ + HNO_3)	73Va	
0.850	(-3)	-----	25	C	μ=1.0($NaClO_4$ + $HClO_4$)	73Va	
0.810	(-3)	-----	25	C	μ=1.0(KNO_3 + HNO_3)	73Va	
1.280	(-3)	-----	25	C	μ=1.50(NaCl + HCl)	73Va	
0.790	(-3)	-----	25	C	μ=1.50($NaClO_4$ + $HClO_4$)	73Va	
0.860	(-3)	-----	25	C	μ=1.50(KNO_3 + HNO_3)	73Va	
1.200	(-3)	-----	25	C	μ=2.0($NaNO_3$ + HNO_3)	73Va	
P94 PHOSPHORIC ACID, dipoly-, imide,			$H_5O_6NP_2$		$(HO)_2OPNHPO(OH)_2$		
-3.2	7.32(25°)	-44.2	25-50	T	μ=0.1	61Ia	c
-6.4	10.22(25°)	-66.2	25-50	T	μ=0.1	61Ia	c
P95 PHOSPHORIC ACID, triethyl ester-			$C_6H_{15}O_4P$		$(CH_3CH_2O)_3PO$		
20.6	-----	-----	25	C	C∿10^{-3}M S: Fluorosulfuric acid	74Ab	
P96 PHOSPHORIC ACID, trimethyl ester			$C_3H_9O_4P$		$(CH_3O)_3PO$		
20.4	-----	-----	25	C	C∿10^{-3} S: Fluorosulfuric acid	74Ab	
P97 PHOSPHORIC ACID, tripoly-			$P_3H_5O_{10}$		$H_5O_{10}P_3$		
-2.5	-0.51(0)	-6	25	C	μ>1	66Ia	
-2.5	1.20(-1)	-14	25	C	μ>1	66Ia	
-2.5	2.30(-2)	-19	25	C	μ>1	66Ia	
-1.3	6.2(-3)	-33	25	C	μ=0	62Cc	
0.37	6.61(-3)	-29.1	25	T	μ=0	65Pe	
2.3	5.55(25°,-3)	-17.8	2-45	T	μ=0.10M(KNO_3)	72Kb	c,q: ±0.5
-1.5	6.50(-3)	-31	25	C	μ=0.4	66Ia	
-3.8	8.9(-4)	-53	25	C	μ=0	62Cc	
2.98	9.261(-4)	-33.6	25	T	μ=0	65Pe	
0.10	8.82(-4)	-40	20	C	0.1M$(CH_3)_4$NOH	65Aa	
0.4	8.10(25°,-4)	-37.9	2-45	T	μ=0.10M(KNO_3)	72Kb	c,q: ±0.1
0.1	9.24(-4)	-42	25	C	μ=0.65	66Ia	
P98 PHOSPHORIC ACID, tripoly-, diimido			$H_7O_8N_2P_3$		$(HO)_2OPNHP(O)(OH)NHPO(OH)_2$		
-4.5	9.84(25°)	-60.3	25-50	T	μ=0.1	61Ia	c
3.2	6.61(25°)	-19.5	25-50	T	μ=0.1	61Ia	c
P99 PHOSPHORYL CHLORIDE			$OPCl_3$		$POCl_3$		
4.8	(+1)	-----	25	C	C∿10^{-3}M S: Fluorosulfuric acid	74Ab	

P100 PHTHALIC ACID $C_8H_6O_4$ $C_6H_4(CO_2H)_2$

ΔH, kcal /mole	pK	ΔS, cal /mole°K	T, °C	M	Conditions	Ref.	Remarks
-0.122	2.925(0)	-13.8	0	T	μ=0	45Hb	
-0.221	2.927(0)	-14.2	5	T	μ=0	45Hb	
-0.323	2.931(0)	-14.6	10	T	μ=0	45Hb	
-0.426	2.937(0)	-14.7	15	T	μ=0	45Hb	
-0.61	2.914(0)	-15.5	15	T	μ=0	75La	
-0.63	2.922(0)	-15.5	20	T	μ=0	75La	
-0.530	2.943(0)	-15.3	20	T	μ=0	45Hb	
-0.637	2.950(0)	-15.6	25	T	μ=0	45Hb	
-0.65	2.930(0)	-15.6	25	T	μ=0	75La	
-0.746	2.958(0)	-16.0	30	T	μ=0	45Hb	
-0.856	2.967(0)	-16.4	35	T	μ=0	45Hb	
-0.70	2.946(0)	-15.8	35	T	μ=0	75La	
-0.968	2.978(0)	-16.7	40	T	μ=0	45Hb	
-1.082	2.988(0)	-17.3	45	T	μ=0	45Hb	
-1.197	3.001(0)	-17.4	50	T	μ=0	45Hb	
-1.315	3.014(0)	-17.8	55	T	μ=0	45Hb	
-1.434	3.028(0)	-18.2	60	T	μ=0	45Hb	
1.183	5.432(-1)	-20.5	0	T	μ=0	45Ha	
0.859	5.418(-1)	-21.7	5	T	μ=0	45Ha	
0.529	5.410(-1)	-22.9	10	T	μ=0	45Ha	
0.194	5.405(-1)	-24.1	15	T	μ=0	45Ha	
-0.50	5.347(-1)	-26.3	15	T	μ=0	75La	
-0.52	5.354(-1)	-26.3	20	T	μ=0	75La	
-0.148	5.405(-1)	-25.2	20	T	μ=0	45Ha	
-0.496	5.408(-1)	-26.4	25	T	μ=0	45Ha	
-0.54	5.360(-1)	-26.3	25	T	μ=0	75La	
-0.849	5.416(-1)	-27.6	30	T	μ=0	45Ha	
-1.208	5.427(-1)	-28.8	35	T	μ=0	45Ha	
-0.57	5.374(-1)	-26.5	35	T	μ=0	75La	
-1.574	5.442(-1)	-29.9	40	T	μ=0	45Ha	
-1.945	5.462(-1)	-31.1	45	T	μ=0	45Ha	
-2.322	5.485(-1)	-32.3	50	T	μ=0	45Ha	
-2.705	5.512(-1)	-33.5	55	T	μ=0	45Ha	
-3.093	5.541(-1)	-34.6	60	T	μ=0	45Ha	
-0.469	4.59(25°,-1)	-21.0	5-45	T	μ=0	68Ba	
-0.591	5.63(25°,-1)	-27.6	5-45	T	S: 10 w % Ethylene Glycol	68Ba	c
-1.053	6.05(25°,-1)	-31.2	5-45	T	S: 30 w % Ethylene Glycol	68Ba	c
-1.674	6.53(25°,-1)	-35.5	5-45	T	S: 50 w % Ethylene Glycol	68Ba	c
-1.735	7.28(25°,-1)	-32.9	5-45	T	S: 70 w % Ethylene Glycol	68Ba	c
-1.844	8.34(25°,-1)	-44.4	5-45	T	S: 90 w % Ethylene Glycol	68Ba	c
0.07	5.00	-23	25	C	μ=0.1(KNO_3) HA=[K(Phthalic acid)]	70Kb	
-0.93	2.316(0)	-13.7	25	C	μ=0 HA=[Co(III)$(NH_3)_5$-(Phthalic acid)]	74La	

P101 PIPERAZINE $C_4H_{10}N_2$

ΔH, kcal /mole	pK	ΔS, cal /mole°K	T, °C	M	Conditions	Ref.	Remarks
7.42	5.333(25°,+2)	0.53	0-50	T	μ=0	68Hb	c
6.9	5.76(25°,+2)	-3.3	25	C	μ=0.1M(KNO_3)	72Ea	
7.12	5.60(30°,+2)	-1.8	25	C	0.1M(KCl)	63Pb	
7.6	5.76(25°,+2)	-1	20-40	T	μ=0.1M(KNO_3)	72Ea	c,q:+1
8.03	6.044(+2)	-0.7	25	C	μ=1M(KNO_3)	72Cd	
7.0	5.205(+2)	0	25	C	μ=0.1M(KNO_3) S: 52 w % Ethanol	71Bf	

ΔH, kcal /mole	pK	ΔS, cal /mole°K	T,°C	M	Conditions	Ref.	Remarks
P101, cont.							
7.817	5.496(+2)	2.46	10	T	μ=0 S: 10 w % Methanol ΔCp=5.5	70Pa	
7.903	5.190(+2)	2.75	25	T	μ=0 S: 10 w % Methanol ΔCp=6.0	70Pa	
7.994	4.911(+2)	3.06	40	T	μ=0 S: 10 w % Methanol ΔCp=6.2	70Pa	
8.348	5.372(+2)	4.90	10	T	μ=0 S: 20 w % Methanol ΔCp=-2.2	70Pa	
8.315	5.048(+2)	4.78	25	T	μ=0 S: 20 w % Methanol ΔCp=-2.2	70Pa	
8.280	4.756(+2)	4.68	40	T	μ=0 S: 20 w % Methanol ΔCp=-2.4	70Pa	
8.663	4.916(+2)	8.10	10	T	μ=0 S: 50 w % Methanol ΔCp=-14.8	70Pa	
8.434	4.583(+2)	7.31	25	T	μ=0 S: 50 w % Methanol ΔCp=-15.8	70Pa	
8.193	4.291(+2)	6.52	40	T	μ=0 S: 50 w % Methanol ΔCp=-16.5	70Pa	
10.24	9.731(25°,+1)	-10.17	0-50	T	μ=0	68Hb	c
10.17	9.72(25°,+1)	-10.3	25	C	0.1M(KCl)	63Pb	
10.4	9.77(25°,+1)	-9.9	25	C	μ=0.1M(KNO_3)	72Ea	q:±2
8.5	9.77(25°,+1)	-16	20-40	T	μ=0.1M(KNO_3)	72Ea	c
10.40	10.007(+1)	-10.9	25	C	μ=1M(KNO_3)	72Cd	
10.5	9.236(+1)	-7.0	25	C	μ=0.1M(KNO_3) S: 52 w % Ethanol	71Bf	
P102 PIPERAZINE, 1,4-dimethyl-			$C_6H_{14}N_2$		$C_4H_8N_2(CH_3)_2$		
4.42	4.630(+2)	-6.3	25	C	μ=1M(KNO_3)	72Cd	
6.21	8.539(+1)	-18.2	25	C	μ=1M(KNO_3)	72Cd	
P103 PIPERAZINE, 1,4-di(2-sulphoethyl)-			$C_8H_{18}O_6N_2S_2$		$C_4H_8N_2(CH_2CH_2SO_3H)_2$		
2.7	(+1)	-----	5	C	μ=0.1M	71Hf	q:±0.3
2.74	6.93	-22.5	25	C	μ=0.1M(NaCl) pH=9.0	71Be	
P104 PIPERAZINE, 1-(2-hydroxyethyl)-4-(2-sulphoethyl)			$C_8H_{18}O_4N_2S$		$(HOCH_2CH_2)C_4H_8N_2(CH_2CH_2SO_3H)$		
4.7	(+1)	-----	5	C	μ=0.1M	71Hf	
5.01	7.23	-16.3	25	C	μ=0.1M(NaCl) pH=9.0	71Be	
P105 PIPERAZINE, 1-(4'-methoxyphenyl)-2-methyl-			$C_{12}H_{18}ON_2$		$C_4H_8(CH_3)(C_6H_5OCH_3)$		
8.4	8.232(+1)	-9.5	25	C	μ=0.1M(KNO_3) S: 52 w % Ethanol	71Bf	

ΔH, kcal /mole	pK	ΔS, cal /mole°K	T, °C	M	Conditions	Ref.	Remarks
P106 PIPERAZINE, 1-methyl-			$C_5H_{12}N_2$		$C_4H_9N_2(CH_3)$		
4.0	4.94(25°,+2)	-9.2	25	C	μ=0.1M(KNO_3)	72Ea	
5.8	4.94(25°,+2)	-3	20-40	T	μ=0.1M(KNO_3)	72Ea	c,q: ±1.5
5.58	5.199(+2)	-5.0	25	C	μ=1M(KNO_3)	72Cd	
8.4	9.09(25°,+1)	-13.5	25	C	μ=0.1M(KNO_3)	72Ea	
7.6	9.09(25°,+1)	-16	20-40	T	μ=0.1M(KNO_3)	72Ea	c,q: ±1.6
9.09	9.317(25°,+1)	-12.1	25	C	μ=1M(KNO_3)	72Cd	
P107 PIPERAZINE, 2-methyl-			$C_5H_{12}N_2$		$C_4H_9N_2(CH_3)$		
6.7	5.62(25°,+2)	-3.3	25	C	μ=0.1M(KNO_3)	72Ea	
7.0	5.62(25°,+2)	-2	20-40	T	μ=0.1M(KNO_3)	72Ea	c,q: ±1.3
6.4	5.117(+2)	-1.9	25	C	μ=0.1M(KNO_3) S: 52 w % Ethanol	71Bf	
10.0	9.60(25°,+1)	-10.4	25	C	μ=0.1M(KNO_3)	72Ea	
7.5	9.60(25°,+1)	-19	20-40	T	μ=0.1M(KNO_3)	72Ea	c,q:±3
10.4	8.958(+1)	-6.1	25	C	μ=0.1M(KNO_3) S: 52 w % Ethanol	71Bf	
P108 PIPERAZINE, 2-methyl-1-(3'-tolyl)-			$C_{12}H_{18}N_2$		$(CH_3)C_4H_7N_2(C_6H_4CH_3)$		
8.8	8.152(+1)	-7.8	25	C	μ=0.1M(KNO_3) S: 52 w % Ethanol	71Bf	
P109 PIPERAZINE, 2-methyl-1-(4'-tolyl)-			$C_{12}H_{18}N_2$		$(CH_3)C_4H_7N_2(C_6H_4CH_3)$		
8.1	8.188(+1)	-10.0	25	C	μ=0.1M(KNO_3) S: 52 w % Ethanol	71Bf	
P110 PIPERAZINE, 1-phenyl-			$C_{10}H_{14}N_2$		$C_4H_9N_2(C_6H_5)$		
9.0	8.71(25°,+1)	-9.7	25	C	μ=0.1M(KNO_3)	72Ea	
9.3	8.71(25°,+1)	-9	20-40	T	μ=0.1M(KNO_3)	72Ea	q:±1.9
P111 PIPERIDINE			$C_5H_{11}N$		$CH_2CH_2CH_2CH_2CH_2NH$ (ring)		
12.26	11.963(+1)	-9.85	0	T	μ=0	56Ba	
12.36	11.786(+1)	-9.51	5	T	μ=0	56Ba	
12.45	11.613(+1)	-9.15	10	T	μ=0	56Ba	
12.55	11.443(+1)	-8.80	15	T	μ=0	56Ba	
12.66	11.280(+1)	-8.46	20	T	μ=0	56Ba	
12.71	11.12(+1)	-8.3	25	C	μ=0	68Ca	
12.66	11.12(+1)	-8.4	25	C	μ=0	71Ca	
12.190	11.13(+1)	-10.0	25	C	μ=0	60Sa	
12.9	11.06(+1)	-7.34	25	T	μ=0.	38Wa	
12.76	11.123(+1)	-8.10	25	T	μ=0	56Ba 66Hc 64Sc	
12.86	10.974(+1)	-7.74	30	T	μ=0	56Ba	
12.97	10.818(+1)	-7.41	35	T	μ=0	56Ba	
13.08	10.670(+1)	-7.05	40	T	μ=0	56Ba	
13.20	10.526(+1)	-6.69	45	T	μ=0	56Ba	
13.31	10.384(+1)	-6.33	50	T	μ=0	56Ba	
13.19	11.26(+1)	-7.34	25	C	μ=0.5M(KNO_3)	74Be	

ΔH, kcal /mole	pK	ΔS, cal /mole°K	T, °C	M	Conditions	Ref.	Remarks
P111, cont.							
12.4	10.96(+1)	-8.56	25	T	μ=0 S: 10 w % Methanol	38Wa	
13.3	10.09(+1)	-1.56	25	T	μ=0 S: 60 w % Methanol	38Wa	
P112 PIPERIDINE, 2-(2-aminoethyl)-			$C_7H_{16}N_2$		$C_5H_{10}N(CH_2CH_2NH_2)$		
3.77	3.94(10°)	-5	10-40	T	μ=0	59Ga	c
12	8.79(10°)	2	10-40	T	μ=0	63Hc	c
11	10.78(10°)	-9	10-40	T	μ=0	63Hc	c
11.4	10.03(10°)	-6	10-40	T	μ=0	59Ga	c
P113 PIPERIDINE, 2-(aminomethyl)-			$C_6H_{14}N_2$		$C_5H_{10}N(CH_2NH_2)$		
10	6.54(10°)	4	10-40	T	μ=0	63Hc	c
11	9.97(10°)	-6	10-40	T	μ=0	63Hc	c
P114 PIPERIDINE, 2,6-dimethyl-			$C_7H_{15}N$		$C_5H_9N(CH_3)_2$		
14.53	11.26(+1)	-2.85	25	C	μ=0.5M(KNO_3)	74Be	
P115 PIPERIDINE, 1-ethyl-			$C_7H_{15}N$		$C_5H_{10}N(C_2H_5)$		
10.39	10.76(+1)	-14.45	25	C	μ=0.5M(KNO_3)	74Be	
P116 PIPERIDINE, 2-ethyl-			$C_7H_{15}N$		$C_5H_{10}N(C_2H_5)$		
14.18	11.25(+1)	-3.99	25	C	μ=0.5M(KNO_3)	74Be	
P117 PIPERIDINE, 1-methyl-			$C_6H_{13}N$		$C_5H_{10}N(CH_3)$		
9.66	10.08(+1)	-13.7	25	C	μ=0	68Ca	
9.441	10.08(+1)	-14.4	25	C	μ=0	71Ca	
9.91	10.33(+1)	-14.09	25	C	μ=0.5M(KNO_3)	74Be	
P118 PIPERIDINE, 2-methyl-			$C_6H_{13}N$		$C_5H_{10}N(CH_3)$		
14.03	11.21(+1)	-4.29	25	C	μ=0.5M(KNO_3)	74Be	
P119 PIPERIDINE, 3-methyl-			$C_6H_{13}N$		$C_5H_{10}N(CH_3)$		
13.75	11.20(+1)	-5.20	25	C	μ=0.5M(KNO_3)	74Be	
P120 PIPERIDINE, 4-methyl-			$C_6H_{13}N$		$C_5H_{10}N(CH_3)$		
13.48	11.23(+1)	-6.24	25	C	μ=0.5M(KNO_3)	74Be	
P121 PIPERIDINE, 2,2,6,6-tetra-methyl-			$C_9H_{19}N$		$C_5H_7N(CH_3)_4$		
14.17	11.24(+1)	-3.96	25	C	μ=0.5M(KNO_3)	74Be	

ΔH, kcal /mole	pK	ΔS, cal /mole°K	T,°C	M	Conditions	Ref.	Re-marks
P122 2-PIPERIDINECARBOXYLIC ACID			$C_6H_{11}O_2N$		$C_5H_{10}N(CO_2H)$		
0.54	2.12(+1)	-7.9	25	C	μ=0	72Ha	e:CO_2H
10.76	10.75(0)	-13.1	25	C	μ=0	72Ha	e:NH_3^+
P123 3-PIPERIDINECARBOXYLIC ACID			$C_6H_{11}O_2N$		$C_5H_{10}N(CO_2H)$		
0.12	3.35(+1)	-14.9	25	C	μ=0	72Ha	e:CO_2H
10.26	10.64(0)	-14.3	25	C	μ=0	72Ha	e:NH_3^+
P124 4-PIPERIDINECARBOXYLIC ACID			$C_6H_{11}O_2N$		$C_5H_{10}N(CO_2H)$		
-0.08	3.73(+1)	-17.3	25	C	μ=0	72Ha	e:CO_2H
11.01	10.72(0)	-12.1	25	C	μ=0	72Ha	e:NH_3^+
P125 POLYACRYLIC ACID			$(C_3H_4O_2)_n$		$[-CH_2CH(CO_2H)-]_n$		
					C=4.3 x 10^{-2} monomoles per liter α=Degree of neutralization		
0.0	4.40	-20.1	25	C	α=0.0	73Cb	
-0.61	5.10	-25.4	25	C	α=0.1	73Cb	
-0.96	5.46	-28.4	25	C	α=0.2	73Cb	
-1.15	5.71	-30.0	25	C	α=0.3	73Cb	
-1.25	5.91	-31.3	25	C	α=0.4	73Cb	
-1.30	6.11	-32.4	25	C	α=0.5	73Cb	
-1.32	6.30	-33.3	25	C	α=0.6	73Cb	
-1.29	6.48	-34.0	25	C	α=0.7	73Cb	
-1.18	6.64	-34.4	25	C	α=0.8	73Cb	
P126 PORCINE ELASTASE							
5.87	7.24(23.6°)	-13.3	10.1-37.6	T	μ=0.1 (Phosphate Buffer)	73Mb	c,e: Kinetic Group on Elastase
P127 PROLINE			$C_5H_9O_2N$		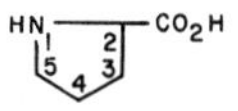		
1.149	2.011(+1)	-5.0	1	T	μ=0	42Sa	
0.780	1.964(+1)	-6.3	12.5	T	μ=0	42Sa	
0.342	1.952(+1)	-7.8	25	T	μ=0	42Sa	
						58Ea	
-0.181	1.950(+1)	-9.5	37.5	T	μ=0	42Sa	
-0.746	1.958(+1)	-11.3	50	T	μ=0	42Sa	
10.360	11.296(0)	-13.9	1	T	μ=0	42Sa	
10.400	10.972(0)	-13.8	12.5	T	μ=0	42Sa	
10.310	10.640(0)	-14.1	25	T	μ=0	42Sa	
10.085	10.342(0)	-14.7	37.5	T	μ=0	42Sa	
9.715	10.064(0)	-16.0	50	T	μ=0	42Sa	
9.99	(0)	-----	20	C	μ≃0.01	71Ma	e:N_1-H
11.35	10.41(0)	-9.4	25	C	μ=0.1M(KNO_3)	73Ia	e:N_1-H
9.85	(0)	-----	25	C	μ=0.10M	71Ba	e:N_1-H

ΔH, kcal /mole	pK	ΔS, cal /mole°K	T, °C	M	Conditions	Ref.	Remarks
P128 PROLINE(L), 1-glycyl-			$C_7H_{12}O_3N_2$		$C_5H_8O_2N[C(O)CH_2NH_2]$		
10.6	8.49(0)	-0.3	25	C	μ=0.1M(KNO_3)	72Bi	
P129 PROLINE, 4-hydroxy-			$C_5H_9O_3N$		$C_5H_8O_2N(OH)$		
1.602	1.900(+1)	-2.8	1	T	μ=0	42Sa	
1.310	1.850(+1)	-3.9	12.5	T	μ=0	42Sa	
0.918	1.818(+1)	-5.2	25	T	μ=0	42Sa 58Ea	
0.446	1.798(+1)	-6.8	37.5	T	μ=0	42Sa	
-0.115	1.796(+1)	-8.5	50	T	μ=0	42Sa	
9.585	10.274(+1)	-12.0	1	T	μ=0	42Sa	
9.543	9.958(0)	-12.2	12.5	T	μ=0	42Sa	
9.385	9.662(0)	-12.7	25	T	μ=0	42Sa	
9.080	9.394(0)	-13.7	37.5	T	μ=0	42Sa	
8.635	9.138(0)	-15.1	50	T	μ=0	42Sa	
10.34	9.46(0)	-8.5	25	C	μ=0.1M(KNO_3)	73Ia	e:NH_2^+
P130 PROPANE, 1-amino-			C_3H_9N		$CH_3CH_2CH_2NH_2$		
13.84	10.568(+1)	-2.0	25	C	μ=0	69Cc	
13.80	10.56(+1)	-2.03	25	C	μ=0	68Ob	
13.7	10.568(+1)	-2.5	25	T	μ=0	61Bb	
13.85	10.047(40°,+1)	-1.7	20-40	T	μ=0	51Ea 51Eb	c
13.670	10.568(+1)	-2.5	0-50	T	μ=0	68Cf	
13.85	10.060(40°,+1)	-1.8	20-40	-	μ=0.05	51Ea	c
13.94	10.062(40°,+1)	-1.5	20-40	-	μ=0.10	51Ea	c
13.95	10.086(40°,+1)	-1.6	20-40	T	μ=0.20	51Ea	c
P131 PROPANE, 2-amino-			C_3H_9N		$CH_3CH(NH_2)CH_3$		
13.97	10.67(+1)	-2.0	25	C	μ=0	69Cc	
13.92	10.59(+1)	-1.86	25	C	μ=0	68Ob	
P132 PROPANE, 1-amino-2-methyl-			$C_4H_{11}N$		$CH_3CH(CH_3)CH_2(NH_2)$		
13.92	10.48(+1)	-1.3	25	C	μ=0	69Cc	
P133 PROPANE, 2-amino-2-methyl-			$C_4H_{11}N$		$(CH_3)_2C(NH_2)CH_3$		
14.29	11.439(+1)	-0.98	5	T	μ=0	62Hd	
14.30	11.240(+1)	-0.91	10	T	μ=0	62Hd	
14.32	11.048(+1)	-0.86	15	T	μ=0	62Hd	
14.34	10.862(+1)	-0.79	20	T	μ=0	62Hd	
14.43	10.685(+1)	-0.5	25	C	μ=0	69Cc	
14.38	10.68(+1)	-0.67	25	C	μ=0	68Ob	
14.36	10.685(+1)	-0.74	25	T	μ=0	62Hd	
14.37	10.511(+1)	-0.67	30	T	μ=0	62Hd	
14.39	10.341(+1)	-0.62	35	T	μ=0	62Hd	
P134 PROPANE, 1,2-diamino-			$C_3H_{10}N_2$		$CH_3CH(NH_2)CH_2(NH_2)$		
9.65	6.607(+2)	2.1	25	C	μ=0	66Pg	
11.3	6.96(25°,+2)	6.2	15-40	T	μ=0.3M($NaClO_4$)	71Mb	c
10.89	7.06(+2)	4.2	25	C	μ=0.5M(KNO_3)	66Va	
10.1	7.13(25°,+2)	1.25	0&25	T	0.50M(KNO_3)	53Ba	b,c
11.92	9.720(+1)	-4.5	25	C	μ=0	66Pg	
12.5	9.92(25°,+1)	-3.5	15-40	T	μ=0.3M($NaClO_4$)	71Mb	c

ΔH, kcal /mole	pK	ΔS, cal /mole°K	T, °C	M	Conditions	Ref.	Remarks
P134, cont.							
12.26	9.91(+1)	-4.2	25	C	μ=0.5M(KNO_3)	66Va	
11.3	10.00(25°,+1)	-7.85	0&25	T	0.50M(KNO_3)	53Ba	b,c
3.46	3.26(+2)	-3.31	25.3	C	μ=0.02 HA=[Re(1,2-Diaminopropane)(O)(OH)]$^{2+}$	63Mb	

P135 PROPANE, 1,3-diamino- $C_3H_{10}N_2$ $H_2NCH_2CH_2CH_2NH_2$

ΔH, kcal /mole	pK	ΔS, cal /mole°K	T, °C	M	Conditions	Ref.	Remarks
12.44	8.49(+2)	2.9	25	C	μ=0	69Cc	
12.03	8.91(+2)	0.3	20	C	μ=0.11	73Fa	
12.11	8.97(+2)	0.2	20	C	μ=0.21	73Fa	
12.97	9.13(+2)	1.7	25	C	0.3M($NaClO_4$)	67Hc	
12.63	8.88(+2)	1.9	25	C	μ=0.5M(KNO_3)	70Ba	
12.20	9.10(+2)	0	20	C	μ=0.51	73Fa	
12.32	9.26(+2)	-0.3	20	C	μ=1.01	73Fa	
12.57	9.46(+2)	-0.4	20	C	μ=2.01	73Fa	
12.61	9.55(+2)	-0.7	20	C	μ=2.51	73Fa	
12.76	9.63(+2)	-0.5	20	C	μ=3.01	73Fa	
13.19	10.47(+1)	-3.7	25	C	μ=0	69Cc	
13.3	10.23(30°,+1)	-3	10-40	T	μ=0	50Ba	c
12.71	10.74(+1)	-5.8	20	C	μ=0.21	73Fa	
13.9	10.62(+1)	-4.6	25	C	0.3M($NaClO_4$)	67Hc	
13.08	10.55(+1)	-4.4	25	C	μ=0.5	67Va	
13.09	10.55(+1)	-4.4	25	C	μ=0.5M(KNO_3)	70Ba	
12.93	10.80(+1)	-5.3	20	C	μ=0.51	73Fa	
13.20	10.93(+1)	-5.0	20	C	μ=1.01	73Fa	
13.59	11.12(+1)	-4.5	20	C	μ=2.01	73Fa	
13.83	11.20(+1)	-4.1	20	C	μ=2.51	73Fa	
13.96	11.30(+1)	-4.1	20	C	μ=3.01	73Fa	
3.44	3.26(+2)	-3.38	25.3	C	μ=0.02 HA=[Re(1,3-diaminopropane)(O)(OH)]$^{2+}$	63Mb	

P136 PROPANE, 1,3-diamino-N,N'-bis(2-aminoethyl)- $C_7H_{20}N_4$ $H_2NCH_2CH_2NHCH_2CH_2CH_2NHCH_2$-$CH_2NH_2$

ΔH, kcal /mole	pK	ΔS, cal /mole°K	T, °C	M	Conditions	Ref.	Remarks
9.18	6.01(+4)	3.3	25	C	μ=0.5M(KCl)	72Fa	
10.03	7.26(+3)	0.4	25	C	μ=0.5M(KCl)	72Fa	
11.33	9.49(+2)	-5.4	25	C	μ=0.5M(KCl)	72Fa	
10.96	10.23(+1)	-10.1	25	C	μ=0.5M(KCl)	72Fa	

P137 PROPANE, 1,2-diamino-2-methyl- $C_4H_{12}N_2$ $(CH_3)_2C(NH_2)CH_2NH_2$

ΔH, kcal /mole	pK	ΔS, cal /mole°K	T, °C	M	Conditions	Ref.	Remarks
9.66	6.178(+2)	4.1	25	C	μ=0	66Pg	
9.2	6.79(25°,+2)	-0.211	0-25	T	0.5M(KNO_3)	53Ba	c
11.78	9.420(+1)	-3.6	25	C	μ=0	66Pg	
11.0	10.00(25°,+1)	-8.86	0-25	T	0.5M(KNO_3)	53Ba	c

P138 PROPANE, 1,2-diamino-N,N,N',N'-tetraacetic acid $C_{11}H_{18}O_8N_2$ $(HO_2CCH_2)_2NCH_2CH[N(CH_2CO_2H)_2]$-$CH_3$

ΔH, kcal /mole	pK	ΔS, cal /mole°K	T, °C	M	Conditions	Ref.	Remarks
6.1	10.46(-3)	-27.1	20	C	μ=1M(KCl)	69Sc	

P139 PROPANE, 1,3-diamino-N,N,N',N'-tetraacetic acid $C_{11}H_{18}O_8N_2$ $(HO_2CCH_2)_2N(CH_2)_3N(CH_2CO_2H)_2$

ΔH, kcal /mole	pK	ΔS, cal /mole°K	T, °C	M	Conditions	Ref.	Remarks
4.43	8.62	-21.6	20	C	0.1M(KNO_3)	64Ab	
5.16	10.45	-30.2	20	C	0.1M(KNO_3)	64Ab	

ΔH, kcal /mole	pK	ΔS, cal /mole°K	T,°C	M	Conditions	Ref.	Remarks
P140 PROPANE, 1,3-diamino-N,N,N',N'-tetra(2-aminoethyl)-			$C_{11}H_{30}N_6$		$(NH_2CH_2CH_2)_2NCH_2CH_2CH_2N-(CH_2CH_2NH_2)_2$		
4.26	2.45(+5)	3.1	25	C	μ=0.1(KNO_3)	71Pc	
13.35	8.44(+4)	2.8	25	C	μ=0.1(KNO_3)	71Pc	
12.68	9.18(+3)	0.5	25	C	μ=0.1(KNO_3)	71Pc	
11.61	9.56(+2)	-4.8	25	C	μ=0.1(KNO_3)	71Pc	
11.00	10.24(+1)	-9.9	25	C	μ=0.1(KNO_3)	71Pc	
7.20	5.73(+3)	-2.0	25	C	μ=0.1(KNO_3) HA=[Ni(II) (1,3-Diamino-N,N,N',N'-tetra(2-aminoethyl)propane]$^{3+}$	71Pc	
11.37	8.21(+3)	0.7	25	C	μ=0.1(KNO_3) HA=[Cu(II) (1,3-Diamino-N,N,N',N'-tetra(2-aminoethyl)propane]$^{3+}$	71Pc	
11.26	7.65(+3)	2.9	25	C	μ=0.1(KNO_3) HA=[Zn(II) (1,3-Diamino-N,N,N',N'-tetra(2-aminoethyl)propane]$^{3+}$	71Pc	
P141 PROPANE, 1,1-dinitro-			$C_3H_6O_4N_2$		$CH_3CH_2CH(NO_2)_2$		
3.675	5.5	-12.83	20	T	--	61Sc	
P142 PROPANE, 1-nitro-			$C_3H_7O_2N$		$CH_3CH_2CH_2NO_2$		
2.58	8.98(0)	-32.4	25	C	μ=0	73Mc	
P143 PROPANE, 2-nitro-			$C_3H_7O_2N$		$CH_3CH(NO_2)CH_3$		
0.08	7.7(0)	-35.1	25	C	μ=0	73Mc	
3.400	7.675(25°,0)	-23.7	25-35	T	μ<0.007	43Ta	c
P144 1,2-PROPANEDIOL			$C_3H_8O_2$		$CH_3CH(OH)CH_2OH$		
11.47	17.81	-40.3	5	T	μ=0 S: Propylene Glycol R: 2HA=H_2A^+ + A^- ΔCp=-28.5	70Kd	
11.32	17.64	-40.8	10	T	μ=0 S: Propylene Glycol R: 2HA=H_2A^+ + A^- ΔCp=-29.0	70Kd	
11.18	17.50	-41.3	15	T	μ=0 S: Propylene Glycol R: 2HA=H_2A^+ + A^- ΔCp=-29.5	70Kd	
11.03	17.35	-41.8	20	T	μ=0 S: Propylene Glycol R: 2HA=H_2A^+ + A^- ΔCp=-30.0	70Kd	
10.88	17.21	-42.3	25	T	μ=0 S: Propylene Glycol R: 2HA=H_2A^+ + A^- ΔCp=-30.5	70Kd	
10.73	17.08	-42.8	30	T	μ=0 S: Propylene Glycol R: 2HA=H_2A^+ + A^- ΔCp=-31.0	70Kd	
10.56	16.96	-43.3	35	T	μ=0 S: Propylene Glycol R: 2HA=H_2A^+ + A^- ΔCp=-31.6	70Kd	
10.41	16.83	-43.8	40	T	μ=0 S: Propylene Glycol R: 2HA=H_2A^+ + A^- ΔCp=-32.1	70Kd	
10.24	16.73	-44.4	45	T	μ=0 S: Propylene Glycol R: 2HA=H_2A^+ + A^- ΔCp=-32.6	70Kd	

ΔH, kcal /mole	pK	ΔS, cal /mole°K	T, °C	M	Conditions	Ref.	Remarks

P145 1,3-PROPANEDIOL, 2-amino-N,N-dihydroxyethyl-2-hydroxymethyl- $C_8H_{19}O_5N$ $(HOH_2C)_3CN(C_2H_4OH)$

ΔH, kcal /mole	pK	ΔS, cal /mole°K	T, °C	M	Conditions	Ref.	Remarks
6.692	6.6536	-9.5	15	T	μ=0 ΔCp=5.5	70Pb	
6.749	6.4835	-7.0	25	T	μ=0 ΔCp=5.7	70Pb	
6.807	6.3212	-6.8	35	T	μ=0 ΔCp=6.0	70Pb	

P146 1,3-PROPANEDIOL, 2-amino-2-(hydroxymethyl)- $C_4H_{11}O_3N$ $(HOCH_2)_3CNH_2$

ΔH, kcal /mole	pK	ΔS, cal /mole°K	T, °C	M	Conditions	Ref.	Remarks
11.95	8.6792(+1)	3.24	5	T	μ=0	63Da	
11.60	8.3602(+1)	1.99	15	T	μ=0	63Da	
11.347	(+1)	-----	25	C	μ=0	69Hc 69Wc	
11.89	8.01(+1)	3.23	25	C	μ=0	61Ta	
11.33	8.0686(+1)	1.08	25	T	μ=0	63Da	
11.377	8.075(+1)	1.2	25	T	μ=0 ΔCp=-15 $\Delta H=13.642 - 2.550 \times 10^{-5} T^2$ $(0<T\ °C<50°C)$	61Bb	
11.16	7.8006(+1)	0.52	35	T	μ=0	63Da	
11.10	7.5515(+1)	0.31	45	T	μ=0	63Da	
11.14	7.3204(+1)	0.46	55	T	μ=0	63Da	
11.319	(+1)	-----	25	C	μ=0.003 - 0.02M	71Hb	
11.343	(+1)	-----	20	C	μ=0.01	72Sa	
11.39	8.030(+1)	1.5	25	C	μ=0.010	68Ce	
11.335	(+1)	-----	25	C	μ=0.01M	72Oa	
10.930	8.08(+1)	0.3	25	C	μ=0.013	55Sa	
11.75	(+1)	-----	5	C	μ=0.01 - 0.02M	70Gb	
11.44	(+1)	-----	20	C	μ=0.01 - 0.02M	70Gb	
11.34	(+1)	-----	25	C	μ=0.01 - 0.02M	70Gb	
11.19	(+1)	-----	35	C	μ=0.01 - 0.02M	70Gb	
11.00	(+1)	-----	50	C	μ=0.01 - 0.02M	70Gb	
11.44	(+1)	-----	5	C	μ=0.1M	71Hf	
11.36	(+1)	-----	25	C	μ=0.1	64Nb	
11.70	8.38(+1)	0.90	25	T	μ=0.65	56Bb	
14.53	10.48(+1)	0.8	25	T	μ=0 S: Ethylene glycol ΔCp=2	70Kc	
11.43	7.818(+1)	2.55	25	T	S: 50 w % Methanol	65Wb	
15.48	9.435(+1)	11.5	10	T	μ=0 S: N-methylpropionamide	73Ea	
15.53	9.227(+1)	11.7	15	T	μ=0 S: N-methylpropionamide	73Ea	
15.58	9.025(+1)	11.9	20	T	μ=0 S: N-methylpropionamide	73Ea	
15.63	8.831(+1)	12.0	25	T	μ=0 S: N-methylpropionamide	73Ea	
15.68	8.642(+1)	12.2	30	T	μ=0 S: N-methylpropionamide	73Ea	
15.74	8.458(+1)	12.4	35	T	μ=0 S: N-methylpropionamide	73Ea	
15.79	8.279(+1)	12.5	40	T	μ=0 S: N-methylpropionamide	73Ea	
15.85	8.105(+1)	12.7	45	T	μ=0 S: N-methylpropionamide	73Ea	
15.90	7.937(+1)	12.9	50	T	μ=0 S: N-methylpropionamide	73Ea	
15.96	7.773(+1)	13.1	55	T	μ=0 S: N-methylpropionamide	73Ea	
10.8	14.03(0)	-28.0	25	C	μ=0	71Oa	

P147 1,3-PROPANEDIOL, 2-amino-2-methyl- $C_4H_{11}O_2N$ $HOCH_2C(NH_2)(CH_3)CH_2OH$

ΔH, kcal /mole	pK	ΔS, cal /mole°K	T,°C	M	Conditions	Ref.	Remarks

P147, cont.

ΔH, kcal /mole	pK	ΔS, cal /mole°K	T,°C	M	Conditions	Ref.	Remarks
12.17	9.6116(+1)	0.60	0	T	μ=0	62Hc	
12.12	9.4328(+1)	0.41	5	T	μ=0	62Hc	
12.07	9.2658(+1)	0.24	10	T	μ=0	62Hc	
12.02	9.1044(+1)	0.05	15	T	μ=0	62Hc	
11.97	8.9508(+1)	-0.12	20	T	μ=0	62Hc	
12.43	8.80(+1)	1.43	25	C	μ=0	61Ta	
11.92	8.8013(+1)	-0.31	25	T	μ=0	62Hc	
11.86	8.6588(+1)	-0.48	30	T	μ=0	62Hc	
11.81	8.5193(+1)	-0.67	35	T	μ=0	62Hc	
11.75	8.3854(+1)	-0.84	40	T	μ=0	62Hc	
11.70	8.2569(+1)	-1.03	45	T	μ=0	62Hc	
11.64	8.1322(+1)	-1.20	50	T	μ=0	62Hc	

P148 2,2-PROPANEDIOL, 1,1,1, 3,3,3-hexafluoro- $C_3H_2O_2F_6$ $CF_3COH(OH)CF_3$

ΔH, kcal /mole	pK	ΔS, cal /mole°K	T,°C	M	Conditions	Ref.	Remarks
6.00	6.65	-10.3	25	C	μ=0	71Wb	
						71Wa	

P149 1,2-PROPANEDIOL, 3-mercapto- $C_3H_8O_2S$ $HSCH_2CH(OH)CH_2OH$

ΔH, kcal /mole	pK	ΔS, cal /mole°K	T,°C	M	Conditions	Ref.	Remarks
6.73	9.43(0)	-20.55	25	C	μ=0.5M(KNO_3)	74Da	

P150 1,2,3-PROPANETRICARBOXYLIC ACID $C_6H_8O_6$ $HO_2CCH_2CH(CO_2H)CH_2CO_2H$

ΔH, kcal /mole	pK	ΔS, cal /mole°K	T,°C	M	Conditions	Ref.	Remarks
0.56	3.67(0)	-14.9	25	T	μ=0 ΔCp=-27	72Pf	
-1.19	4.87(-1)	-26.3	25	T	μ=0 ΔCp=-49	72Pf	
-1.81	6.38(-2)	-35.2	25	T	μ=0 ΔCp=-52	72Pf	

P151 PROPANOIC ACID $C_3H_6O_2$ $CH_3CH_2CO_2H$

ΔH, kcal /mole	pK	ΔS, cal /mole°K	T,°C	M	Conditions	Ref.	Remarks
0.737	4.895(0)	-19.7	0	T	μ=0	33Hc	
0.562	4.884(0)	-20.3	5	T	μ=0	33Hc	
0.384	4.877(0)	-21.0	10	T	μ=0	33Hc	
						34Hc	
0.203	4.874(0)	-21.6	15	T	μ=0	33Hc	
0.182	4.873(0)	-21.58	15	T	μ=0 ΔCp=-36.17	69Gc	
-0.001	4.870(0)	-22.21	20	T	μ=0 ΔCp=-36.80	69Gc	
0.019	4.873(0)	-22.2	20	T	μ=0	33Hc	
-0.08	4.88(0)	-22.6	25	C	μ=0	58Ca	a:13.5
-0.18	4.875(0)	-22.8	25	C	μ=0	66Ae	
-0.14	4.874(0)	-22.8	25	C	μ=0	67Ca	
-0.14	4.87(0)	-22.7	25	C	μ=0	74Md	
-0.163	4.874(0)	-22.8	25	T	μ=0	41Ha	
-0.180	4.875(0)	-22.57	25	T	μ=0	52Ea	
-0.187	4.874(0)	-22.84	25	T	μ=0 ΔCp=-37.43	69Gc	
-0.167	4.874(0)	-22.9	25	T	μ=0 ΔCp=-41.8	39Hb	
						33Hc	
						34Hc	
-0.16	4.865(0)	-22.8	25	T	μ=0	62Aa	
-0.358	4.877(0)	-23.5	30	T	μ=0	33Hc	
-0.375	4.880(0)	-23.46	30	T	μ=0 ΔCp=-38.05	69Gc	
-0.551	4.883(0)	-24.1	35	T	μ=0	33Hc	
-0.567	4.891(0)	-24.00	35	T	μ=0 ΔCp=-38.68	69Gc	
-0.746	4.891(0)	-24.8	40	T	μ=0	33Hc	
-0.767	4.901(0)	-24.72	40	T	μ=0 ΔCp=-39.31	69Gc	
-0.860	(0)	-----	40	T	μ=0	66Aa	
-0.945	4.901(0)	-25.4	45	T	μ=0	33Hc	
-0.966	4.912(0)	-25.35	45	T	μ=0 ΔCp=-39.94	69Gc	

ΔH, kcal /mole	pK	ΔS, cal /mole°K	T, °C	M	Conditions	Ref.	Remarks
P151, cont.							
-1.147	4.910(0)	-26.0	50	T	μ=0	33Hc	
-1.351	4.923(0)	-26.6	55	T	μ=0	33Hc	
-1.559	4.936(0)	-27.3	60	T	μ=0	33Hc	
0.107	4.860(0)	-22	0&35	T	--	39Wa	
-0.148	5.467(0)	-25.5	25	T	S: 20 w % Dioxane	41Ha	
-0.05	5.536(0)	-25.5	25	T	S: 20 w % Dioxane	62Aa	
-0.206	6.554(0)	-30.7	25	T	S: 45 w % Dioxane	41Ha	
-0.21	6.555(0)	-30.7	25	T	S: 45 w % Dioxane	62Aa	
-0.201	8.614(0)	-40.1	25	T	S: 70 w % Dioxane	41Ha	
-0.20	8.617(0)	-40.1	25	T	S: 70 w % Dioxane	62Aa	
-1.064	10.415(0)	-51.2	25	T	S: 82 w % Dioxane	41Ha	
-1.06	10.412(0)	-51.2	25	T	S: 82 w % Dioxane	62Aa	
-0.14	5.01(0)	-23.4	25	C	μ=0 S: 5 w % tert-Butanol	74Md	
-0.17	5.11(0)	-23.9	25	C	μ=0 S: 10 w % tert-Butanol	74Md	
-0.096	5.26(0)	-24.4	25	C	μ=0 S: 15 w % tert-Butanol	74Md	
-0.14	5.32(0)	-24.8	25	C	μ=0 S: 17.5 w % tert-Butanol	74Md	
-0.14	5.36(0)	-25.1	25	C	μ=0 S: 20 w % tert-Butanol	74Md	
-0.048	5.56(0)	-25.6	25	C	μ=0 S: 25 w % tert-Butanol	74Md	
0.072	5.73(0)	-26.0	25	C	μ=0 S: 30 w % tert-Butanol	74Md	
0.22	5.91(0)	-26.3	25	C	μ=0 S: 35 w % tert-Butanol	74Md	
0.26	6.07(0)	-26.8	25	C	μ=0 S: 40 w % tert-Butanol	74Md	
0.6721	4.9924(0)	-20.38	0	T	S: 5 w % Isopropanol	47Ma	
-0.2424	4.9766(0)	-23.58	25	T	S: 5 w % Isopropanol	47Ma	
-0.8294	4.9953(0)	-25.50	40	T	S: 5 w % Isopropanol	47Ma	
0.8227	5.1064(0)	-20.35	0	T	S: 10 w % Isopropanol	47Ma	
-0.2493	5.0857(0)	-24.10	25	T	S: 10 w % Isopropanol	47Ma	
-0.9376	5.1062(0)	-26.35	40	T	S: 10 w % Isopropanol	47Ma	
1.1199	5.3633(0)	-20.44	0	T	S: 20 w % Isopropanol	47Ma	
-0.2168	5.3310(0)	-25.12	25	T	S: 20 w % Isopropanol	47Ma	
-1.0750	5.3533(0)	-27.92	40	T	S: 20 w % Isopropanol		

P152 PROPANOIC ACID, 2-amino-N-carbamoyl-2-methyl- $C_5H_{10}O_3N_2$ $NH_2CONHC(CH_3)_2CO_2H$

ΔH, kcal /mole	pK	ΔS, cal /mole°K	T, °C	M	Conditions	Ref.	Remarks
0.217	4.4627	-19.69	25	T	μ=0	56Kb	

P153 PROPANOIC ACID, 2-amino-2-methyl- $C_4H_9O_2N$ $(CH_3)_2C(NH_2)CO_2H$

ΔH, kcal /mole	pK	ΔS, cal /mole°K	T, °C	M	Conditions	Ref.	Remarks
1.300	2.419(+1)	-6.2	1	T	μ=0	37Sa	
0.980	2.380(+1)	-7.1	12.5	T	μ=0	37Sa	
1.41	2.36(+1)	-6	25	C	μ=0	64Ic	
0.560	2.357(+1)	-8.9	25	T	μ=0 ΔCp=-36.9	37Sa 39Hb	
0.492	2.357(+1)	-9.1	25	T	μ=0	58Ea	
0.060	2.351(+1)	-10.6	37.5	T	μ=0	37Sa	
-0.540	2.356(+1)	-12.4	50	T	μ=0	37Sa	
11.480	10.960(0)	-8.2	1	T	μ=0	37Sa	
11.600	10.580(0)	-7.8	12.5	T	μ=0	37Sa	
11.57	10.205(0)	-7.9	25	C	μ=0	69Cc	
11.610	10.205(0)	-7.7	25	T	μ=0	37Sa	
11.510	9.872(0)	-8.1	37.5	T	μ=0	37Sa	
11.260	9.561(0)	-8.9	50	T	μ=0	37Sa	

ΔH, kcal /mole	pK	ΔS, cal /mole°K	T,°C	M	Conditions	Ref.	Remarks

P154 PROPANOIC ACID, 2-amino-3-methyl- $C_4H_9O_2N$ $CH_3CH_2CH(NH_2)CO_2H$

| 11.090 | 10.205($\underline{0}$) | -9.29 | 25 | C | -- | 67Ac | |

P155 PROPANOIC ACID, 2-bromo- $C_3H_5O_2Br$ $CH_3CH(Br)CO_2H$

| -1.31 | 2.971($\underline{0}$) | -18.0 | 25 | C | $\mu=0$ | 67Ca | |
| -1.28 | 2.98($\underline{0}$) | -17.9 | 25 | C | $\mu=0$ | 66Ae | |

P156 PROPANOIC ACID, 3-bromo- $C_3H_5O_2Br$ $BrCH_2CH_2CO_2H$

-0.17	3.992($\underline{0}$)	-18.9	25	C	$\mu=0$	68Cd	
-0.32	3.992($\underline{0}$)	-19.3	25	C	$\mu=0$	67Ca	
-1.72	3.98($\underline{0}$)	-23.9	25	C	$\mu=0$	66Ae	
-1.426	3.026($\underline{0}$)	-18	0&35	T	--	39Wa	

P157 PROPANOIC ACID, 2-chloro- $C_3H_5O_2Cl$ $CH_3CH(Cl)CO_2H$

| -0.51 | 3.992($\underline{0}$) | -20.0 | 25 | C | $\mu=0$ | 68Cd | |
| -0.94 | 2.84($\underline{0}$) | -16.1 | 25 | C | $\mu=0$ | 66Ae | |

P158 PROPANOIC ACID, 3-chloro $C_3H_5O_2Cl$ $ClCH_2CH_2CO_2H$

-0.457	($\underline{0}$)	-----	20	C	$\mu=0$	48Ca	
-0.32	3.992($\underline{0}$)	-19.3	25	C	$\mu=0$	67Ca	
-0.593	3.983($\underline{0}$)	-20.2	25	C	$\mu=0$	48Ca	
-0.78	4.08($\underline{0}$)	-21.2	25	C	$\mu=0$	66Ae	
-0.729	($\underline{0}$)	-----	30	C	$\mu=0$	48Ca	

P159 PROPANOIC ACID, 2-cyano-2-methyl- $C_5H_7O_2N$ $(CH_3)_2C(CN)CO_2H$

-1.207	2.3420($\underline{0}$)	-15.06	5	T	$\mu=0$ $\Delta Cp=-29.7$	70Ia	
-1.360	2.3597($\underline{0}$)	-15.06	10	T	$\mu=0$ $\Delta Cp=-31.4$	70Ia	
-1.521	2.3786($\underline{0}$)	-16.16	15	T	$\mu=0$ $\Delta Cp=-33.1$	70Ia	
-1.691	2.3998($\underline{0}$)	-16.75	20	T	$\mu=0$ $\Delta Cp=-34.9$	70Ia	
-1.870	2.4218($\underline{0}$)	-17.35	25	T	$\mu=0$ $\Delta Cp=-36.7$	70Ia	
-2.058	2.4458($\underline{0}$)	-17.98	30	T	$\mu=0$ $\Delta Cp=-38.6$	70Ia	
-2.256	2.4709($\underline{0}$)	-18.63	35	T	$\mu=0$ $\Delta Cp=-40.5$	70Ia	
-2.463	2.4982($\underline{0}$)	-19.29	40	T	$\mu=0$ $\Delta Cp=-42.4$	70Ia	
-2.680	2.5253($\underline{0}$)	-19.98	45	T	$\mu=0$ $\Delta Cp=-44.4$	70Ia	

P160 PROPANOIC ACID, 2,3-diamino- $C_3H_8O_2N_2$ $H_2NCH_2CH(NH_2)CO_2H$

| 9.56 | 6.674(25°,$\underline{+1}$) | 1.5 | 25-50 | T | $\mu=0.1M$ | 72Ha 71He | c,e: α-NH$_3^+$, q:±0.61 |
| 9.40 | 9.623(25°,$\underline{0}$) | -12.5 | 25-50 | T | $\mu=0.1M$ | 72Ha 71He | c,e: β-NH$_3^+$ |

P161 PROPANOIC ACID, 2,3-diamino-, methylester $C_4H_{10}O_2N_2$ $H_2NCH_2CH(NH_2)CO_2CH_3$

| 9.08 | 4.412(25°,$\underline{+1}$) | 10.3 | 25-50 | T | $\mu=0.1M$ | 72Ha 71He | c,e: α-NH$_3^+$, q:±0.76 |
| 10.90 | 8.250(25°,$\underline{0}$) | -1.2 | 25-50 | T | $\mu=0.1M$ | 72Ha 71He | c,e: β-NH$_3^+$ |

ΔH, kcal /mole	pK	ΔS, cal /mole°K	T,°C	M	Conditions	Ref.	Remarks
P162 PROPANOIC ACID, 2,3-dibromo-			$C_3H_4O_2Br_2$		$BrCH_2CH(Br)CO_2H$		
-1.89	2.33(0)	-17.0	25	C	μ=0	66Ae	
P163 PROPANOIC ACID, 2,2-dichloro-			$C_3H_4O_2Cl_2$		$CH_3C(Cl)_2CO_2H$		
-0.45	2.06(0)	-10.9	25	C	μ=0	66Ae	
P164 PROPANOIC ACID, 2,3-dichloro-			$C_3H_4O_2Cl_2$		$ClCH_2CH(Cl)CO_2H$		
-0.17	2.85(0)	-13.6	25	C	μ=0	66Ae	
P165 PROPANOIC ACID, 2,2-di (hydroxymethyl)-3-hydroxy-			$C_5H_{12}O_5$		$HOCH_2C(CH_3OH)_2CO_2H$		
1.860	4.5026(0)	-14.2	15	T	μ=0 ΔCp=-32.0	71Gf	
1.535	4.4595(0)	-15.3	25	T	μ=0 ΔCp=-33.1	71Gf	
1.199	4.4275(0)	-16.4	35	T	μ=0 ΔCp=-34.2	71Gf	
3.622	5.5331(0)	-12.7	15	T	μ=0 S: 50 w % Methanol ΔCp=-54.3	71Gf	
3.070	5.4479(0)	-14.6	25	T	μ=0 S: 50 w % Methanol ΔCp=-56.1	71Gf	
2.499	5.3809(0)	-16.5	35	T	μ=0 S: 50 w % Methanol ΔCp=-58.0	71Gf	
P166 PROPANOIC ACID, 2,2-dimethyl-			$C_5H_{10}O_2$		$(CH_3)_3CCOOH$		
-0.21	5.014(0)	-23.7	10	C	μ=0	70Ce	q:+0.07
-0.67	5.03(0)	-25.3	25	C	μ=0	74Md	
-0.69	5.03(0)	-25.3	25	C	μ=0	70Ce	
-0.724	5.031(0)	-25.45	25	T	μ=0	52Ea	
-1.19	5.067(0)	-27.0	40	C	μ=0	70Ce	
-----	(0)	-----	10-40	C	μ=0 ΔCp=-33	70Ce	
-0.57	5.22(0)	-25.8	25	C	μ=0 S: 5 w % tert-Butanol	74Md	
-0.64	5.39(0)	-26.8	25	C	μ=0 S: 10 w % tert-Butanol	74Md	
-1.2	5.59(0)	-29.6	25	C	μ=0 S: 15 w % tert-Butanol	74Md	
-1.7	5.75(0)	-32.0	25	C	μ=0 S: 17.5 w % tert-Butanol	74Md	
-1.1	5.82(0)	-30.4	25	C	μ=0 S: 20 w % tert-Butanol	74Md	
0.14	6.11(0)	-27.5	25	C	μ=0 S: 25 w % tert-Butanol	74Md	
0.93	6.37(0)	-26.0	25	C	μ=0 S: 30 w % tert-Butanol	74Md	
1.4	6.56(0)	-25.3	25	C	μ=0 S: 35 w % tert-Butanol	74Md	
1.2	6.76(0)	-27.0	25	C	μ=0 S: 40 w % tert-Butanol	74Md	
P167 PROPANOIC ACID, 2,2-dicyano-			$C_4H_4O_2N_2$		$CH_3C(CN)_2CO_2H$		
-4.5	-2.8(0)	-2	25	T	μ=0	65Bb	

P168 PROPANOIC ACID, 2-hydroxy- $C_3H_6O_3$ $CH_3CH(OH)CO_2H$

ΔH, kcal /mole	pK	ΔS, cal /mole°K	T, °C	M	Conditions	Ref.	Remarks
0.768	3.880(0)	-14.9	0	T	μ=0	37Ma	
0.795	3.890(0)	-14.10	0.3	T	μ=0	36Nb	
0.619	3.873(0)	-15.5	5	T	μ=0	37Ma	
0.458	3.868(0)	-16.1	10	T	μ=0	37Ma	
0.460	3.866(0)	-16.25	12.5	T	μ=0	36Nb	
0.285	3.861(0)	-16.7	15	T	μ=0	37Ma	
0.098	3.857(0)	-17.3	20	T	μ=0	37Ma	
-0.061	3.862(0)	-17.9	25	T	μ=0 ΔCp=-41.1	36Nb	
						39Hb	
-0.102	3.858(0)	-18.0	25	T	μ=0	37Ma	
-0.315	3.861(0)	-18.7	30	T	μ=0	37Ma	
-0.543	3.867(0)	-19.45	35	T	μ=0	37Ma	
-0.620	3.874(0)	-19.70	37.5	T	μ=0	36Nb	
-0.686	3.870(0)	-19.9	38	T	μ=0	37Ma	
-0.785	3.873(0)	-20.2	40	T	μ=0	37Ma	
-1.041	3.883(0)	-21.0	45	T	μ=0	37Ma	
-1.313	3.895(0)	-21.9	50	T	μ=0	37Ma	
-1.265	3.896(0)	-21.75	50	T	μ=0	36Nb	

P169 PROPANOIC ACID, 2-hydroxy-2-methyl- $C_4H_8O_3$ $(CH_3)_2C(OH)CO_2H$

ΔH, kcal /mole	pK	ΔS, cal /mole°K	T, °C	M	Conditions	Ref.	Remarks
0.3	3.717(0)	-16	25	C	μ=0.10M	71Bc	
0.518	3.971(35°,0)	-16	0&35	T	--	39Wa	b,c
1.8	4.598(0)	-15	25	C	μ=0.10M S: 40% DMSO	71Bc	
7.5	8.558(0)	-14	25	C	μ=0.10M S: DMSO	71Bc	
1.4	7.583(0)	-30	25	C	μ=0.10M S: Ethanol	71Bc	
2.2	6.858(0)	-24	25	C	μ=0.10M S: Ethylene glycol	71Bc	

P170 PROPANOIC ACID, 3(4-imidazole) $C_6H_8O_2N_2$

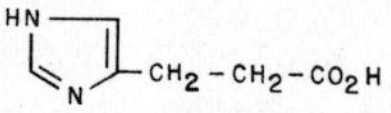

ΔH, kcal /mole	pK	ΔS, cal /mole°K	T, °C	M	Conditions	Ref.	Remarks
-0.01	3.96(+1)	18.2	25	C	μ=0.16M(KNO_3)	70Me	e:CO_2H
8.82	7.57(0)	5.1	25	C	μ=0.16M(KNO_3)	70Me	e:Imidazole group

P171 PROPANOIC ACID, 3-iodo- $C_3H_5O_2I$ $ICH_2CH_2CO_2H$

ΔH, kcal /mole	pK	ΔS, cal /mole°K	T, °C	M	Conditions	Ref.	Remarks
-1.38	4.08(0)	-23.2	25	C	μ=0	66Ae	
-0.450	4.031(35°,0)	-20	0&35	T	--	39Wa	b,c

P172 PROPANOIC ACID, 2-mercapto- $C_3H_6O_2S$ $CH_3CH(SH)CO_2H$

ΔH, kcal /mole	pK	ΔS, cal /mole°K	T, °C	M	Conditions	Ref.	Remarks
-4.75	3.66(30°,0)	-32.4	20-40	T	μ=0.1M($NaClO_4$)	71Sa	c

P173 PROPANOIC ACID, 3-mercapto- $C_3H_6O_2S$ $HSCH_2CH_2CO_2H$

ΔH, kcal /mole	pK	ΔS, cal /mole°K	T, °C	M	Conditions	Ref.	Remarks
6.10	10.84(-1)	-29.1	25	C	μ=0.05	64Ia	

P174 PROPANOIC ACID, 2-methyl $C_4H_8O_2$ $(CH_3)_2CHCO_2H$

ΔH, kcal /mole	pK	ΔS, cal /mole°K	T, °C	M	Conditions	Ref.	Remarks
-0.14	4.827(0)	-22.6	10	C	μ=0	70Ce	q:±0.08
-0.345	(0)	-----	10	C	μ=0	48Ca	

ΔH, kcal /mole	pK	ΔS, cal /mole°K	T, °C	M	Conditions	Ref.	Remarks
P174, cont.							
-0.623	(0)	-----	20	C	μ=0	52Ea	
-0.775	4.853(0)	-24.8	25	C	μ=0	68Cd	
-0.74	4.85(0)	-24.6	25	C	μ=0	74Md	
-1.01	4.86(0)	-25.6	25	C	μ=0	58Ca	a:13.5
-0.799	4.8486(0)	-24.86	25	T	μ=0	52Ea	
-0.75	4.853(0)	-24.7	25	C	μ=0	70Ce	
-0.921	(0)	-----	30	C	μ=0	48Ca	
-1.22	4.918(0)	-26.4	40	C	μ=0	70Ce	
-----	(0)	-----	10-40	C	μ=0 ΔCp=-36	70Ce	
-0.6	5.467(0)	-23	25	C	μ=0.10M	71Bc	
-0.419	4.848(0)	-24	0&35	T	--	39Wa	b,c
-0.69	4.98(0)	-25.1	25	C	μ=0 S: 5 w % tert-Butanol	74Md	
-0.74	5.14(0)	-26.0	25	C	μ=0 S: 10 w % tert-Butanol	74Md	
-1.0	5.28(0)	-27.5	25	C	μ=0 S: 15 w % tert-Butanol	74Md	
-1.2	5.44(0)	-28.9	25	C	μ=0 S: 17.5 w % tert-Butanol	74Md	
-0.98	5.47(0)	-28.4	25	C	μ=0 S: 20 w % tert-Butanol	74Md	
-0.48	5.70(0)	-27.7	25	C	μ=0 S: 25 w % tert-Butanol	74Md	
-0.096	5.95(0)	-27.5	25	C	μ=0 S: 30 w % tert-Butanol	74Md	
0.14	6.16(0)	-27.7	25	C	μ=0 S: 35 w % tert-Butanol	74Md	
0.024	6.34(0)	-28.9	25	C	μ=0 S: 40 w % tert-Butanol	74Md	
1.9	5.54(0)	-19	25	C	μ=0.01M S: 40% DMSO	71Bc	
8.6	10.68(0)	-20	25	C	μ=0.01M S: DMSO	71Bc	
2.1	8.97(0)	-34	25	C	μ=0.01M S: Ethanol	71Bc	
0.5	7.80(0)	-34	25	C	μ=0.01M S: Ethylene glycol	71Bc	
1.2	3.72(0)	-13	25	C	μ=0.01M S: Formamide	71Bc	

P175 PROPANOIC ACID, 2-oxo- $C_3H_4O_3$ CH_3COCO_2H

2.90	2.49(0)	-1.7	25	C	μ=0	67Ob	
3.011	2.60(0)	-1.80	25	C	μ=0.05	67Ob	

P176 PROPANOIC ACID, 3-phenyl- $C_9H_{10}O_2$ $C_6H_5CH_2CH_2CO_2H$

-0.066	4.664(35°,0)	-22	0&35	T	--	39Wa	b,c

P177 1-PROPANOL, 1-amino- C_3H_9ON $CH_3CH_2CH(NH_2)OH$

14.4	9.96(+1)	3	25	T	μ=0	59Sc	

P178 1-PROPANOL, 2-amino- C_3H_9ON $CH_3CH(NH_2)CH_2OH$

12.042	9.469(25°,+1)	-2.9	0-60	T	μ=0	68Tb	c

P179 1-PROPANOL, 3-amino- C_3H_9ON $H_2NCH_2CH_2CH_2OH$

12.74	9.96(+1)	-2.9	25	C	μ=0	69Cc	

ΔH, kcal /mole	pK	ΔS, cal /mole°K	T, °C	M	Conditions	Ref.	Remarks
P180 1-PROPANOL, 2-amino-2-methyl-			$C_4H_{11}ON$		$(CH_3)_2C(NH_2)CH_2OH$		
13.38	9.71(+1)	0.45	25	C	μ=0	61Ta	
12.93	9.69(+1)	-0.98	25	C	μ=0	68Ob	
12.894	9.694(25°,+1)	-1.1	0-60	T	μ=0	68Tb	c
P181 2-PROPANOL, 1,3-diamino-			$C_3H_{10}ON_2$		$H_2NCH_2CH(OH)CH_2NH_2$		
10.4	7.68(30°,+2)	-1	10-40	T	μ=0	50Ba	c
11.2	9.42(30°,+1)	-6	10-40	T	μ=0	50Ba	c
P182 2-PROPANOL, 1,1,1,3,3,3-hexafluoro-			$C_3H_2OF_6$		$CF_3CH(OH)CF_3$		
6.37	9.42(0)	-21.7	25	C	μ=0	71Wb	
P183 1-PROPANONE, 1,3-diphenyl-3-mercapto-			$C_{15}H_{12}OS$		$H_5C_6C(O)CH_2CH(SH)C_6H_5$		
2.5	11.07(25°,0)	42.3	15-40	T	μ=0.02M[$N(CH_3)_4Cl$] S: 75 v % Dioxane	74Sb	c
P184 PROPENE, 3-amino-			C_3H_7N		$H_2NCH_2CH:CH_2$		
13.06	9.52(+1)	0.2	25	C	μ=0	69Cc	
12.89	9.691(+1)	-1.1	25	T	μ=0	62Hd	
P185 PROPENOIC ACID			$C_3H_4O_2$		$CH_2:CHCO_2H$		
0.115	4.267	-19.03	15	T	μ=0 ΔCp=-55.09	69Gc	
-0.163	4.250	-19.99	20	T	μ=0 ΔCp=-56.05	69Gc	
-0.445	4.247	-20.94	25	T	μ=0 ΔCp=-57.00	69Gc	
-0.673	4.249	-21.90	30	T	μ=0 ΔCp=-57.96	69Gc	
-1.025	4.267	-22.86	35	T	μ=0 ΔCp=-58.92	69Gc	
-1.322	4.301	-23.81	40	T	μ=0 ΔCp=-59.81	69Gc	
-1.623	4.311	-24.76	45	T	μ=0 ΔCp=-60.82	69Gc	
P185a PROPYNOIC ACID			$C_3H_2O_2$		HC⋮CCO_2H		
-3.362	1.791	-20.04	10	T	μ=0 ΔCp=56.43	69Gc	
-3.077	1.829	-19.04	15	T	μ=0 ΔCp=57.43	69Gc	
-2.787	1.867	-18.04	20	T	μ=0 ΔCp=58.43	69Gc	
-2.493	1.887	-17.05	25	T	μ=0 ΔCp=59.42	69Gc	
-2.193	1.940	-16.05	30	T	μ=0 ΔCp=60.42	69Gc	
-1.888	1.932	-15.05	35	T	μ=0 ΔCp=61.42	69Gc	
-1.579	1.963	-14.06	40	T	μ=0 ΔCp=62.41	69Gc	
-1.263	1.978	-13.06	45	T	μ=0 ΔCp=63.41	69Gc	
P186 PROPYNOIC ACID, bromo-			C_3HO_2Br		BrC⋮CCO_2H		
-2.075	1.786(0)	-15.50	10	T	μ=0 ΔCp=21.46	69Gc	
-1.967	1.814(0)	-15.12	15	T	μ=0 ΔCp=21.84	69Gc	
-1.857	1.839(0)	-14.75	20	T	μ=0 ΔCp=22.21	69Gc	
-1.715	1.855(0)	-14.37	25	T	μ=0 ΔCp=22.59	69Gc	
-1.631	1.879(0)	-13.99	30	T	μ=0 ΔCp=22.97	69Gc	
-1.515	1.900(0)	-13.61	35	T	μ=0 ΔCp=23.35	69Gc	
-1.398	1.919(0)	-13.23	40	T	μ=0 ΔCp=23.73	69Gc	
-1.278	1.924(0)	-12.85	45	T	μ=0 ΔCp=24.11	69Gc	

ΔH, kcal /mole	pK	ΔS, cal /mole°K	T,°C	M	Conditions	Ref.	Remarks
P187 PROPYNOIC ACID, chloro-			C_3HO_2Cl		ClC:CCO$_2$H		
-2.137	1.766(0)	-15.64	10	T	μ=0 ΔCp=28.48	69Gc	
-1.993	1.796(0)	-15.14	15	T	μ=0 ΔCp=28.98	69Gc	
-1.847	1.820(0)	-14.63	20	T	μ=0 ΔCp=29.49	69Gc	
-1.698	1.845(0)	-14.13	25	T	μ=0 ΔCp=29.99	69Gc	
-1.547	1.864(0)	-13.63	30	T	μ=0 ΔCp=30.49	69Gc	
-1.393	1.879(0)	-13.12	35	T	μ=0 ΔCp=31.00	69Gc	
-1.237	1.893(0)	-12.61	40	T	μ=0 ΔCp=31.51	69Gc	
-1.078	1.901(0)	-12.10	45	T	μ=0 ΔCp=32.02	69Gc	
P188 PROPYNOIC ACID, phenyl-			$C_9H_6O_2$		HO$_2$CC:CC$_6$H$_5$		
-0.792	2.269(35°,0)	-13	0&35	T	--	39Wa	b,c
P189 PSEUDOISOCYANINE			$C_{23}H_{26}N_2$		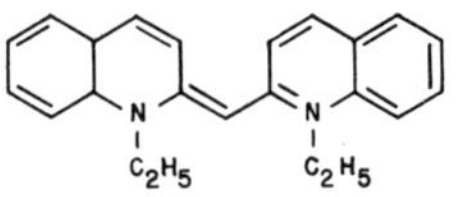		
6.32	4.01(25°,+2)	2.85	3.8-25	T	μ=0.20	72Dd	c
6.95	4.59(25°,+2)	2.31	3.8-25	T	μ=0.20 R: $DA^{2+} = D^+ + A^+$	72Dd	c
P190 PURINE			$C_5H_4N_4$		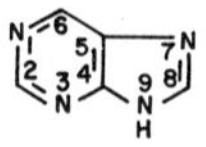		
8.2	8.74(30°,0)	-14.0	20-50	T	μ=0	66Lb	c
P191 PURINE, 6-amino-			$C_5H_5N_5$		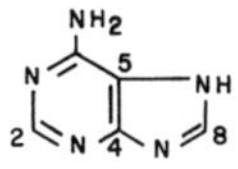		
4.9	4.2(+1)	-2.7	25	C	μ=0	62Cc	
3.8	4.22(20°,+1)	-5	20&30	T	μ=0	62La	b,c,e: NH$_2$
2.7	4.12(30°,+1)	-10	30&40	T	μ=0	62La	b,c,e: NH$_2$
4.2	4.18(25°,+1)	-5	10-40	T	μ=0.005	58Ha	
3.8	4.12(+1)	-5	25	T	μ=0.006 - 0.012	62La	
2.7	4.06(+1)	-10	35	T	μ=0.01	62La	
5.10	(+1)	-----	20	C	μ=0.025	72Za	e:N$_1$-H
5.06	4.19(+1)	-2.4	25	C	μ=0.025 ΔCp=-11	72Za	
4.87	(+1)	-----	40	C	μ=0.025	72Za	
4.72	(+1)	-----	57	C	μ=0.025	72Za	
4.72	(+1)	-----	20	C	μ=0.05	72Za	
4.21	(+1)	-----	20	C	μ=0.10	72Za	
4.2	4.12(+1)	-4.7	25	T	μ=0.1	64Sd	
3.99	4.1(+1)	-5.4	25	C	μ=0.1M(NaCl)	60Ra	d
4.81	4.20(+1)	-3.17	25	C	--	70Cb	e:N$_1$-H
9.1	9.8(0)	-14	25	C	μ=0	62Cc	
11.0	9.96(20°,0)	-5	20&30	T	μ=0	62La	b,c,e: N$_9$-H
9.1	9.75(30°,0)	-14.8	20-50	T	μ=0	66Lb	

ΔH, kcal /mole	pK	ΔS, cal /mole°K	T, °C	M	Conditions	Ref.	Remarks
P191, cont.							
8.2	9.67(30°,0)	-14	30&40	T	μ=0	62La	b,c,e: N_9-H
11.0	9.67(0)	-10	25	T	μ=0.001 - 0.012	62La	
8.2	9.49(0)	-14	35	T	μ=0.01	62La	
9.13	9.92(0)	-15.1	25	C	μ=0.025 ΔCp=-17	72Za	e:N_9-H
8.88	(0)	-----	40	C	μ=0.025	72Za	
9.5	9.72(0)	-12.4	25	T	μ=0.1	64Sd	
9.65	9.87(0)	-12.88	25	C	--	70Cb	e:N_9-H

P192 PYRAZINE, 2,3,5,6-tetra methyl-, 1,4-dioxide $C_8H_{12}O_2N_2$

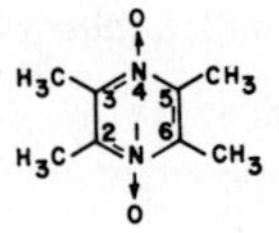

ΔH, kcal /mole	pK	ΔS, cal /mole°K	T, °C	M	Conditions	Ref.	Remarks
5.095	4.562(25°)	3.574	25-90	T	Aqueous H_2SO_4 solution	74Cb	c

P193 PYRAZOLE $C_3H_4N_2$

ΔH, kcal /mole	pK	ΔS, cal /mole°K	T, °C	M	Conditions	Ref.	Remarks
3.52	2.61(+1)	-0.2	25	C	μ=0	70Hb	

P194 PYRENE, 3-amino $C_{16}H_{11}N$

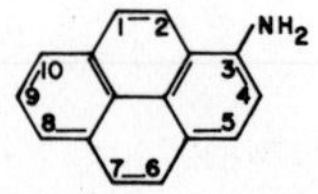

ΔH, kcal /mole	pK	ΔS, cal /mole°K	T, °C	M	Conditions	Ref.	Remarks
3.9	2.91(20°,+1)	-0.1	0-20	T	C=0.005 S: 50 w % Ethanol	59Eb	c

P195 PYRIDINE C_5H_5N

ΔH, kcal /mole	pK	ΔS, cal /mole°K	T, °C	M	Conditions	Ref.	Remarks
4.80	5.21(+1)	-7.75	25	C	μ=0	74Lb	
4.92	5.17(+1)	-7.2	25	C	μ=0	69Cc	
5.70	5.22(+1)	-4.76	25	C	μ=0	59Mb	
4.773	5.158(+1)	-7.6	25	C	μ=0	65Ca	
4.795	5.17(+1)	-7.55	25	C	μ=0	60Sa	
4.67	(+1)	-----	25	C	μ=0	61Ta	
4.020	5.21(+1)	10.4	25	T	μ=0	69Bb	
5.40	5.19(+1)	-5.6	25	T	μ=0	68Pa	
4.721	5.22(+1)	-8.0	25	T	μ=0	54Aa	
4.37	5.27(20°,+1)	-9.21	15-36	T	μ=0	61Eb	
5.057	5.158(25°,+1)	-7.6	5-45	T	μ=0	65Ca	
3.13	5.42(+1)	-13.6	5	T	$\mu=10^{-4}$	69Gb	
4.28	5.22(+1)	-9.6	25	T	$\mu=10^{-4}$	69Gb	
5.83	4.94(+1)	-4.3	50	T	$\mu=10^{-4}$	69Gb	
4.97	(+1)	-----	25	C	μ=0.01M(Na_2SO_4)	71Ha	

ΔH, kcal /mole	pK	ΔS, cal /mole°K	T,°C	M	Conditions	Ref.	Remarks
P195, cont.							
4.98	5.168(+1)	-7.0	25	C	μ=0.012	68Ce	
4.130	(+1)	-----	10	C	C=0.05	49La	
4.640	(+1)	-----	20	C	C=0.05	49La	
4.721	5.30(+1)	-8.4	25	C	C=0.05	49La	
4.800	(+1)	-----	30	C	C=0.05	49La	
5.41	5.49(+1)	-6.98	25	C	μ=0.5M(KNO_3)	73Be	
4.00	5.31(+1)	-10.39	25	T	μ=1.0M	73Ca	
4.795	5.17(+1)	-7.55	25	C	--	60Sa	
5.230	4.38(+1)	-2.50	25	C	S: 50 w % Ethanol	60Sa	

P196 PYRIDINE, 2-acetyl- C_7H_7ON $C_5H_4N(OCCH_3)$

ΔH	pK	ΔS	T,°C	M	Conditions	Ref.	Remarks
3.073	2.64(+1)	-1.7	25	C	μ=0	65Ca	
3.381	2.643(+1)	-0.75	25	C	μ=0	65Ca	
3.073	2.643(25°,+1)	-0.75	5-45	T	μ=0	65Ca	c

P197 PYRIDINE, 3-acetyl- C_7H_7ON $C_5H_4N(OCCH_3)$

ΔH	pK	ΔS	T,°C	M	Conditions	Ref.	Remarks
3.127	3.256(+1)	-4.4	25	C	μ=0	65Ca	
3.136	3.256(25°,+1)	-4.4	5-45	T	μ=0	65Ca	c
2.40	3.26(25°,+1)	-6.99	25	T	μ=1.0M	73Ca	q:±0.20

P198 PYRIDINE, 4-acetyl- C_7H_7ON $C_5H_4N(COCH_3)$

ΔH	pK	ΔS	T,°C	M	Conditions	Ref.	Remarks
3.681	3.505(+1)	-3.7	25	C	μ=0	65Ca	
3.775	3.505(25°,+1)	-3.7	5-45	T	μ=0	65Ca	c
3.71	3.62(+1)	-4.03	25	T	μ=1.0M	73Ca	

P199 PYRIDINE, 2-aldehyde-2'-pyridylhydrazone- $C_{11}H_{10}N_4$ $(C_5H_4N)CH:NNH(C_5H_4N)$

ΔH	pK	ΔS	T,°C	M	Conditions	Ref.	Remarks
5.3	2.91(25°)	4	5-60	T	μ=0	68Gc	c
6.9	5.62(25°)	-3	5-60	T	μ=0	68Gc	c

P200 PYRIDINE, 2-aldoxime- $C_6H_6ON_2$

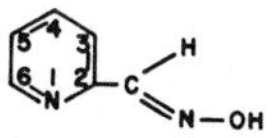

ΔH	pK	ΔS	T,°C	M	Conditions	Ref.	Remarks
3.32	3.42	-4.5	25	T	μ=0	62Ha	
7.7	3.88	9.9	5	T	$C=10^{-4}M$	61Ga	
6.3	3.70	4.9	15	T	$C=10^{-4}M$	61Ga	
4.8	3.56	0.0	25	T	$C=10^{-4}M$	61Ga	
4.1	3.51	-2.5	30	T	$C=10^{-4}M$	61Ga	
2.6	3.42	-7.5	40	T	$C=10^{-4}M$	61Ga	
1.0	3.39	-12.5	50	T	$C=10^{-4}M$	61Ga	
-0.7	3.38	-17.5	60	T	$C=10^{-4}M$	61Ga	
6.8	10.22	-24.0	25	T	μ=0	62Ha	
0.7	10.25	-44.3	5	T	$C=10^{-4}M$	61Ga	
1.4	10.21	-41.8	15	T	$C=10^{-4}M$	61Ga	
2.1	10.17	-39.3	25	T	$C=10^{-4}M$	61Ga	
2.5	10.13	-38.1	30	T	$C=10^{-4}M$	61Ga	
3.3	10.08	-35.7	40	T	$C=10^{-4}M$	61Ga	
4.0	10.00	-33.2	50	T	$C=10^{-4}M$	61Ga	
4.9	9.91	-30.7	60	T	$C=10^{-4}M$	61Ga	

ΔH, kcal /mole	pK	ΔS, cal /mole°K	T, °C	M	Conditions	Ref.	Remarks
P201 PYRIDINE, 2-amino-			$C_5H_6N_2$		$C_5H_4N(NH_2)$		
8.43	6.25(+1)	-1.9	10	C	μ=0	72Cc	
8.40	6.71(+1)	-2.5	25	C	μ=0	72Cc	
8.170	6.71(+1)	-3.49	25	T	μ=0 ΔCp=-63.38	73Bd	
8.38	6.43(+1)	-2.6	40	C	μ=0	72Cc	
-----	(+1)	-----	10-40	C	μ=0 ΔCp=-2	72Cc	
8.4	6.86(20°,+1)	-2.9	5&35	T	C=0.02M	59Eb	b,c
8.39	7.04(+1)	-4.06	25	C	μ=0.5M(KNO_3)	73Be	
P202 PYRIDINE, 3-amino-			$C_5H_6N_2$		$C_5H_4N(NH_2)$		
6.46	6.25(+1)	-5.7	10	C	μ=0	72Cc	
6.43	6.03(+1)	-6.0	25	C	μ=0	72Cc	
6.41	6.04(+1)	-6.14	25	C	μ=0	74Lb	
6.315	6.03(+1)	-6.64	25	T	μ=0 ΔCp=-13.63	73Bd	
6.50	5.80(+1)	-5.7	40	C	μ=0	72Cc	
-----	-----	-----	10-40	C	μ=0 ΔCp=1	72Cc	
5.1	5.98(20°,+1)	-7.4	5&35	T	C=0.02M	59Eb	b,c
6.66	6.34(+1)	-6.67	25	C	μ=0.5M(KNO_3)	73Be	
P203 PYRIDINE, 4-amino-			$C_5H_6N_2$		$C_5H_4N(NH_2)$		
11.338	9.8731(+1)	-3.65	0	T	μ=0	60Bb	
11.321	9.7043(+1)	-3.70	5	T	μ=0	60Bb	
11.33	9.549(+1)	-3.6	10	C	μ=0	72Cc	
11.305	9.5486(+1)	-3.78	10	T	μ=0	60Bb	
11.288	9.3979(+1)	-3.82	15	T	μ=0	60Bb	
11.271	9.2524(+1)	-3.90	20	T	μ=0	60Bb	
11.27	9.12(+1)	-3.92	25	C	μ=0	74Lb	
11.31	9.114(+1)	-3.7	25	C	μ=0	72Cc	
11.254	9.1141(+1)	-3.94	25	T	μ=0	60Bb	
11.235	8.9783(+1)	-4.02	30	T	μ=0	60Bb	
11.216	8.8455(+1)	-4.06	35	T	μ=0	60Bb	
10.88	9.29(20°,+1)	-5.36	5-35	T	μ=0	61Eb	
11.200	8.7170(+1)	-4.13	40	T	μ=0	60Bb	
11.27	8.717(+1)	-3.9	40	C	μ=0	72Cc	
-----	(+1)	-----	10-40	C	μ=0 ΔCp=-2	72Cc	
11.180	8.5941(+1)	-4.18	45	T	μ=0	60Bb	
11.161	8.4768(+1)	-4.25	50	T	μ=0	60Bb	
10.8	9.17(20°,+1)	-5.8	5-35	T	μ=0.02	59Eb	c
11.34	9.40(+1)	-5.00	25	C	μ=0.5M(KNO_3)	73Be	
5.23	9.68(+1)	-24.30	25	T	μ=1.0M	73Ca	
11.133	8.942(+1)	-1.67	10	T	μ=0 S: 50 w % Methanol	66Pc	
10.739	8.520(+1)	-2.96	25	T	μ=0 S: 50 w % Methanol	66Pc	
10.346	8.149(+1)	-4.24	40	T	μ=0 S: 50 w % Methanol	66Pc	
P204 PYRIDINE, 4-amino-3-bromo-			$C_5H_5N_2Br$		$C_5H_3N(NH_2)(Br)$		
7.69	7.04(+1)	-5.97	20	T	μ=0	61Eb	
P205 PYRIDINE, 4-amino-3-bromo-1-methyl-			$C_6H_8N_2Br$		$C_5H_3N(NH_2)(Br)(CH_3)$		
9.01	7.47(+1)	-3.45	20	T	μ=0	61Eb	

ΔH, kcal /mole	pK	ΔS, cal /mole°K	T, °C	M	Conditions	Ref.	Remarks
P206 PYRIDINE, 4-amino-3,5-dimethyl-			$C_7H_{10}N_2$		$C_5H_2N(NH_2)(CH_3)_2$		
10.48	9.54(+1)	-7.88	20	T	μ=0	61Eb	
P207 PYRIDINE, 2-(2'-aminoethyl)-			$C_7H_{10}N_2$		$C_5H_4N(CH_2CH_2NH_2)$		
3.77	5.94(10°,+2)	-5	10-40	T	μ=0	59Ga	c
6.27	3.96(+2)	-2.9	25	C	0.3M($NaClO_4$)	67Hc	
4.60	4.24(+2)	-4.0	25	C	μ=0.5M(KNO_3)	71Ge	
11.4	10.03(10°,+1)	-6	10-40	T	μ=0	59Ga	c
11.60	9.46(+1)	-4.4	25	C	0.3M($NaClO_4$)	67Hc	
12.12	9.78(+1)	-4.2	25	C	μ=0.5M(KNO_3)	71Ge	
P208 PYRIDINE, 4-amino-3-ethyl-			$C_7H_{10}N_2$		$C_5H_3N(NH_2)(C_2H_5)$		
10.89	9.51(+1)	-6.34	20	T	μ=0	61Eb	
P209 PYRIDINE, 4-amino-3-isopropyl-			$C_8H_{12}N_2$		$C_5H_3N(NH_2)(C_3H_7)$		
11.28	9.54(+1)	-5.15	20	T	μ=0	61Eb	
P210 PYRIDINE, 2-aminomethyl-			$C_6H_8N_2$		$C_5H_4N(CH_2NH_2)$		
6.31	2.33(+2)	-10.5	25	C	0.3M($NaClO_4$)	67Hc	
2.95	2.31(+2)	-0.7	25	C	μ=0.5M(KNO_3)	71Ge	
10.3	9.09(10°,+1)	-5	10-40	T	μ=0	59Ga	c
10.57	8.55(+1)	-3.7	25	C	0.3M($NaClO_4$)	67Hc	
10.98	8.79(+1)	-3.4	25	C	μ=0.5M(KNO_3)	71Ge	
P211 PYRIDINE, 4-aminomethyl-			$C_6H_8N_2$		$C_5H_4N(CH_2NH_2)$		
-0.31	4.41	-20.93	25	T	μ=1.0M	73Ca	q: ±0.04
P212 PYRIDINE, 2-amino-4-methyl-			$C_6H_8N_2$		$C_5H_3N(NH_2)(CH_3)$		
5.78	3.885(25°)	1.60	15&35	T	μ=0 S: Dimethylformamide	67Ra	b,c
5.5	3.77(25°)	1.2	0.1&25	T	μ=10^{-3} S: Methanol	65Ra	b,c
P213 PYRIDINE, 4-amino-3-methyl-			$C_6H_8N_2$		$C_5H_3N(NH_2)(CH_3)$		
11.66	9.43	-3.34	20	T	μ=0	61Eb	
P214 PYRIDINE, 2-(aminomethyl)-6-methyl-			$C_7H_{10}N_2$		$C_5H_3N(CH_2NH_2)(CH_3)$		
9.3	8.70(30°,+1)	-9.1	0-40	T	--	60Wa	c
P215 PYRIDINE, 4-amino-2,3-5,6-tetramethyl-			$C_9H_{14}N_2$		$C_5N(NH_2)(CH_3)_4$		
10.35	10.58	-13.10	20	T	μ=0	61Eb	
P216 PYRIDINE, 4-benzyl-, 1-oxide			$C_{12}H_{11}ON$		$C_5H_4N(CH_2C_6H_5)(\rightarrow O)$		
1.237	-1.018(+1)	8.80	25	T	μ=0	73Kb	

ΔH, kcal /mole	pK	ΔS, cal /mole°K	T, °C	M	Conditions	Ref.	Remarks
P217 PYRIDINE, 2-bromo-			C_5H_4NBr		$C_5H_4N(Br)$		
-0.01	0.72(+1)	-3.2	10	C	μ=0	72Cc	q:±0.01
0.08	0.71(+1)	-2.9	25	C	μ=0	72Cc	q:±0.02
0.05	0.66(+1)	-2.8	40	C	μ=0	72Cc	q:±0.03
-----	(+1)	-----	10-40	C	μ=0 ΔCp=2	72Cc	
30.2	(+1)	-----	25	C	$C\sim10^{-3}M$ S: Fluorosulfuric acid	74Ab	
P218 PYRIDINE, 3-bromo-			C_5H_4NBr		$C_5H_4N(Br)$		
1.32	2.95(+1)	-8.8	10	C	μ=0	72Cc	
1.85	2.91(+1)	-7.03	20	T	μ=0	61Eb	
1.35	2.89(+1)	-8.6	25	C	μ=0	72Cc	
2.78	2.85(+1)	-3.72	25	C	μ=0	74Lb	
1.797	2.72(+1)	-6.44	25	T	μ=0 ΔCp=-16.49	73Bd	
1.36	2.80(+1)	-8.4	40	C	μ=0	72Cc	
-----	(+1)	-----	10-40	C	μ=0 ΔCp=2	72Cc	
P219 PYRIDINE, 4-bromo-			C_5H_4NBr		$C_5H_4N(Br)$		
3.51	3.75(+1)	-5.40	25	C	μ=0	74Lb	
3.120	3.68(+1)	-6.39	25	T	μ=0 ΔCp=-17.45	73Bd	
2.38	4.05(+1)	-10.0	25	T	μ=1.0M	73Ca	
P220 PYRIDINE, 3-bromo-4-(dimethylamino)-			$C_7H_9N_2Br$		$C_5H_3N(Br)[N(CH_3)_2]$		
6.36	6.52	-8.15	20	T	μ=0	61Eb	
P221 PYRIDINE, 2-chloro-			C_5H_4NCl		$C_5H_4N(Cl)$		
0.00	0.49(+1)	-2.2	10	C	μ=0	72Cc	q:±0.01
0.02	0.49(+1)	-2.1	25	C	μ=0	72Cc	q:±0.01
0.07	0.48(+1)	-2.0	40	C	μ=0	72Cc	q:±0.03
-----	(+1)	-----	10-40	C	μ=0 ΔCp=2	72Cc	
-0.775	-0.750(25°,+1)	-0.83	25-90	T	S: Aqueous H_2SO_4 Solution	74Cb	c
31.7	(+1)	-----	25	C	$C\sim10^{-3}M$ S: Fluorosulfuric acid	74Ab	
P222 PYRIDINE, 3-chloro-			C_5H_4NCl		$C_5H_4N(Cl)$		
1.86	2.92(+1)	-6.7	10	C	μ=0	72Cc	
2.60	2.81(+1)	-4.13	25	C	μ=0	74Lb	
2.11	2.84(+1)	-5.9	25	C	μ=0	72Cc	
2.30	2.74(+1)	-5.1	40	C	μ=0	72Cc	
-----	(+1)	-----	10-40	C	μ=0 ΔCp=14	72Cc	
1.63	2.88(+1)	-7.82	25	T	μ=1.0M	73Ca	q:±0.17
-2.480	-2.698(25°,+1)	-4.70	25-90	T	S: Aqueous H_2SO_4 Solution	74Cb	c
P223 PYRIDINE, 4-chloro-			C_5H_4NCl		$C_5H_4N(Cl)$		
3.58	3.83(+1)	-5.53	25	C	μ=0	74Lb	
2.57	4.11(+1)	-9.97	25	T	μ=1.0M	73Ca	

ΔH, kcal /mole	pK	ΔS, cal /mole°K	T, °C	M	Conditions	Ref.	Remarks
P224 PYRIDINE, 2-chloro-6-fluoro-			C_5H_3NClF		$C_5H_3N(Cl)(F)$		
3.923	3.569(25°,+1)	-3.78	25-90	T	Aqueous H_2SO_4 Solution	74Cb	c
P225 PYRIDINE, deuterated							
4.740	5.72(+1)	10.3	25	T	μ=0 R: $C_5H_5ND^+ = C_5H_5N + D^+$	69Bb	
4.160	5.33(+1)	10.4	25	T	μ=0 R: $C_5D_5NH^+ = C_5D_5N + H^+$	69Bb	
4.910	5.83(+1)	10.2	25	T	μ=0 R: $C_5D_5ND^+ = C_5D_5N + D^+$	69Bb	
P226 PYRIDINE, 2,4-dichloro-			$C_5H_3NCl_2$		$C_5H_3N(Cl)_2$		
2.611	1.250(25°,+1)	-3.05	25-90	T	Aqueous H_2SO_4 Solution	74Cb	c
P227 PYRIDINE, 2,6-dichloro-			$C_5H_3NCl_2$		$C_5H_3NCl_2$		
3.405	3.888(25°,+1)	-0.50	25-90	T	Aqueous H_2SO_4 Solution	74Cb	c
P228 PYRIDINE, 3,5-dichloro-			$C_5H_3NCl_2$		$C_5H_3N(Cl)_2$		
30.7	(+1)	-----	25	C	C∿10^{-3}M S: Fluorosulfuric acid	74Ab	
P229 PYRIDINE, 2,6-dichloro-, 1-oxide			$C_5H_3ONCl_2$		$C_5H_3N(Cl)_2(\rightarrow O)$		
3.505	2.112(25°,+1)	-2.124	25-90	T	Aqueous H_2SO_4 Solution	74Cb	c
P230 PYRIDINE, 3,5-dichloro-, 1-oxide			$C_5H_3ONCl_2$		$C_5H_3N(Cl)_2(\rightarrow O)$		
1.045	0.836(25°,+1)	0.206	25-90	T	Aqueous H_2SO_4 Solution	74Cb	c
P231 PYRIDINE, 2,6-di(isonitroso-methyl)-			$C_7H_7O_2N_3$		$C_5H_3N(CH:NOH)_2$		
0	7.40	-34	25	T	μ=0 HA={Fe(II)[2,6-di(isonitrosomethyl)pyridine]$_2$}$^{-1}$	65Ha	
P232 PYRIDINE, 2,6-di(1-isonitrosoethyl)-			$C_9H_{11}O_2N_3$		$C_5H_3N[C(CH_3):NOH]_2$		
5.3	10.08	-28	25	T	μ=0	65Ha	
6.7	10.88	-27	25	T	μ=0	65Ha	
P233 PYRIDINE, 3,5-dimethoxy-2-nitro-, 1-oxide			$C_7H_8O_5N_2$		$C_5H_2N(OCH_3)_2(NO_2)(\rightarrow O)$		
2.764	1.690(25°)	-1.790	25-90	T	Aqueous H_2SO_4 Solution	74Cb	c

ΔH, kcal /mole	pK	ΔS, cal /mole°K	T, °C	M	Conditions	Ref.	Remarks
P234 PYRIDINE, 2,3-dimethyl-			C_7H_9N		$C_5H_3N(CH_3)_2$		
8.51	6.59(+1)	-1.6	25	T	μ=0	68Pa	
7.46	6.91(+1)	-6.61	25	C	μ=0.5M(KNO_3)	73Be	
P235 PYRIDINE, 2,4-dimethyl-			C_7H_9N		$C_5H_3N(CH_3)_2$		
8.14	6.74(+1)	-3.5	25	T	μ=0	68Pa	
8.26	6.98(+1)	-4.23	25	C	μ=0.5M(KNO_3)	73Be	
6.75	6.46(37°,+1)	-7.93	0-70	T	μ=0.5M	72Mb	c
6.91	6.68(+1)	-6.79	25	T	μ=1.0M	73Ca	
7.165	6.79(+1)	-7.00	25	C	--	60Sa	
3.05	5.070(25°,+1)	-13.0	15&35	T	μ=0 S: Dimethylformamide	67Ra	b,c,e: NH
6.9	6.090(25°,+1)	-4.8	0.1&25	T	$μ=10^{-3}$ S: Methanol	65Ra	b,c
P236 PYRIDINE, 2,5-dimethyl-			C_7H_9N		$C_5H_3N(CH_3)_2$		
8.64	6.43(+1)	-0.5	25	T	μ=0	68Pa	
7.12	6.74(+1)	-6.94	25	C	μ=0.5M(KNO_3)	73Be	
6.28	6.18(37°,+1)	-7.57	0-70	T	μ=0.5M	72Mb	c
5.59	6.25(+1)	-10.11	25	T	μ=1.0M	73Ca	q:+0.80
6.815	6.51(+1)	-6.90	25	C	--	60Sa	
P237 PYRIDINE, 2,6-dimethyl-			C_7H_9N		$C_5H_3N(CH_3)_2$		
6.15	6.72(+1)	-10.11	25	C	μ=0	59Mb	
8.25	6.71(+1)	-3.0	25	T	μ=0	68Pa	
8.36	7.06(+1)	-4.26	25	C	μ=0.5M(KNO_3)	73Be	
8.65	6.67(37°,+1)	-2.19	0-70	T	μ=0.5M	72Mb	c
7.93	6.81(+1)	-4.49	25	T	μ=1.0M	73Ca	q:+0.70
7.235	6.75(+1)	-6.60	25	C	--	60Sa	
P238 PYRIDINE, 3,4-dimethyl-			C_7H_9N		$C_5H_3N(CH_3)_2$		
8.15	6.47(+1)	-2.3	25	T	μ=0	68Pa	
7.48	6.81(+1)	-6.07	25	C	μ=0.5M(KNO_3)	73Be	
5.75	6.28(37°,+1)	-9.23	0-70	T	μ=0.5M	72Mb	c
4.11	6.23(+1)	-15.00	25	T	μ=1.0M	73Ca	
P239 PYRIDINE, 3,5-dimethyl-			C_7H_9N		$C_5H_3N(CH_3)_2$		
5.29	6.23(+1)	-10.44	20	T	μ=0	61Eb	
7.13	6.09(+1)	-3.9	25	T	μ=0	68Pa	
6.10	5.85(37°,+1)	-5.52	0-70	T	μ=0.5M	72Mb	c
7.82	6.03(+1)	-1.41	25	T	μ=1.0M	73Ca	q:+0.51
6.365	6.18(+1)	-6.95	25	C	--	60Sa	
P240 PYRIDINE, 4-(dimethylamino)			$C_7H_{10}N_2$		$C_5H_4N[N(CH_3)_2]$		
10.75	9.71(+1)	-7.71	20	T	μ=0	61Eb	
9.53	9.47(+1)	-1.23	25	T	μ=1.0M	73Ca	
P241 PYRIDINE, 4(dimethylamino)-3,5-dimethyl-			$C_9H_{14}N_2$		$C_5H_2N[(CH_3)_2N](CH_3)_2$		
9.83	8.15	-3.75	20	T	μ=0	61Eb	

ΔH, kcal /mole	pK	ΔS, cal /mole°K	T,°C	M	Conditions	Ref.	Remarks
P242 PYRIDINE, 4(dimethylamino)-3-ethyl-			$C_9H_{14}N_2$		$C_5H_3N[N(CH_3)_2](C_2H_5)$		
9.15	8.66	-8.43	20	T	μ=0	61Eb	
P243 PYRIDINE, 4(dimethylamino)-3-isopropyl-			$C_{10}H_{16}N_2$		$C_5H_3N[N(CH_3)_2][CH(CH_3)_2]$		
8.76	8.27	-7.95	20	T	μ=0	61Eb	
P244 PYRIDINE, 4(dimethylamino)-3-methyl-			$C_9H_{14}N_2$		$C_5H_3N[N(CH_3)_2](C_2H_5)$		
9.02	8.68	-8.94	20	T	μ=0	61Eb	
P245 PYRIDINE, 3,5-dimethyl-4(methylamino)-			$C_8H_{12}N_2$		$C_5H_2N(CH_3)_2(NHCH_3)$		
12.06	9.43	-2.01	20	T	μ=0	61Eb	
P246 PYRIDINE, 3,5-dimethyl-4-nitro-, 1-oxide			$C_7H_8O_3N_2$		$C_5H_2N(CH_3)_2(NO_2)(\rightarrow O)$		
0.797	0.518(25°)	-0.366	25-90	T	Aqueous H_2SO_4 Solution	74Cb	c
P247 PYRIDINE, 2,4-dimethyl-, 1-oxide			C_7H_9ON		$C_5H_3N(CH_3)_2(\rightarrow O)$		
1.002	1.627(+1)	10.83	25	T	μ=0	73Ka	
P248 PYRIDINE, 2,5-dimethyl-, 1-oxide			C_7H_9ON		$C_5H_3N(CH_3)_2(\rightarrow O)$		
0.825	1.208(+1)	8.29	25	T	μ=0	73Ka	
P249 PYRIDINE, 2,6-dimethyl-, 1-oxide			C_7H_9ON		$C_5H_3N(CH_3)_2(\rightarrow O)$		
1.051	1.366(+1)	9.77	25	T	μ=0	73Ka	
P250 PYRIDINE, 3,4-dimethyl-, 1-oxide			C_7H_9ON		$C_5H_3N(CH_3)_2(\rightarrow O)$		
0.935	1.493(+1)	9.97	25	T	μ=0	73Ka	
P251 PYRIDINE, 3,5-dimethyl-, 1-oxide			C_7H_9ON		$C_5H_3N(CH_3)_2(\rightarrow O)$		
1.086	1.181(+1)	9.02	25	T	μ=0	73Ka	
P252 PYRIDINE, 2-ethyl-			C_7H_9N		$C_5H_4N(C_2H_5)$		
6.48	6.23(+1)	-6.77	25	C	μ=0.5M(KNO_3)	73Be	

ΔH, kcal /mole	pK	ΔS, cal /mole°K	T, °C	M	Conditions	Ref.	Remarks
P253 PYRIDINE, 3-ethyl-			C_7H_9N		$C_5H_4N(C_2H_5)$		
5.30	5.80(+1)	-8.43	20	T	μ=0	61Eb	
P254 PYRIDINE, 4-ethyl-			C_7H_9N		$C_5H_4N(C_2H_5)$		
6.67	6.49(+1)	-7.31	25	C	μ=0.5M(KNO_3)	73Be	
P255 PYRIDINE, 2-ethyl-4-methyl-			$C_8H_{11}N$		$C_5H_3N(C_2H_5)(CH_3)$		
3.62	5.085(25°,+1)	-11.1	15&35	T	μ=0 S: Dimethylformamide	67Ka	b,c
6.9	6.073(25°,+1)	-4.7	0.1&25	T	$μ=10^{-3}$ S: Methanol	65Ka	b,c
P256 PYRIDINE, 3-ethyl-4 (methylamino)-			$C_8H_{12}N_2$		$C_5H_3N(C_2H_5)(NHCH_3)$		
11.54	9.90	-5.90	20	T	μ=0	61Eb	
P257 PYRIDINE, 3-ethyl-4-methyl-, 1-oxide			$C_8H_{11}ON$		$C_5H_3N(C_2H_5)(CH_3)(\rightarrow O)$		
0.683	-1.534(+1)	9.31	25	T	μ=0	73Kb	
P258 PYRIDINE, 5-ethyl-2-methyl-, 1-oxide			$C_8H_{11}ON$		$C_5H_3N(C_2H_5)(CH_3)(\rightarrow O)$		
0.987	-1.288(+1)	9.21	25	T	μ=0	73Kb	
P259 PYRIDINE, 2-ethyl-, 1-oxide			C_7H_9ON		$C_5H_4N(C_2H_5)(\rightarrow O)$		
1.331	-1.191(+1)	9.92	25	T	μ=0	73Kb	
P260 PYRIDINE, 3-ethyl-, 1-oxide			C_7H_9ON		$C_5H_4N(C_2H_5)(\rightarrow O)$		
1.222	-0.965(+1)	8.51	25	T	μ=0	73Kb	
P261 PYRIDINE, 6-fluoro-2,3,4,5-tetrachloro-			C_5NCl_4F		$C_5N(F)(Cl)_4$		
10.808	7.707(25°)	-0.80	25-90	T	Aqueous H_2SO_4 Solution	74Cb	c
P262 PYRIDINE, 2-hydroxy-			C_5H_5ON		$C_5H_4N(OH)$		
-0.01	1.26(+1)	-5.8	10	C	μ=0	72Cc	q:±0.05
-0.07	1.25(+1)	-5.9	25	C	μ=0	72Cc	q:±0.04
-0.05	1.22(+1)	-5.7	40	C	μ=0	72Cc	q:±0.05
-----	(+1)	-----	10-40	C	μ=0 ΔCp=-2	72Cc	
P263 PYRIDINE, 3-hydroxy-			C_5H_5ON		$C_5H_4N(OH)$		
3.92	4.95(+1)	-8.8	10	C	μ=0	72Cc	
4.01	4.80(+1)	-8.5	25	C	μ=0	72Cc	
3.87	4.68(+1)	-9.0	40	C	μ=0	72Cc	
-----	(+1)	-----	10-40	C	μ=0 ΔCp=-5	72Cc	

ΔH, kcal /mole	pK	ΔS, cal /mole°K	T,°C	M	Conditions	Ref.	Remarks
P264 PYRIDINE, 4-hydroxy-			C_5H_5ON		$C_5H_4N(OH)$		
1.57	3.29(+1)	-9.5	10	C	μ=0	72Cc	
1.49	3.23(+1)	-9.7	25	C	μ=0	72Cc	
1.28	3.18(+1)	-10.4	40	C	μ=0	72Cc	
-----	(+1)	-----	10-40	C	μ=0 ΔCp=-9	72Cc	
P265 PYRIDINE, 2-hydroxy-4-methyl-			C_6H_7ON		$C_5H_3N(OH)(CH_3)$		
5.96	4.529(25°,+1)	-7.3	15&35	T	μ=0 S: Dimethylformamide	67Ra	b,c
7.3	5.846(25°,+1)	-2.3	0.1&25	T	$\mu=10^{-3}$ S: Methanol	65Ra	b,c
P266 PYRIDINE, 3-isopropyl-			$C_8H_{11}N$		$C_5H_4N(C_3H_7)$		
5.57	5.88(+1)	-7.91	20	T	μ=0	61Eb	
P267 PYRIDINE, 3-isopropyl-4 (methylamino)-			$C_9H_{14}N_2$		$C_5H_3N(C_3H_7)(NHCH_3)$		
11.93	9.96	-4.88	20	T	μ=0	61Eb	
P268 PYRIDINE, 4-mercapto-			C_5H_5NS		$C_5H_4N(SH)$		
-3.01	1.56	-17.01	25	T	μ=1.0M	73Ca	
P269 PYRIDINE, 4-methoxy-			C_6H_7ON		$C_5H_4N(OCH_3)$		
6.85	6.58(+1)	-7.14	25	C	μ=0	74Lb	
P270 PYRIDINE, 2-methyl-			C_6H_7N		$C_5H_4N(CH_3)$		
5.94	6.19(+1)	-7.3	10	C	μ=0	72Cc	
6.22	5.96(+1)	-6.4	25	C	μ=0	72Cc	
6.95	5.96(+1)	-3.95	25	C	μ=0	59Mb	
6.94	5.95(+1)	-3.9	25	T	μ=0	68Pa	
6.095	5.96(+1)	-6.8	25	T	μ=0	54Aa	
6.48	5.74(+1)	-5.5	40	C	μ=0	72Cc	
-----	(+1)	-----	10-40	C	μ=0 ΔCp=17	72Cc	
6.29	(+1)	-----	25	C	$\mu=0.1(KNO_3)$	71Fb	
5.930	(+1)	-----	10	C	C=0.5M	49La	
6.075	(+1)	-----	20	C	C=0.5M	49La	
6.55	6.14(+1)	-6.11	25	C	$\mu=0.5M(KNO_3)$	73Be	
6.095	6.02(+1)	-7.0	25	C	C=0.5M	49La	
6.115	(+1)	-----	30	C	C=0.5M	49La	
6.57	5.65(37°,+1)	-4.65	0-70	T	μ=0.5M	72Mb	c
7.14	5.91(+1)	-2.40	25	T	μ=1.0M	73Ca	
5.990	5.97(+1)	-7.20	25	C	--	60Sa	
6.26	(+1)	-----	25	C	--	71Fb	
P271 PYRIDINE, 3-methyl-			C_6H_7N		$C_5H_4N(CH_3)$		
5.33	5.76(+1)	-7.5	10	C	μ=0	72Cc	
5.87	5.67(+1)	-6.27	25	C	μ=0	74Lb	
6.70	5.63(+1)	-3.28	25	C	μ=0	59Mb	
5.71	5.63(+1)	-6.2	25	C	μ=0	72Cc	
6.39	5.66(+1)	-4.5	25	T	μ=0	68Pa	
4.64	5.79(20°,+1)	-10.64	5-35	T	μ=0	61Eb	
5.96	5.33(+1)	-5.3	40	C	μ=0	72Cc	

ΔH, kcal /mole	pK	ΔS, cal /mole°K	T,°C	M	Conditions	Ref.	Remarks
P271, cont.							
-----	(+1)	-----	10-40	C	μ=0 ΔCp=21	72Cc	
6.04	6.02(+1)	-7.31	25	C	μ=0.5M(KNO_3)	73Be	
4.85	5.52(37°,+1)	-9.56	0-70	T	μ=0.5M	72Mb	c
4.87	5.68(+1)	-9.84	25	T	μ=1.0M	73Ca	
5.640	5.68(+1)	-7.05	25	C	--	60Sa	
P272 PYRIDINE, 4-methyl-		C_6H_7N			$C_5H_4N(CH_3)$		
5.96	6.12(+1)	-7.9	10	C	μ=0	72Cc	
6.13	6.03(+1)	-7.04	25	C	μ=0	74Lb	
7.03	5.97(+1)	-3.77	25	C	μ=0	59Mb	
6.10	5.98(+1)	-6.5	25	C	μ=0	72Cc	
7.56	6.00(+1)	-2.1	25	T	μ=0	68Pa	
6.45	5.67(+1)	-5.3	40	C	μ=0	72Cc	
-----	(+1)	-----	10-40	C	μ=0 ΔCp=25	72Cc	
6.44	6.18(+1)	-6.67	25	C	μ=0.5M(KNO_3)	73Be	
6.89	5.88(37°,+1)	-4.60	0-70	T	μ=0.5M	72Mb	c
6.54	6.04(+1)	-5.91	25	T	μ=1.0M	73Ca	q:±0.43
6.020	6.02(+1)	-7.35	25	C	--	60Sa	
2.93	5.055(25°,+1)	-13.0	15&35	T	μ=0 S: Dimethylformamide	67Ra	b,c,e: NH
39.0	(+1)	-----	25	C	C∿10^{-3}M S: Fluorosulfuric acid	74Ab	
6.7	6.090(25°,+1)	-5.2	0.1&25	T	μ=10^{-3} S: Methanol	65Ra	a,b
P273 PYRIDINE, 3-methylamino		$C_6H_8N_2$			$C_5H_4N(NHCH_3)$		
9.3	8.70(30°)	-9.1	10-40	T	μ=0	60Wa	c
P274 PYRIDINE, 4-methylamino		$C_6H_8N_2$			$C_5H_4N(NHCH_3)$		
11.02	9.66(30°)	-6.62	20	T	μ=0	61Eb	
P275 PYRIDINE, 2-(methyl-aminoethyl)-		$C_8H_{12}N_2$			$C_5H_4N(CH_2CH_2NHCH_3)$		
2.4	3.58(30°,+2)	-8	10-40	T	μ=0	61Ra	c
4.41	4.10(+2)	-4.0	25	C	μ=0.5M(KNO_3)	71Ge	
10.1	9.65(30°,+1)	-11	10-40	T	μ=0	61Ra	c
10.84	10.04(+1)	-9.6	25	C	μ=0.5M(KNO_3)	71Ge	
P276 PYRIDINE, 2-(methyl-aminomethyl)-		$C_7H_{10}N_2$			$C_5H_4N(CH_2NHCH_3)$		
2.34	1.99(+2)	-1.3	25	C	μ=0.5M(KNO_3)	71Ge	
9.43	9.30(10°,+1)	-9(10°)	10-40	T	μ=0	59Ga	c
9.43	8.82(30°,+1)	-9(10°)	10-40	T	μ=0	59Ga	c
9.88	9.09(+1)	-8.5	25	C	μ=0.5M(KNO_3)	71Ge	
P277 PYRIDINE, 2-(methyl-aminomethyl)-6-methyl-		$C_8H_{12}N_2$			$C_5H_3N(CH_2NHCH_3)(CH_3)$		
3.93	3.03(+2)	-0.7	25	C	μ=0.5M(KNO_3)	71Ge	
4.21	8.79(30°,+1)	-3(30°)	10-40	T	μ=0	61Ra	c
9.70	9.15(+1)	-9.4	25	C	μ=0.5M(KNO_3)	71Ge	

ΔH, kcal /mole	pK	ΔS, cal /mole°K	T, °C	M	Conditions	Ref.	Remarks
P278 PYRIDINE, 4(methylamino)-3-methyl-			$C_7H_{10}N_2$		$C_5H_3N(NHCH_3)(CH_3)$		
10.87	9.83	-7.88	20	T	μ=0	61Eb	
P279 PYRIDINE, 4(methylamino)-2,3,5,6-tetramethyl-			$C_{10}H_{16}N_2$		$C_5N(NHCH_3)(CH_3)_4$		
9.96	10.06(20°)	-12.08	5-35	T	μ=0	61Eb	
P280 PYRIDINE, 2-methyl-, 1-oxide			C_6H_7ON		$C_5H_4N(CH_3)(\rightarrow O)$		
1.458	1.029(+1)	9.59	25	T	μ=0	73Ka	
P281 PYRIDINE, 3-methyl-, 1-oxide			C_6H_7ON		$C_5H_4N(CH_3)(\rightarrow O)$		
1.276	0.921(+1)	8.49	25	T	μ=0	73Ka	
P282 PYRIDINE, 4-methyl-, 1-oxide			C_6H_7ON		$C_5H_4N(CH_3)(\rightarrow O)$		
1.250	1.258(+1)	9.94	25	T	μ=0	73Ka	
P283 PYRIDINE, 4-methyl-2-phenyl-			$C_{12}H_{11}N$		$C_5H_3N(CH_3)(C_6H_5)$		
2.91	4.854(25°,+1)	-12.5	15&35	T	μ=0 S: Dimethylformamide	67Ka	b,c
6.5	5.794(25°,+1)	-4.7	0.1&25	T	$\mu=10^{-3}$ S: Methanol	65Ra	b,c
P284 PYRIDINE, 2-nitro-			$C_5H_4O_2N_2$		$C_5H_4N(NO_2)$		
-5.720	-2.06(+1)	-9.76	25	T	μ=0.02M	72Bb	
2.083	2.629(25°,+1)	-3.29	25-90	T	Aqueous H_2SO_4 Solution	74Cb	c
P285 PYRIDINE, 3-nitro-			$C_5H_4O_2N_2$		$C_5H_4N(NO_2)$		
-1.844	0.79(+1)	-9.80	25	T	μ=0.02M	72Bb	
P286 PYRIDINE, 4-nitro-			$C_5H_4O_2N_2$		$C_5H_4N(NO_2)$		
-1.100	1.23(+1)	-9.32	25	T	μ=0.02M	72Bb	
P287 PYRIDINE, 1-oxide			C_5H_5ON		$C_5H_5N(\rightarrow O)$		
1.786	0.686(+1)	9.17	25	T	μ=0	73Ka	
33.4	(+1)	-----	25	C	$C\sim10^{-3}M$ S: Fluorosulfuric acid	74Ab	
0.793	-0.792(25°,+1)	-6.302	25-90	T	Aqueous H_2SO_4 Solution	74Cb	c
P288 PYRIDINE, 2,3,4,5,6-pentachloro-			C_5NCl_5		$C_5N(Cl)_5$		
7.653	6.088(25°,+1)	-2.17	25-90	T	Aqueous H_2SO_4 Solution	74Cb	c

ΔH, kcal /mole	pK	ΔS, cal /mole°K	T,°C	M	Conditions	Ref.	Remarks
P289 PYRIDINE, 2,3,4,5,6-pentachloro-, 1-oxide		C_5ONCl_5			$C_5N(Cl)_5(\to O)$		
3.900	2.591(25°,+1)	-1.200	25-90	T	Aqueous H_2SO_4 Solution	74Cb	c
P290 PYRIDINE, 2-propyl-		$C_8H_{11}N$			$C_5H_4N(C_3H_7)$		
6.83	6.30(+1)	-5.90	25	C	μ=0.5M(KNO_3)	73Be	
P291 PYRIDINE, 2,3,4,5-tetrachloro-		C_5HNCl_4			$C_5HN(Cl)_4$		
2.580	3.211(25°,+1)	-5.57	25-90	T	Aqueous H_2SO_4 Solution	74Cb	c
P292 PYRIDINE, 2,3,5,6-tetramethyl-		$C_9H_{13}N$			$C_5HN(CH_3)_4$		
8.08	7.88(20°,+1)	-8.49	5-35	T	μ=0	61Eb	c
P293 PYRIDINE, 2,3,6-trimethyl-		$C_8H_{11}N$			$C_5H_2N(CH_3)_3$		
8.33	7.60(+1)	-6.84	25	C	μ=0.5M(KNO_3)	73Be	
P294 PYRIDINE, 2,4,6-trimethyl-		$C_8H_{11}N$			$C_5H_2N(CH_3)_3$		
42.7	(+1)	-----	25	C	S: Fluorosulfuric C_L∿10^{-3}M	74Ab	
P295 PYRIDINE, 2,4,6-trimethyl-, 1-oxide		C_8H_8ON			$C_5H_2N(CH_3)_3(\to O)$		
0.456	1.990(+1)	10.64	25	T	μ=0	73Ka	
P296 2-PYRIDINECARBOXALDEHYDE		C_6H_5ON			$C_5H_4N(CHO)$		
3.84	4.24(+1)	-6.5	25	C	μ=0 ΔCp=3 ΔH=3.805 + 0.0031(t-15) (5<t °C<25)	69Ca	
5.0	4.13(+1)	-0.7	5	T	C=10^{-4}M	61Ga	
5.7	4.00(+1)	1.7	15	T	C=10^{-4}M	61Ga	
6.5	3.84(+1)	4.1	25	T	C=10^{-4}M	61Ga	
6.8	3.76(+1)	5.3	30	T	C=10^{-4}M	61Ga	
7.6	3.57(+1)	7.7	40	T	C=10^{-4}M	61Ga	
8.3	3.42(+1)	10.1	50	T	C=10^{-4}M	61Ga	
9.1	3.25(+1)	12.6	60	T	C=10^{-4}M	61Ga	
P297 4-PYRIDINECARBOXALDEHYDE		C_6H_5ON			$C_5H_4N(CHO)$		
4.92	4.74(+1)	-5.2	25	C	μ=0 ΔCp=24 ΔH=4.673 + 0.0243(t-15) (5<t °C<25)	69Ca	
4.82	4.86(+1)	-6.02	25	T	μ=1.0M	73Ca	
P298 2-PYRIDINECARBOXYLIC ACID		$C_6H_5O_2N$			$C_5H_4N(CO_2H)$		
0.52	1.01(+1)	-2.9	25	C	μ=0.13-0.27	69Cc	
2.35	5.32(0)	-16.5	25	C	μ=0	69Cc	
2.48	5.29(0)	-15.9	25	C	μ=0	71Hc	g:69Cc

ΔH, kcal /mole	pK	ΔS, cal /mole$^\circ K$	T,$^\circ$C	M	Conditions	Ref.	Re-marks

P298, cont.

2.45	5.29(0)	−16.0	25	C	μ=0	64Ma	
2.49	(0)	-----	25	C	μ=0.019	69Cc	
2.51	(0)	-----	25	C	μ=0.036	69Cc	
2.72	(0)	-----	25	C	μ=0.190	69Cc	

P299   3-PYRIDINECARBOXYLIC ACID $C_6H_5O_2N$ $C_5H_4N(CO_2H)$

0.75	2.07(+1)	−6.9	25	C	μ=0.01−0.09	69Cc	
2.71	4.81(0)	−12.9	25	C	μ=0	69Cc	
3.08	4.75(0)	−11.4	25	C	μ=0	71Hc	g:69Cc, o
2.57	4.77(0)	−13.2	25	C	μ=0	64Ma	
2.76	(0)	-----	25	C	μ=0.008	69Cc	
2.81	(0)	-----	25	C	μ=0.017	69Cc	
2.97	(0)	-----	25	C	μ=0.168	69Cc	
3.34	3.98(0)	−7.07	25	T	μ=1.0M	73Ca	q:+−0.30

P300   4-PYRIDINECARBOXYLIC ACID $C_6H_5O_2N$ $C_5H_4N(CO_2H)$

0.48	1.84(+1)	−6.8	25	C	μ=0.03−0.14	69Cc	
3.02	4.86(0)	−12.1	25	C	μ=0	69Cc	
3.16	4.84(0)	−11.5	25	C	μ=0	71Hc	g:69Cc, o
3.00	4.84(0)	−12.1	25	C	μ=0	64Ma	
2.99	(0)	-----	25	C	μ=0.008	69Cc	
3.05	(0)	-----	25	C	μ=0.017	69Cc	
3.19	(0)	-----	25	C	μ=0.026	69Cc	
3.20	(0)	-----	25	C	μ=0.168	69Cc	
8.14	4.61(0)	0.60	25	T	μ=1.0M	73Ca	

P301   2-PYRIDINECARBOXYLIC ACID, 4-methyl-, nitrile $C_7H_6N_2$ $C_5H_3N(CN)(CH_3)$

0.77	3.711(25°,+1)	−14.4	15&35	T	μ=0 S: Dimethylformamide	67Ra	b,c
6.5	4.565(25°,+1)	1.0	0.1&25	T	μ=10^{-3} S: Methanol	65Ra	b,c

P302   3-PYRIDINECARBOXYLIC ACID, nitrile $C_6H_4N_2$ $C_5H_4N(CN)$

0.88	1.35(+1)	−3.22	25	C	μ=0	74Lb	
−0.596	1.17(+1)	−7.35	25	T	μ=0 ΔCp=1.74	73Bd	
5.59	1.59(+1)	1.13	25	T	μ=1.0M	73Ca	

P303   4-PYRIDINECARBOXYLIC ACID, nitrile $C_6H_4N_2$ $C_5H_4N(CN)$

1.26	1.86(+1)	−4.29	25	C	μ=0	74Lb	
1.26	1.86(+1)	−4.29	25	C	μ=0	74Lb	
−0.123	1.48(+1)	−7.17	25	T	μ=0 ΔCp=−0.13	73Bd	
3.04	2.26(+1)	0.30	25	T	μ=1.0M	73Ca	

P304   3-PYRROLECARBOXYLIC ACID $C_5H_5O_2N$

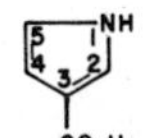

ΔH, kcal /mole	pK	ΔS, cal /mole°K	T, °C	M	Conditions	Ref.	Remarks
P304, cont.							
0.45	4.453	-18.9	25	C	μ=0	69Cc	
0.47	-----	-----	25	C	μ=0.002	69Cc	
0.53	-----	-----	25	C	μ=0.017	69Cc	
0.68	-----	-----	25	C	μ=0.172	69Cc	
P305 PYRROLIDINE			C_4H_9N		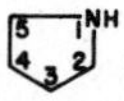		
12.63	12.17(+1)	-9.42	0	T	μ=0	63He	
12.71	11.98(+1)	-9.15	5	T	μ=0	63He	
12.78	11.81(+1)	-8.87	10	T	μ=0	63He	
12.86	11.63(+1)	-8.60	15	T	μ=0	63He	
12.94	11.43(+1)	-8.34	20	T	μ=0	63He	
12.82	11.30(+1)	-8.7	25	C	μ=0	68Ca 71Ca	
13.02	11.305(+1)	-8.05	25	T	μ=0	63He	
13.10	11.15(+1)	-7.79	30	T	μ=0	63He	
13.18	10.99(+1)	-7.53	35	T	μ=0	63He	
13.27	10.84(+1)	-7.24	40	T	μ=0	63He	
13.35	10.70(+1)	-6.98	45	T	μ=0	63He	
13.44	10.56(+1)	-6.72	50	T	μ=0	63He	
12.370	11.11(+1)	-9.30	25	C	--	60Sa	
P306 PYRROLIDINE, 1-methyl			$C_5H_{11}N$		$C_4H_8N(CH_3)$		
9.78	10.45(+1)	-15.0	25	C	μ=0	68Ca	
9.05	10.46(+1)	-17.5	25	C	μ=0	71Ca	
P307 2-PYRROLIDONE, 1-methyl-			C_5H_9ON		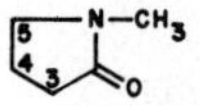		
31.3	-----	-----	25	C	C∿10^{-3}M S: Fluorosulfuric acid	74Ab	

ΔH, kcal /mole	pK	ΔS, cal /mole$^\circ K$	T,$^\circ$C	M	Conditions	Ref.	Re- marks

Q

Q1 QUINAZOLINE, 4-methyl-, 3-oxide $C_9H_8ON_2$ $C_6H_4C_2HN_2(O)(CH_3)$

| 3.960 | 3.728(25°,+$\underline{1}$) | 3.780 | 25-90 | T | Aqueous H_2SO_4 Solution | 74Cb | c |

Q2 QUINOLINE C_9H_7N

| 5.365 | 4.80(+$\underline{1}$) | -3.95 | 25 | C | -- | 60Sa | |
| 37.0 | (+$\underline{1}$) | ----- | 25 | C | C$\sim$10^{-3}M S: Fluorosulfuric acid | 74Ab | |

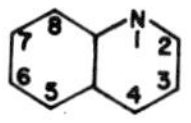

Q3 QUINOLINE, 2-amino- $C_9H_8N_2$ $C_9H_6N(NH_2)$

| 9.8 | 7.34(20°) | -0.5 | 5-35 | T | C=0.01 | 59Eb | c |

Q4 QUINOLINE, 3-amino- $C_9H_8N_2$ $C_9H_6N(NH_2)$

| 5.3 | 4.95(20°) | -4.6 | 5-35 | T | C=0.01 | 59Eb | c |

Q5 QUINOLINE, 4-amino- $C_9H_8N_2$ $C_9H_6N(NH_2)$

| 11.9 | 9.17(20°) | -1.2 | 5-35 | T | C=0.01 | 59Eb | c |

Q6 QUINOLINE, 5-amino- $C_9H_8N_2$ $C_9H_6N(NH_2)$

| 5.6 | 5.46(20°) | -6.2 | 5-35 | T | C=0.01 | 59Eb | c |

Q7 QUINOLINE, 6-amino- $C_9H_8N_2$ $C_9H_6N(NH_2)$

| 6.1 | 5.63(20°) | -5.3 | 5-35 | T | C=0.01 | 59Eb | c |

Q8 QUINOLINE, 8-amino- $C_9H_8N_2$ $C_9H_6N(NH_2)$

| 5.0 | 3.99(20°) | 1.3 | 5-35 | T | C=0.01 | 59Eb | c |

Q9 QUINOLINE, 8-hydroxy- C_9H_7ON $C_9H_6N(OH)$

5.7	(+$\underline{1}$)	-4	25	C	μ=0.1M	65Lb	
5.83	4.13(+$\underline{1}$)	-0.7	25	C	μ=0.1(NaClO$_4$) S: 50 v % Dioxane	68Gd	
6.3	9.76(+$\underline{1}$)	-23	25	T	μ=0.1M	65Lb	
6.41	10.95(+$\underline{1}$)	-28.6	25	C	μ=0.1(NaClO$_4$) S: 50 v % Dioxane	68Gd	

Q10 QUINOLINE, 8-hydroxy-2-methyl $C_{10}H_9ON$ $C_9H_5N(OH)(CH_3)$

| 6.96 | 4.72(+$\underline{1}$) | -1.7 | 25 | C | μ=0.1(NaClO$_4$) S: 50 v % Dioxane | 68Gd | |
| 6.92 | 11.31($\underline{0}$) | -28.5 | 25 | C | μ=0.1(NaClO$_4$) S: 50 v % Dioxane | 68Gd | |

ΔH, kcal /mole	pK	ΔS, cal /mole°K	T, °C	M	Conditions	Ref.	Remarks
Q11 QUINOLINE, 8-hydroxy-4-methyl-			$C_{10}H_9ON$		$C_9H_5N(OH)(CH_3)$		
6.74	4.83(+1)	-0.5	25	C	μ=0.1($NaClO_4$) S: 50 v % Dioxane	68Gd	
6.18	11.10(0)	-30.0	25	C	μ=0.1($NaClO_4$) S: 50 v % Dioxane	68Gd	
Q12 QUINOLINE, 8-mercapto-			C_9H_7NS		$C_9H_6N(SH)$		
1.73	0.77(+1)	2.3	25	C	μ=0.1 S: 50 v % Dioxane	68Gd	
3.02	4.80(0)	-32.1	25	C	μ=0.1 S: 50 v % Dioxane	68Gd	
Q13 QUINOLINE, 8-mercapto-2-methyl-			$C_{10}H_9NS$		$C_9H_5N(SH)(CH_3)$		
2.3	1.96(+1)	1	25	C	μ=0.1($NaClO_4$) S: 50 v % Dioxane	68Gd	
3.58	9.76(0)	-33	25	C	μ=0.1($NaClO_4$) S: 50 v % Dioxane	68Gd	
Q14 QUINOLINE, 2-methyl-8-hydroxy			$C_{10}H_9ON$		$C_9H_5N(CH_3)(OH)$		
6.9	4.58(+1)	2	25	T	μ=0.005	54Ja	
7.5	11.71(0)	-28	25	T	μ=0.005	54Ja	
Q15 QUINOLINE, 4-methyl-8-hydroxy-			$C_{10}H_9ON$		$C_9H_5N(CH_3)(OH)$		
7.1	4.67(+1)	2	25	T	μ=0.005	54Ja	
9.1	11.62(0)	-23	25	T	μ=0.005	54Ja	
Q16 5-QUINOLINESULFONIC ACID, 8-hydroxy-			$C_9H_7O_4NS$		$C_9H_5N(OH)(SO_3H)$		
4.2	4.405(0)	-4.8	0	T	μ=0	57Ua	
4.5	4.104(0)	-3.8	25	T	μ=0	57Ua	
2.8	4.01(0)	-9	25	T	μ=0	58Fa	
5.2	3.813(0)	-2.0	50	T	μ=0	57Ua	
4.27	(0)	-----	25	C	μ=0.1M(KNO_3)	71Gd	
9.2	3.97(0)	-25.8	25	C	μ=0.1(KNO_3)	70Kb	
4.37	3.88(0)	3.1	25	C	μ=0.1($NaClO_4$)	68Gd	
4.04	3.56(0)	2.7	25	C	μ=0.1($NaClO_4$) S: 50 v % Dioxane	68Gd	
5.9	9.125(-1)	-20	0	T	μ=0	57Ua	
5.0	8.750(-1)	-23	25	T	μ=0	57Ua	
5.8	8.53(-1)	-20	25	T	μ=0	58Fa	
3.8	8.491(-1)	-27	50	T	μ=0	57Ua	
4.00	8.43(-1)	25.2	25	C	μ=0.1($NaClO_4$)	68Gd	
4.95	9.85(-1)	28.5	25	C	μ=0.1($NaClO_4$) S: 50 v % Dioxane	68Gd	
Q17 5-QUINOLINESULFONIC ACID, 8-hydroxy-7-iodo-			$C_9H_6O_4NIS$		$C_9H_4N(OH)(I)(SO_3H)$		
1.3	2.61(0)	-7.2	0	T	μ=0	57Ua	
1.5	2.51(0)	-6.4	25	T	μ=0	57Ua	
1.8	2.42(0)	-6.0	50	T	μ=0	57Ua	
6.1	7.768(-1)	-13.5	0	T	μ=0	57Ua	
4.1	7.417(-1)	-21	25	T	μ=0	57Ua	
1.0	7.267(-1)	-29	50	T	μ=0	57Ua	

ΔH, kcal /mole	pK	ΔS, cal /mole°K	T,°C	M	Conditions	Ref.	Remarks
Q17, cont.							
7.14	7.22(25°,-1)	-9.10	25-45	T	μ=0.1M(KNO_3)	73La	c
5.72	7.93(25°,-1)	-17.20	25-45	T	μ=0.1M(KNO_3) S: 30 v % Acetone	73La	c
5.80	7.96(25°,-1)	-17.10	25-45	T	μ=0.1M(KNO_3) S: 30 v % 1,4-Dioxane	73La	c
6.71	7.61(25°,-1)	-12.10	25-45	T	μ=0.1M(KNO_3) S: 30 v % Ethanol	73La	c
3.51	7.05(25°,-1)	-18.40	25-45	T	μ=0.1M(KNO_3) S: 30 v % Glycerol	73La	c
3.09	7.09(25°,-1)	-22.70	25-45	T	μ=0.1M(KNO_3) S: 30 v % Glycol	73La	c
5.86	7.90(25°,-1)	-16.40	25-45	T	μ=0.1M(KNO_3) S: 30 v % Isopropanol	73La	c
6.76	7.49(25°,-1)	-11.40	25-45	T	μ=0.1M(KNO_3) S: 30 v % Methanol	73La	c

Q18 5-QUINOLINESULFONIC ACID, 8-hydroxy-7-nitro- $C_9H_6O_6N_2S$ $C_9H_4N(OH)(NO_2)(SO_3H)$

ΔH	pK	ΔS	T	M	Conditions	Ref.	Remarks
-0.2	1.93(0)	-9.3	0	T	μ=0	57Ua	
-0.2	1.94(0)	-9.6	25	T	μ=0	57Ua	
-0.4	1.94(0)	-10.0	50	T	μ=0	57Ua	
2.5	5.919(-1)	-18	0	T	μ=0	57Ua	
2.6	5.750(-1)	-18	25	T	μ=0	57Ua	
2.6	5.605(-1)	-18	50	T	μ=0	57Ua	

Q19 5-QUINOLINESULFONIC ACID, 7 (4-nitrophenylazo)-8-hydroxy- $C_{15}H_{10}O_6N_4S$ $C_9H_4N(N{:}NC_6H_4NO_2)(OH)(SO_3H)$

ΔH	pK	ΔS	T	M	Conditions	Ref.	Remarks
4.5	3.41(0)	0.9	0	T	μ=0	57Ua	
3.5	3.14(0)	-2.7	25	T	μ=0	57Ua	
2.1	2.98(0)	-7.2	50	T	μ=0	57Ua	
4.9	7.811(-1)	-18	0	T	μ=0	57Ua	
4.5	7.495(-1)	-19	25	T	μ=0	57Ua	
3.7	7.262(-1)	-22	50	T	μ=0	57Ua	

Q20 5-QUINOLINESULFONIC ACID, 7-phenylazo-8-hydroxy- $C_{15}H_{11}O_4N_3S$ $C_9H_4N(N{:}NC_6H_5)(OH)(SO_3H)$

ΔH	pK	ΔS	T	M	Conditions	Ref.	Remarks
4.0	3.65(0)	-2.1	0	T	μ=0	57Ua	
3.1	3.41(0)	-5.2	25	T	μ=0	57Ua	
1.1	3.27(0)	-10	50	T	μ=0	57Ua	
5.2	8.183(-1)	-18	0	T	μ=0	57Ua	
4.6	7.850(-1)	-21	25	T	μ=0	57Ua	
3.4	7.620(-1)	-25	50	T	μ=0	57Ua	

Q21 QUINUCLIDINE $C_7H_{13}N$

ΔH	pK	ΔS	T	M	Conditions	Ref.	Remarks
45.8	(+1)	-----	25	C	C∿10^{-3}M S: Fluorosulfuric acid	74Ab	

ΔH, kcal /mole	pK	ΔS, cal /mole°K	T, °C	M	Conditions	Ref.	Remarks
			R				
R1 RIBONUCLEASE							
7.0	9.92(25°)	-22	6&25	T	μ=0.15	55Ta	b,c,e: phenolic group
R2 RIBOSE			$C_5H_{10}O_5$		$CH_2CH(OH)CH(OH)CH(OH)CHOH$ (ring via O)		
9.06	12.54(0)	-25.4	10	C	μ=0	70Cd	
8.62	12.11(0)	-26.5	25	C	μ=0	70Cd	
8.1	12.22(0)	-28.7	25	C	μ=0	66Ib	
R3 RIBOSE, 2-deoxy-			$C_5H_{10}O_4$		$CH_2CH(OH)CH(OH)CH_2CHOH$ (ring via O)		
9.0	12.98(0)	-27.6	10	C	μ=0	70Cd	
7.7	12.67(0)	-32.1	25	C	μ=0	66Ib	
8.2	12.61(0)	-30.3	25	C	μ=0	70Cd	
R4 RIBOSE, 5-phosphoric acid			$C_5H_{11}O_8P$		see R2		
-2.7	6.70(-1)	-40	25	C	μ=0	62Cc	
6.1	13.05(-2)	-39.4	25	C	μ=0	65Ib	
R5 RIBOFLAVIN			$C_{17}H_{20}O_6N_4$				
8.3	9.69(25°)	-16	10-40	T	μ=0.01	59Ha	

ΔH, kcal /mole	pK	ΔS, cal /mole°K	T, °C	M	Conditions	Ref.	Remarks

S

S1 SARCOSINE $C_3H_7O_2N$ $CH_3NH_2CH_2CO_2H$

ΔH, kcal /mole	pK	ΔS, cal /mole°K	T, °C	M	Conditions	Ref.	Remarks
1.72	2.12(+1)	-4	25	C	μ=0	64Ic	
9.74	10.71(0)	-14.0	5	T	μ=0	58Da	
9.71	10.45(0)	-14.1	15	C	μ=0	58Da	
9.75	10.20(0)	-13	25	C	μ=0	64Ic	
9.68	10.20(0)	-14.2	25	C	μ=0	58Da	
9.65	9.97(0)	-14.3	35	C	μ=0	58Da	
9.62	9.76(0)	-14.4	45	C	μ=0	58Da	
9.740	10.18(0)	-13.72	25	C	--	67Ac	

S2 SARCOSINE, N-glycyl- $C_5H_{11}O_3N_2$ $H_2NCH_2CONHCH_2(CH_3)CO_2H$

ΔH, kcal /mole	pK	ΔS, cal /mole°K	T, °C	M	Conditions	Ref.	Remarks
10.2	8.55(0)	-4.9	25	C	μ=0.1M(KNO_3)	72Bi	

S3 SELENIC ACID H_2O_4Se H_2SeO_4

ΔH, kcal /mole	pK	ΔS, cal /mole°K	T, °C	M	Conditions	Ref.	Remarks
5.57	-1.65(-1)	26.3	25	T	μ=0 Sign on ΔH, pK and ΔS changed.	64Na	
5.7	-1.74(25°,-1)	27.1	0-45	T	μ=0	70Ga	c

S4 SELENIOUS ACID H_2O_3Se H_2SeO_3

ΔH, kcal /mole	pK	ΔS, cal /mole°K	T, °C	M	Conditions	Ref.	Remarks
-1.50	2.27(0)	-15.4	25	C	μ=1M($NaClO_4$)	72Aa	
-1.26	2.61(0)	-16.2	25	C	μ=3M($NaClO_4$)	72Aa	
-0.74	2.52(0)	-14.1	25	C	μ=3M($LiClO_4$)	72Aa	
2.8	8.39(-1)	-29	25	T	μ=0	70Ta	q:±0.2
1.20	7.78(-1)	-31.6	25	C	μ=1M($NaClO_4$)	72Aa	
1.26	8.05(-1)	-32.6	25	C	μ=3M($NaClO_4$)	72Aa	
1.85	7.66(-1)	-28.8	25	C	μ=3M($LiClO_4$)	72Aa	
0.62	8.01	-33.2	25	C	μ=1M($NaClO_4$) HA=$[H_2(SeO_3)_2]^{2-}$	72Aa	q:±0.29
1.14	7.79	-31.4	25	C	μ=3M($NaClO_4$) HA=$[H_2(SeO_3)_2]^{2-}$	72Aa	q:±0.33
-1.53	2.11	-14.8	25	C	μ=1M($NaClO_4$) HA=$[H_4(SeO_3)_2]$	72Aa	q:±0.29
0.10	1.89	-8.3	25	C	μ=3M($NaClO_4$) HA=$[H_4(SeO_3)_2]$	72Aa	q:±0.5
-1.92	1.83	-14.8	25	C	μ=3M($LiClO_4$) HA=$[H_4(SeO_3)_2]$	72Aa	q:±0.68

S5 SEMICARBAZIDE CH_5ON_3 $H_2NCONHNH_2$

ΔH, kcal /mole	pK	ΔS, cal /mole°K	T, °C	M	Conditions	Ref.	Remarks
6.08	3.53	3.9	25	C	μ=0.1	69Ga	

S6 SEMICARBAZIDE, 1,1-diacetic acid acid $C_5H_9O_5N_3$ $H_2NC(O)NHN(CO_2H)_2$

ΔH, kcal /mole	pK	ΔS, cal /mole°K	T, °C	M	Conditions	Ref.	Remarks
-0.2	2.96	-14	30	C	μ=0.1(KNO_3)	67Gb	
-0.6	4.04	-20	30	C	μ=0.1(KNO_3)	67Gb	

S7 SEMICARBAZIDE, 1,1-diacetic acid, 3-thio- $C_5H_9O_4N_3S$ $H_2NC(S)NHN(CH_2CO_2H)_2$

ΔH, kcal /mole	pK	ΔS, cal /mole°K	T, °C	M	Conditions	Ref.	Remarks
-0.04	2.94	-14	30	C	μ=0.1(KNO_3)	67Gb	
-0.2	4.07	-19	30	C	μ=0.1(KNO_3)	67Gb	

ΔH, kcal /mole	pK	ΔS, cal /mole°K	T,°C	M	Conditions	Ref.	Remarks
S8 SEMICARBAZIDE, 3-seleno			CH_5N_3Se		$H_2NNHC(Se)NH_2$		
4.5	0.8	11.2	25	C	μ=0.1	69Ga	
S9 SEMICARBAZIDE, 3-thio-			CH_6N_3S		$H_2NNHC(SH)NH_2$		
4.53	1.5	8.1	25	C	μ=0.1	69Ga	
S10 SERINE			$C_3H_7O_3N$		$HOCH_2CH(NH_2)CO_2H$		
1.981	2.296(+1)	-3.3	1	T	μ=0	42Sa	
1.721	2.232(+1)	-4.2	12.5	T	μ=0	42Sa	
1.320	2.187(+1)	-5.57	25	T	μ=0	60Kb	
1.366	2.186(+1)	-5.4	25.0	T	μ=0	42Sa	
0.932	2.154(+1)	-6.8	37.5	T	μ=0	42Sa	
0.411	2.132(+1)	-8.5	50.0	T	μ=0	42Sa	
1.2	2.15(+1)	-5.81	25	C	μ=0.05(KCl)	72Gc	
10.450	9.880(0)	-7.0	1	T	μ=0	42Sa	
10.490	9.542(0)	-6.9	12.5	T	μ=0	42Sa	
10.350	9.209(0)	-7.41	25	T	μ=0	60Kb	
10.405	9.208(0)	-7.2	25.0	T	μ=0	42Sa	
10.200	8.904(0)	-7.9	37.5	T	μ=0	42Sa	
9.840	8.628(0)	-9.1	50.0	T	μ=0	42Sa	
9.92	(0)	-----	20	C	μ≈0.01	71Ma	e: $\alpha\text{-}NH_3^+$
10.2	9.10(0)	-7.42	25	C	μ=0.05(KCl)	72Gc	
10.46	9.14(0)	-6.3	25	C	μ=0.1M(KNO_3)	72Ia	
10.14	9.260(25°,0)	-8.4	25-50	T	μ=0.1M(KCl)	70Ha	c
10.05	9.18(0)	8.3	25	C	μ=0.16	70Lb	
S11 SERINE, glycyl-			$C_5H_{10}O_4N_2$		$H_2NCH_2CONHCH(CH_2OH)CO_2H$		
0.189	2.9808	-13.0	25	T	μ=0	57Ka	
S12 SERINE, methyl ester			$C_4H_9O_3N$		$HOCH_2CH(NH_2)CO_2CH_3$		
10.90	7.030(25°,0)	4.4	25-50	T	μ=0.1M(KCl)	70Ha	c,e:NH_3^+
S13 SPERM WHALE FERRIMYOGLOBIN							
6.20	8.94(25°, -0.3)	-20	25-34	T	μ=0	67Ha	c
S14 SPINACEAMINE			$C_6H_9N_3$		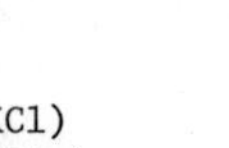		
6.12	4.895(+2)	-1.9	25	T	μ=0.1M(KCl)	73Bg	
9.97	8.904(+1)	-7.3	25	T	μ=0.1M(KCl)	73Bg	
S15 SPINACINE			$C_7H_9O_2N_3$				

ΔH, kcal /mole	pK	ΔS, cal /mole°K	T, °C	M	Conditions	Ref.	Remarks
S15, cont.							
-1.55	1.649(+2)	-12.7	25	T	μ=0.1M(KCl)	73Bg	
5.30	4.936(+1)	-4.83	25	T	μ=0.1M(KCl)	73Bg	
8.29	8.663(0)	-11.8	25	T	μ=0.1M(KCl)	73Bg	
S16 SUCCINIC ACID			$C_4H_6O_4$		$HO_2CCH_2CH_2CO_2H$		
1.526	4.2845(0)	-14.0	0	T	μ=0	50Pb	
1.378	4.2631(0)	-14.6	5	T	μ=0	50Pb	
1.228	4.2449(0)	-15.1	10	T	μ=0	50Pb	
1.075	4.2316(0)	-15.6	15	T	μ=0	50Pb	
0.920	4.2176(0)	-16.2	20	T	μ=0	50Pb	
0.80	4.207(0)	-16.6	25	C	μ=0	67Ca	
0.602	4.163(0)	-17.0	25	C	μ=0	48Cb	
0.761	4.2066(0)	-16.7	25	T	μ=0	50Pb	
0.601	4.1980(0)	-17.2	30	T	μ=0	50Pb	
0.437	4.1914(0)	-17.8	35	T	μ=0	50Pb	
0.270	4.1878(0)	-18.3	40	T	μ=0	50Pb	
0.101	4.1869(0)	-18.8	45	T	μ=0	50Pb	
-0.070	4.1863(0)	-19.4	50	T	μ=0	50Pb	
6.11	6.00(0)	-5.39	25	T	μ=0 S: Formamide	75Db	
1.144	5.674(-1)	-21.8	0	T	μ=0	50Pa	
0.902	5.660(-1)	-22.6	5	T	μ=0	50Pa	
0.656	5.649(-1)	-23.5	10	T	μ=0	50Pa	
0.406	5.642(-1)	-24.4	15	T	μ=0	50Pa	
0.151	5.639(-1)	-25.3	20	T	μ=0	50Pa	
0.042	5.607(-1)	-25.5	25	C	μ=0	48Cb	
0.06	5.635(-1)	-25.6	25	C	μ=0	67Ca	
-0.107	5.635(-1)	-26.1	25	T	μ=0	50Pa	
-0.371	5.641(-1)	-27.0	30	T	μ=0	50Pa	
-0.639	5.647(-1)	-27.9	35	T	μ=0	50Pa	
-0.911	5.654(-1)	-28.8	40	T	μ=0	50Pa	
-1.188	5.669(-1)	-29.6	45	T	μ=0	50Pa	
-1.469	5.680(-1)	-30.5	50	T	μ=0	50Pa	
7.09	8.03(-1)	-12.96	25	T	μ=0 S: Formamide	75Db	
0.275	3.817(-1)	-16.5	25	C	μ=0 HA=[Co(III)$(NH_3)_5$ (Succinic Acid)]	74La	
S17 SUCCINIC ACID (DL), imide-			$C_4H_5O_2N$		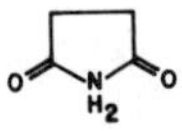		
6.510	9.623	-22.0	25	T	μ=0	52Wa	
S18 SUCCINIC ACID (DL), rac-2,3-di-tert-butyl-			$C_{12}H_{24}O_4$		$HO_2CCH_2[C(CH_3)_3]CH_2[C(CH_3)_3]$-$CO_2H$		
-0.94	3.58(0)	-19	25	C	μ=0.1	63Ea	
4.0	10.2(-1)	-33	25	C	μ=0.1	63Ea	
S19 SUCCINIC ACID, 2,3-dimethyl			$C_6H_{10}O_4$		$HO_2CCH(CH_3)CH(CH_3)CO_2H$		
-0.30	3.67(0)	-17.8	25	T	μ=0 HA=meso isomer of 2,3-dimethylsuccinic acid	73Pb	
-0.35	3.82(0)	-18.7	25	T	μ=0 HA=dl isomer of 2,3-dimethylsuccinic acid	73Pb	
-3.77	5.30(-1)	-36.9	25	T	μ=0 HA=meso isomer of 2,3-dimethylsuccinic acid	73Pb	

ΔH, kcal /mole	pK	ΔS, cal /mole°K	T, °C	M	Conditions	Ref.	Remarks
S19, cont.							
-2.25	5.93(-1)	-34.7	25	T	μ=0 HA=dl isomer of 2,3-dimethylsuccinic acid	73Pb	
S20 SUCCINIC ACID (DL), 2-hydroxy- $C_4H_6O_5$					$HO_2CCH(OH)CH_2CO_2H$		
1.592	3.537(0)	-10.4	0	T	μ=0	59Ea	
1.421	3.520(0)	-11.0	5	T	μ=0	59Ea	
1.247	3.494(0)	-11.6	10	T	μ=0	59Ea	
1.070	3.482(0)	-12.2	15	T	μ=0	59Ea	
0.890	3.472(0)	-12.8	20	T	μ=0	59Ea	
0.706	3.458(0)	-13.4	25	T	μ=0	59Ea	
0.520	3.452(0)	-14.1	30	T	μ=0	59Ea	
0.330	3.446(0)	-14.7	35	T	μ=0	59Ea	
0.138	3.444(0)	-15.3	40	T	μ=0	59Ea	
-0.058	3.446(0)	-15.9	45	T	μ=0	59Ea	
-0.257	3.445(0)	-16.6	50	T	μ=0	59Ea	
0.983	5.119(-1)	-19.8	0	T	μ=0	59Ea	
0.738	5.108(-1)	-20.7	5	T	μ=0	59Ea	
0.490	5.098(-1)	-21.6	10	T	μ=0	59Ea	
0.237	5.096(-1)	-22.5	15	T	μ=0	59Ea	
-0.066	5.096(-1)	-23.4	20	T	μ=0	59Ea	
-0.282	5.097(-1)	-24.2	25	T	μ=0	59Ea	
-0.549	5.099(-1)	-25.1	30	T	μ=0	59Ea	
-0.819	5.104(-1)	-26.0	35	T	μ=0	59Ea	
-1.094	5.117(-1)	-26.9	40	T	μ=0	59Ea	
-1.374	5.133(-1)	-27.8	45	T	μ=0	59Ea	
-1.659	5.149(-1)	-28.7	50	T	μ=0	59Ea	
S21 SULFAMIC ACID			H_3O_3NS		H_2NSO_3H		
0.41	0.998	-3.2	25	T	μ=0 ΔCp=-80	69Kb	
S22 SULFIDE, di-n-butyl-			$C_8H_{18}S$		$(CH_3CH_2CH_2CH_2)_2S$		
18.5	-----	-----	25	C	C∼10^{-3}M S: Fluorosulfuric acid	74Ab	
S23 SULFIDE, diethyl-			$C_4H_{10}S$		$(CH_3CH_2)_2S$		
19.0	-----	-----	25	C	C∼10^{-3}M S: Fluorosulfuric acid	74Ab	
S24 SULFIDE, diethyl-, 2,2'-diamino-N,N,N',N'-tetraacetic acid			$C_{12}H_{20}O_8N_2S$		$(HO_2CH_2C)_2N(CH_2)_2S(CH_2)_2$-$N(CH_2CO_2H)_2$		
6.59	8.47(-2)	-16.3	20	C	0.1M(KNO_3)	64Ab	
6.69	9.42(-3)	-20.3	20	C	0.1M(KNO_3)	64Ab	
S25 SULFIDE, dimethyl-, 1-chloro-			C_2H_5ClS		H_3CSCH_2Cl		
11.9	-----	-----	25	C	C∼10^{-3}M S: Fluorosulfuric acid	74Ab	

ΔH, kcal /mole	pK	ΔS, cal /mole°K	T, °C	M	Conditions	Ref.	Remarks
S26 SULFIDE, diphenyl-			$C_{12}H_{10}S$		$(C_6H_5)_2S$		
7.6	-----	-----	25	C	C∿10^{-3}M S: Fluorosulfuric acid	74Ab	
S27 SULFIDE, methyl phenyl-			C_7H_8S		$H_3CSC_6H_5$		
13.2	-----	-----	25	C	C∿10^{-3}M S: Fluorosulfuric acid	74Ab	
S28 SULFONE, 4-aminophenyl (4-chlorophenyl)			$C_{12}H_{10}O_2NCl$		$H_2NC_6H_4SO_2C_6H_4Cl$		
5.01	1.38	10.5	25	T	μ=0	57Sa	
S29 SULFONE, tetramethylene-			$C_4H_8O_2S$		$\underline{CH_2CH_2CH_2CH_2}SO_2$ (ring)		
11.8	-----	-----	25	C	C∿10^{-3}M S: Fluorosulfuric acid	74Ab	
S30 SULFOXIDE, di-n-butyl-			$C_8H_{18}OS$		$CH_3CH_2CH_2CH_2SOCH_2CH_2CH_2CH_3$		
29.5	-----	-----	25	C	C∿10^{-3}M S: Fluorosulfuric acid	74Ab	
S31 SULFOXIDE, dimethyl-			C_2H_6OS		CH_3SOCH_3		
28.6	-----	-----	25	C	C∿10^{-3}M S: Fluorosulfuric acid	74Ab	
S32 SULFOXIDE, methyl phenyl-			C_7H_8OS		$CH_3SOC_6H_5$		
25.2	-----	-----	25	C	C∿10^{-3}M S: Fluorosulfuric acid	74Ab	
S33 SULFOXIDE, tetramethylene-			C_4H_8OS		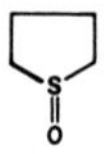		
29.5	-----	-----	25	C	C∿10^{-3}M S: Fluorosulfuric acid	74Ab	
S34 SULFURIC ACID			H_2O_4S		H_2SO_4		
-2.44	1.778(-1)	-17.1	0	T	μ=0 ΔCp=-56	66Ma	
-2.290	1.812(-1)	-16.54	4.34	C	μ=0 ΔCp=-80.1	69Rb	
-1.400	(-1)	-----	10	T	μ=0	34Hc	
-3.150	1.894(-1)	-19.60	15.0	C	μ=0 ΔCp=-88.1	69Rb	
-4.120	1.987(-1)	-22.91	24.99	C	μ=0 ΔCp=-85.2	69Rb	
-4.86	2.00(-1)	-25.4	25	C	μ=0	69Ia	
-5.6	1.91(-1)	-27.5	25	C	μ=0	66Ca	
-5.200	1.920(-1)	-26.3	25	C	μ=0	37Pa	
-5.41	1.930(-1)	-27.0	25	C	μ=0	69Wb	
-4.911	1.987(-1)	-25.6	25	T	μ=0	61Lb	
-3.85	1.988(-1)	-22.0	25	T	μ=0 ΔCp=-57	66Ma	

ΔH, kcal /mole	pK	ΔS, cal /mole°K	T, °C	M	Conditions	Ref.	Re-marks
S34, cont.							
-2.229	1.914(-1)	-16.2	25	T	μ=0 ΔCp=-96.9	39Hb	
-2.150	(-1)	-----	25	T	μ=0	34Hc	
-4.922	2.094(-1)	-25.56	34.96	C	μ=0 ΔCp=-78.7	69Rb	
-5.20	1.98(25°,-1)	-26.5	20-35	T	μ=0	52Da	
-3.040	(-1)	-----	40	T	μ=0	34Hc	
-5.238	1.959(25°,-1)	-26.5	0-45	T	μ=0	69Sa	
-5.6	1.96(25°,-1)	-27.7	0-45	T	μ=0	58Na	
-5.30	2.246(-1)	-26.7	50	T	μ=0 ΔCp=-59	66Ma	
-6.214	2.301(-1)	-29.8	50	T	μ=0	61Lb	
-6.637	2.404(-1)	-30.92	60.03	C	μ=0 ΔCp=-62.8	69Rb	
-7.623	2.636(-1)	-34.0	75	T	μ=0	61Lb	
-6.79	2.539(-1)	-31.1	75	T	μ=0 ΔCp=-60	66Ma	
-8.325	2.754(-1)	-35.81	85.62	C	μ=0 ΔCp=-55.1	69Rb	
-8.808	2.885(-1)	-37.13	95.0	C	μ=0 ΔCp=-53.9	69Rb	
-8.31	2.855(-1)	-35.3	100	T	μ=0 ΔCp=-62	66Ma	
-9.136	2.987(-1)	-38.2	100	T	μ=0	61Lb	
-10.750	3.352(-1)	-42.4	125	T	μ=0	61Lb	
-11.5	3.534(-1)	-43.3	150	T	μ=0 ΔCp=-65	66Ma	
-12.480	3.728(-1)	-46.5	150	T	μ=0	61Lb	
-14.310	4.113(-1)	-50.7	175	T	μ=0	61Lb	
-16.240	4.506(-1)	-54.9	200	T	μ=0	61Lb	
-14.8	4.246(-1)	-50.6	200	T	μ=0 ΔCp=-67	66Ma	
-18.280	4.905(-1)	-59.1	225	T	μ=0	61Lb	
-18.2	4.971(-1)	-57.6	250	T	μ=0 ΔCp=-70	66Ma	
-21.8	5.698(-1)	-64.1	300	T	μ=0 ΔCp=-73	66Ma	
-25.6	6.421(-1)	-70.4	350	T	μ=0 ΔCp=-76	66Ma	
-27.1	6.708(-1)	-72.8	370	T	μ=0 ΔCp=-78	66Ma	
-5.61	1.06(-1)	-23.7	25	C	μ=1.00M($NaClO_4$)	71Ab	
-5.39	(-1)	-----	25	T	μ=1.0(HBr + NaBr)	64Fa	
-5.54	1.08(-1)	-23.5	25	C	μ=2.0	59Zc	
-6.15	2.86(25°,-1)	-34.2	20-37	T	μ=0 S: 20 w % Dioxane	71Cb	c
-7.1	4.03(25°,-1)	-42.6	20-37	T	μ=0 S: 40 w % Dioxane	71Cb	c
-8.0	4.62(27°,-1)	-48.0	20-37	T	μ=0 S: 60 w % Dioxane	71Cb	c
S35 SULFURIC ACID, monamide-			H_4O_3NS		H_2NSO_3H		
0.47	0.988(0)	-3.0	25	C	μ=0	69Cc	
0.25	1.01(0)	-3.8	25	C	μ=0	65Hc	
0.460	0.988(25°,0)	-3.0	0-50	T	μ=0	52Kb	c
S36 SULFUROUS ACID			H_2O_3S		H_2SO_3		
-3.860	1.997(0)	-22.1	25	T	μ=0	34Ja	
-2.91	7.29(-1)	-42.2	25	T	μ=0	69La	
-2.7	7.20(-1)	-42	25	T	μ=0	72Hb	q:±0.3
S37 SULFUROUS ACID, dimethyl ester			$C_2H_6O_3S$		$CH_3OS(O)OCH_3$		
16.0	-----	-----	25	C	C∿10^{-3}M S: Fluorosulfuric acid	74Ab	

ΔH, kcal /mole	pK	ΔS, cal /mole°K	T,°C	M	Conditions	Ref.	Remarks

T

T1 TARTARIC ACID $C_4H_6O_6$ $HO_2CCH(OH)CH(OH)CO_2H$

ΔH, kcal /mole	pK	ΔS, cal /mole°K	T,°C	M	Conditions	Ref.	Remarks
-0.76	3.07(0)	-17	25	T	μ=0	72Dc	
0.741	3.036(0)	-11.4	25	T	μ=0	51Bb	
0.81	3.17(0)	-11.8	25	T	μ=0	73Pb	
-0.76	4.18(-1)	-16.7	15-35	T	μ=0	72Dc	c
1.01	4.39(-1)	-16.5	10	T	μ=0	51Bb	
0.753	4.381(-1)	-17.4	15	T	μ=0	51Bb	
0.497	4.372(-1)	-18.3	20	T	μ=0	51Bb	
0.237	4.366(-1)	-19.2	25	T	μ=0	51Bb	
0.14	4.26(-1)	-19	25	T	μ=0	72Dc	
1.48	4.91(-1)	-17.5	25	T	μ=0	73Pb	
-0.029	4.365(-1)	-20.1	30	T	μ=0	51Bb	
-0.299	4.367(-1)	-21.0	35	T	μ=0	51Bb	
0.14	5.81(-1)	-18.9	15-35	T	μ=0	72Dc	c
-0.571	4.372(-1)	-21.8	40	T	μ=0	51Bb	
5.4	7.17(25°)	-14.7	15-35	T	μ=0.01M R: $H_2A=2H^+ + A^{2-}$	73Kc	c

T2 TAURINE $C_2H_7O_3NS$ $H_2NCH_2CH_2SO_3H$

ΔH, kcal /mole	pK	ΔS, cal /mole°K	T,°C	M	Conditions	Ref.	Remarks
9.99	9.06	-7.95	25	C	μ=0	65Hc	
10.000	9.0614	7.91	25	T	μ=0	54Ka	
11.10	9.18(25°)	-5.10	5-45	T	μ=0	62Da	c

T3 TELLURIC ACID H_6O_6Te H_6TeO_6

ΔH, kcal /mole	pK	ΔS, cal /mole°K	T,°C	M	Conditions	Ref.	Remarks
5.8	7.70(25°,0)	-15.9	5-61	T	μ=0	62Ea	c
9.6	7.62(25°,0)	-2.7	25	T	μ=0.1KCl	56Ab	
9.4	10.99(25°,-1)	-18.6	5-61	T	μ=0	62Ea	c

T4 TETRACYCLINE HYDROCHLORIDE $C_{22}H_{25}O_8N_2Cl$

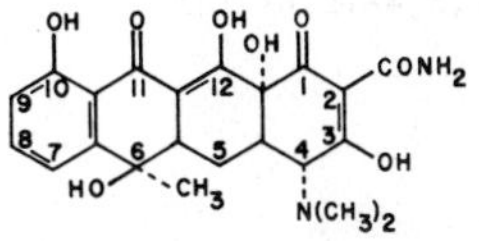

ΔH, kcal /mole	pK	ΔS, cal /mole°K	T,°C	M	Conditions	Ref.	Remarks
2.05	3.34	-8.4	25	C	μ=0.01	66Ba	

T5 TETRACYCLINE HYDROCHLORIDE, chloro- $C_{22}H_{24}O_8N_2Cl_2$ see T4

ΔH, kcal /mole	pK	ΔS, cal /mole°K	T,°C	M	Conditions	Ref.	Remarks
2.35	3.27	-7.1	25	C	μ=0.01	66Ba	

T6 TETRACYCLINE HYDROCHLORIDE, dimethylchloro- $C_{24}H_{28}O_8N_2Cl_2$ see T4

ΔH, kcal /mole	pK	ΔS, cal /mole°K	T, °C	M	Conditions	Ref.	Remarks
T6, cont.							
2.35	3.30	-7.2	25	C	μ=0.01	66Ba	
T7 TETRACYCLINE HYDROCHLORIDE, 4-epi-anhydro-			$C_{22}H_{25}O_8N_2Cl$		see T4		
1.95	3.48	-9.4	25	C	μ=0.01	66Ba	
T8 TETRACYCLINE HYDROCHLORIDE, 4-epi-chloro-			$C_{22}H_{24}O_8N_2Cl_2$		see T4		
2.25	3.59	-8.9	25	C	μ=0.01	66Ba	
T9 TETRAETHYLENEPENTAMINE			$C_8H_{23}N_5$		$H_2NCH_2(CH_2NHCH_2)_3CH_2NH_2$		
6.83	2.98(+5)	9.3	25	C	0.1M(KCl)	64Pa	
7.89	4.72(+4)	4.9	25	C	0.1M(KCl)	64Pa	
10.71	8.08(+3)	-1.0	25	C	0.1M(KCl)	64Pa	
11.32	9.10(+2)	-3.7	25	C	0.1M(KCl)	64Pa	
10.76	9.67(+1)	-8.2	25	C	0.1M(KCl)	64Pa	
T10 1,2,3,4-TETRAZOLE			CH_2N_4		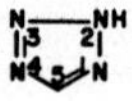		
3.09	4.90	-12.1	25	C	μ=0	70Hb	
T11 1,2,3,4-TETRAZOLE, 5-cyclopropyl-			$C_4H_6N_4$		$CHN_4(C_3H_5)$		
3.66	5.41	-12.5	25	C	μ=0	70Hb	
T12 1,2,3,4-TETRAZOLE, 5-hydroxy-			CH_2ON_4		$CHN_4(OH)$		
3.87	5.40	-11.7	25	C	μ=0	70Hb	
6.19	10.26	-26.3	25	C	μ=0	70Hb	
T13 1,2,3,4-TETRAZOLE, 5-methyl-			$C_2H_4N_4$		$CHN_4(CH_3)$		
3.32	5.63	-14.6	25	C	μ=0	70Hb	
T14 1,2,3,4-TETRAZOLE, 5-methylthio-			$C_2H_4N_4S$		$CHN_4(SCH_3)$		
2.71	4.00	-9.2	25	C	μ=0	70Hb	

ΔH, kcal /mole	pK	ΔS, cal /mole°K	T, °C	M	Conditions	Ref.	Remarks
T15 1,2,3,4-TETRAZOLE, 5-phenoxy-			$C_7H_6ON_4$		$CHN_4(OC_6H_5)$		
2.59	3.49	-7.3	25	C	μ=0	70Hb	
T16 1,2,3,4-TETRAZOLE, 5-phenyl-			$C_7H_6N_4$		$CHN_4(C_6H_5)$		
3.20	4.38	-9.3	25	C	μ=0	70Hb	
T17 1,2,3,4-TETRAZOLE, 5-trifluoromethyl-			$C_2HN_4F_3$		$CHN_4(CF_3)$		
-1.13	1.70	-11.2	25	C	μ=0	70Hb	
T18 THIAZOLE			C_3H_3NS				
2.02	2.55	-4.9	25	C	μ=0.1	66Ga	
T19 THIAZOLE, 2,4-dimethyl-			C_5H_7NS		$C_3HNS(CH_3)_2$		
4.21	3.98	-4.1	25	C	μ=0.1	66Ga	
T20 THIAZOLE, 2,5-dimethyl-			C_5H_7NS		$C_3HNS(CH_3)_2$		
3.99	3.91	-4.5	25	C	μ=0.1	66Ga	
T21 THIAZOLE, 4,5-dimethyl-			C_5H_7NS		$C_3HNS(CH_3)_2$		
3.97	3.73	-4.0	25	C	μ=0.1	66Ga	
T22 THIAZOLE, 2-methyl-			C_4H_5NS		$C_3H_2NS(CH_3)$		
3.21	3.40	-4.8	25	C	μ=0.1	66Ga	
T23 THIAZOLE, 4-methyl-			C_4H_5NS		$C_3H_2NS(CH_3)$		
2.99	3.16	-4.4	25	C	μ=0.1	66Ga	
T24 THIAZOLE, 5-methyl-			C_4H_5NS		$C_3H_2NS(CH_3)$		
2.88	3.03	-4.2	25	C	μ=0.1	66Ga	
T25 THIAZOLE, 2-tert-butyl-			$C_7H_{11}NS$		$C_3H_2NS[C(CH_3)_3]$		
3.71	3.00	-1.3	25	C	μ=0.1	66Ga	
T26 THIAZOLE, 4-tert-butyl-			$C_7H_{11}NS$		$C_3H_2NS[C(CH_3)_3]$		
3.86	3.04	-1.0	25	C	μ=0.1	66Ga	

ΔH, kcal /mole	pK	ΔS, cal /mole°K	T,°C	M	Conditions	Ref.	Remarks
T27 THIAZOLE, 2,4,5-trimethyl-			C_6H_9NS		$C_3NS(CH_3)_3$		
4.95	4.55	-4.3	25	C	μ=0.1	66Ga	
T28 THIOCYANIC ACID			CHNS		HSCN		
13	0.95(25°)	40	20-33	T	μ=0	66Bh	c
T29 THIODIGLYCOLIC ACID			$C_4H_6O_4S$		$S(CH_2CO_2H)_2$		
0.10	3.13(0)	-14.0	25	C	μ=1.00M($NaClO_4$)	73Db	
-0.44	3.99(-1)	-19.7	25	C	μ=1.00M($NaClO_4$)	73Db	
T30 THIOPHENE, 2-acetyl-			C_6H_6OS		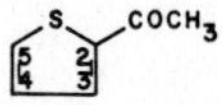		
9.0	5.06(40°)	5.0	25-60	T	S: Aqueous Solution of Sulfuric Acid	73Ke	c,q: ±9.0
T31 THIOPHENE, 2-acetyl-4-methyl-			C_7H_8OS		$C_4H_2S[C(O)CH_3](CH_3)$		
8.2	4.43(40°)	7.5	25-60	T	S: Aqueous Solution of Sulfuric Acid	73Ke	c,q: ±3.0
T32 THIOPHENE, tetrahydro-			C_4H_8S				
19.7	-----	-----	25	C	C∿10^{-3}M S: Fluorosulfuric acid	74Ab	
T33 2-THIOPHENECARBOXYLIC ACID			$C_5H_4O_2S$		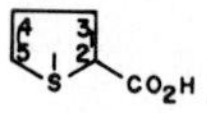		
-1.3	3.529	-20.6	30	T	μ=0	65La	
-1.4	3.546	-20.7	35	T	μ=0	65La	
-1.4	3.561	-20.9	40	T	μ=0	65La	
T34 THREONINE			$C_4H_9O_3N$		$H_3CCH(OH)CH(NH_2)CO_2H$		
1.823	2.200(+1)	-3.4	1	T	μ=0	42Sa	
1.549	2.132(+1)	-4.4	12.5	T	μ=0	42Sa	
1.36	2.09(+1)	-5	25	C	μ=0	64Ic	
1.18	2.006(+1)	-3.6	25	T	μ=0	58Ea	
1.180	2.088(+1)	-5.6	25	T	μ=0	42Sa	
0.728	2.070(+1)	-7.1	37.5	T	μ=0	42Sa	
0.191	2.055(+1)	-8.8	50	T	μ=0	42Sa	
1.2	2.24(+1)	-6.22	25	C	μ=0.05(KCl)	72Gc	
10.075	9.748(0)	-8.0	1	T	μ=0	42Sa	
10.085	9.420(0)	-7.8	12.5	T	μ=0	42Sa	
10.04	9.10(0)	-8	25	C	μ=0	64Ic	
9.960	9.100(0)	-8.2	25	T	μ=0	42Sa	

ΔH, kcal /mole	pK	ΔS, cal /mole°K	T, °C	M	Conditions	Ref.	Remarks
T34, cont.							
9.712	8.812(0)	-9.0	37.5	T	μ=0	42Sa	
9.320	8.548(0)	-10.3	50	T	μ=0	42Sa	
9.75	(0)	-----	20	C	μ≃0.01	71Ma	e:α-NH_3^+
10.0	8.98(0)	-7.55	25	C	μ=0.05(KCl)	72Gc	
9.78	9.07(0)	8.7	25	C	μ=0.16	70Lb	e:α-NH_3^+

T35 THREONINE, allo- $C_4H_9O_3N$ $H_3CCH(OH)CH(NH_2)CO_2H$

ΔH, kcal /mole	pK	ΔS, cal /mole°K	T, °C	M	Conditions	Ref.	Remarks
1.478	2.178(+1)	-4.6	1	T	μ=0	42Sa	
1.176	2.138(+1)	-5.7	12.5	T	μ=0	42Sa	
0.772	2.108(+1)	-7.0	25	T	μ=0	42Sa	
0.287	2.090(+1)	-8.6	37.5	T	μ=0	42Sa	
-0.287	2.086(+1)	-10.4	50	T	μ=0	42Sa	
10.430	9.774(0)	-6.6	1	T	μ=0	42Sa	
10.465	9.432(0)	-7.2	12.5	T	μ=0	42Sa	
10.380	9.096(0)	-6.8	25	T	μ=0	42Sa	
10.150	8.796(0)	-7.5	37.5	T	μ=0	42Sa	
9.800	8.520(0)	-8.7	50	T	μ=0	42Sa	

T36 THYMIDINE $C_{10}H_{14}O_5N_2$

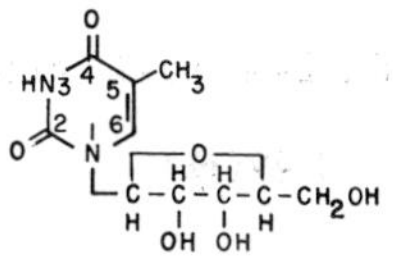

ΔH, kcal /mole	pK	ΔS, cal /mole°K	T, °C	M	Conditions	Ref.	Remarks
7.87	10.14(0)	-18.6	10	C	μ=0	70Cc	e: N_3-C_4O
7.75	9.79(0)	-18.8	25	C	μ=0	67Ia	
7.32	9.79(0)	-20.2	25	C	μ=0	70Cc	e: N_3-C_4O
6.92	9.57(0)	-21.7	40	C	μ=0	70Cc	e: N_3-C_4O
-----	-----	-----	10-40	C	μ=0 ΔCp=-30	70Cc	
12.4	12.85(-1)	-17.4	25	C	μ=0	67Ia	

T37 TOLUENE, α-amino- C_7H_9N $C_6H_5(CH_2NH_2)$

ΔH, kcal /mole	pK	ΔS, cal /mole°K	T, °C	M	Conditions	Ref.	Remarks
12.98	9.35(+1)	0.7	25	C	μ=0	69Cc	
13.160	(+1)	-----	10	C	C=0.06M	49La	
13.180	(+1)	-----	20	C	C=0.06M	49La	
13.087	9.37(+1)	1.0	25	C	C=0.06M	49La	
12.995	(+1)	-----	30	C	C=0.06M	49La	

T38 TOLUENE, 2-amino- C_7H_9N $C_6H_4CH_3(NH_2)$

ΔH, kcal /mole	pK	ΔS, cal /mole°K	T, °C	M	Conditions	Ref.	Remarks
5.43	4.58(+1)	-2.11	15	T	μ=0	71Ga	
5.75	4.495(+1)	-1.01	20	T	μ=0	71Ga	
7.22	4.45(+1)	3.9	25	C	μ=0	68Oa	
8.25	4.39(+1)	7.60	25	C	μ=0	59Za	
7.025	4.45(+1)	3.2	25	C	μ≃0	71Va	
6.08	4.45(+1)	0.12	25	T	μ=0 ΔCp=67.9	71Ga	
6.43	4.345(+1)	1.27	30	T	μ=0	71Ga	
6.78	4.28(+1)	2.44	35	T	μ=0	71Ga	
7.16	4.20(+1)	3.64	40	T	μ=0	71Ga	
7.54	4.12(+1)	4.86	45	T	μ=0	71Ga	
7.367	4.447(25°,+1)	4.35	10-50	T	μ=0	67Ba	
8.163	4.22(+1)	8.1	25	C	μ≃0 S: 0.10 mole fraction Ethanol	71Va	

ΔH, kcal /mole	pK	ΔS, cal /mole°K	T, °C	M	Conditions	Ref.	Remarks
T39 TOLUENE, 3-amino-			C_7H_9N		$C_6H_4CH_3(NH_2)$		
7.37	4.72(+1)	3.1	25	C	μ=0	68Oa	
7.58	4.71(+1)	3.86	25	C	μ=0	73Lb	
8.00	4.66(+1)	5.53	25	C	μ=0	59Za	
7.370	4.726(+1)	3.1	25	C	μ≃0	70Va	
6.84	4.72(+1)	1.36	25	T	μ=0	73Pa	
7.467	4.712(25°,+1)	3.48	10-50	T	μ=0	67Ba	c
8.225	4.526(+1)	6.9	25	C	μ≃0 S: 0.10 mole fraction Ethanol	70Va	
T40 TOLUENE, 4-amino-			C_7H_9N		$C_6H_4CH_3(NH_2)$		
7.86	5.09(+1)	3.09	25	C	μ=0	73Lb	
7.60	5.08(+1)	2.2	25	C	μ=0	68Oa	
4.98	5.07(+1)	-6.48	25	C	μ=0	59Za	
7.592	5.080(+1)	2.2	25	C	μ≃0	70Va	
6.64	5.09(+1)	-1.00	25	T	μ=0	73Pa	
8.057	5.083(25°,+1)	3.75	10-50	T	μ=0	67Ba	c
8.555	4.866(+1)	6.4	25	C	μ≃0 S: 0.10 mole fraction Ethanol	70Va	
T41 TOLUENE, 2-amino-3-chloro-			C_7H_8NCl		$C_6H_3CH_3(NH_2)(Cl)$		
2.62	2.49(+1)	2.48	25	T	μ=0	71Ga	
T42 TOLUENE, 2-amino-4-chloro-			C_7H_8NCl		$C_6H_3CH_3(NH_2)(Cl)$		
6.12	3.55(+1)	4.99	15	T	μ=0	71Ga	
5.98	3.48(+1)	4.52	20	T	μ=0	71Ga	
5.82	3.385(+1)	3.99	25	T	μ=0 ΔCp=-33.3	71Ga	
5.65	3.34(+1)	3.40	30	T	μ=0	71Ga	
5.45	3.265(+1)	2.75	35	T	μ=0	71Ga	
5.23	3.20(+1)	2.04	40	T	μ=0	71Ga	
4.98	3.15(+1)	1.26	45	T	μ=0	71Ga	
T43 TOLUENE, 2-amino-5-chloro-			C_7H_8NCl		$C_6H_3CH_3(NH_2)(Cl)$		
5.18	3.98(+1)	-0.25	15	T	μ=0	71Ga	
5.43	3.92(+1)	0.60	20	T	μ=0	71Ga	
5.68	3.85(+1)	1.47	25	T	μ=0 ΔCp=51.7	71Ga	
5.94	3.78(+1)	2.34	30	T	μ=0	71Ga	
6.22	3.705(+1)	3.22	35	T	μ=0	71Ga	
6.49	3.62(+1)	4.12	40	T	μ=0	71Ga	
6.78	3.57(+1)	5.02	45	T	μ=0	71Ga	
T44 TOLUENE, 2-amino-6-chloro-			C_7H_8NCl		$C_6H_3CH_3(NH_2)(Cl)$		
5.56	3.745(+1)	2.16	15	T	μ=0	71Ga	
5.72	3.67(+1)	2.70	20	T	μ=0	71Ga	
5.88	3.62(+1)	3.24	25	T	μ=0 ΔCp=31.7	71Ga	
6.04	3.52(+1)	3.76	30	T	μ=0	71Ga	
6.20	3.455(+1)	4.28	35	T	μ=0	71Ga	
6.36	3.385(+1)	4.80	40	T	μ=0	71Ga	
6.52	3.32(+1)	5.31	45	T	μ=0	71Ga	
T45 TOLUENE, 3-amino-2,4,6-trinitro-			$C_7H_6O_6N_4$		$C_6HCH_3(NH_2)(NO_2)_3$		
-7.576	-8.447	12.5	25	T	μ=0	70Be	g:69Ja

ΔH, kcal /mole	pK	ΔS, cal /mole°K	T, °C	M	Conditions	Ref.	Remarks
T46 TOLUENE, α(diethylamino)-			$C_{11}H_{17}N$		$C_6H_5[CH_2N(C_2H_5)_2]$		
9.92	9.44(+1)	-9.8	25	C	μ=0	69Cc	
T47 TOLUENE, 2-hydroxy-			C_7H_8O		$C_6H_4CH_3(OH)$		
7.13	10.28(0)	-23.2	25	C	μ=0	59Pa	h
5.73	10.33(0)	-28.1	25	T	μ=0 ΔCp=-28.5	62Cb	
T48 TOLUENE, 3-hydroxy-			C_7H_8O		$C_7H_7(OH)$		
4.90	10.08(0)	-30.7	25	C	μ=0	59Pa	h
5.52	10.10(0)	-27.7	25	T	μ=0 ΔCp=-28.3	62Cb	
T49 TOLUENE, 4-hydroxy-			C_7H_8O		$C_7H_7(OH)$		
5.50	10.281(0)	-28.6	25	C	μ=0	75Pa	
4.23	10.26(0)	-32.8	25	C	μ=0	59Pa	h
5.50	10.276(0)	-28.6	25	T	μ=0	62Cb	
5.50	10.28(0)	-28.6	25	T	μ=0 ΔCp=-28.4	62Cb	
5.486	10.276(0)	-28.6	25	T	μ=0 ΔCp=-53	70Bc	g:62Cb
5.97	10.460(0)	-27.8	25	C	μ=0 S: 0.059 mole fraction Methanol	75Pa	
5.55	10.572(0)	-29.8	25	C	μ=0 S: 0.123 mole fraction Methanol	75Pa	
5.31	10.807(0)	-31.6	25	C	μ=0 S: 0.194 mole fraction Methanol	75Pa	
5.46	11.013(0)	-32.1	25	C	μ=0 S: 0.273 mole fraction Methanol	75Pa	
5.30	11.189(0)	-33.4	25	C	μ=0 S: 0.360 mole fraction Methanol	75Pa	
5.22	11.399(0)	-34.6	25	C	μ=0 S: 0.458 mole fraction Methanol	75Pa	
4.84	11.600(0)	-36.8	25	C	μ=0 S: 0.567 mole fraction Methanol	75Pa	
4.47	11.921(0)	-39.6	25	C	μ=0 S: 0.692 mole fraction Methanol	75Pa	
4.42	13.447(0)	-41.9	25	C	μ=0 S: 0.835 mole fraction Methanol	75Pa	
8.48	14.651(0)	-38.6	25	C	μ=0 S: Methanol	75Pa	
T50 TOLUENE, 3-hydroxy-4-nitro-			$C_7H_7O_3N$		$C_6H_3CH_3(OH)(NO_2)$		
4.58	7.41	-18.6	25	T	μ=0.1	63Ra	
T51 TOLUENE, 4-hydroxy-3-(2'-thiazolyazo)-			$C_{10}H_9ON_3S$		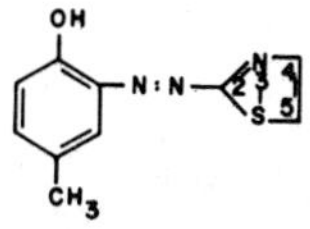		
3.25	8.36(25°)	-27.34	15-35	T	μ=0	74Kb	c
T52 TOLUENE, 3-hydroxy-α,α,α-trifluoro-			$C_7H_5OF_3$		$C_6H_4(OH)(CF_3)$		
5.24	8.950(0)	-23.3	25	C	μ=0	72Lc	

ΔH, kcal /mole	pK	ΔS, cal /mole°K	T, °C	M	Conditions	Ref.	Remarks
T53 TOLUENE, 4-hydroxy-α,α,α-trifluoro-			$C_7H_5OF_3$		$C_6H_4(OH)(CF_3)$		
4.99	8.675(0)	-23.0	25	C	μ=0	72Lc	
T54 1,2,3-TRIAZOLE			$C_2H_3N_3$				
8.88	9.26	-12.6	25	C	μ=0	68Ha	
T55 1,2,4-TRIAZOLE			$C_2H_3N_3$		see T54		
2.6	2.451(+1)	-2.3	15	T	μ=0	74Ld	
2.5	2.418(+1)	-2.4	20	T	μ=0	74Ld	
2.30	2.45(+1)	-3.5	25	C	μ=0	70Hb	
2.5	2.386(+1)	-2.5	25	T	μ=0	74Ld	
2.4	2.327(+1)	-2.7	35	T	μ=0	74Ld	
9.6	10.205(0)	-13.3	15	T	μ=0	74Ld	
9.2	10.083(0)	-14.7	20	T	μ=0	74Ld	
7.90	10.04(0)	-19.4	25	C	μ=0	70Hb	
8.8	9.972(0)	-16.0	25	T	μ=0	74Ld	
8.0	9.768(0)	-18.7	35	T	μ=0	74Ld	
T56 1,2,3-TRIAZOLE, 4,5-dibromo-			$C_2H_3N_3$		see T54		
4.24	5.37	-10.4	25	C	μ=0	68Ha	
T57 1,2,3-TRIAZOLE-4-carboxylic acid			$C_3H_3O_2N_3$		see T54		
0.84	3.22	-11.5	25	C	μ=0	68Ha	
5.89	8.73	-20.2	25	C	μ=0	68Ha	
T58 1,2,3-TRIAZOLE-4-carboxylic acid, 1-phenyl-			$C_9H_7O_2N_3$		see T54		
0.76	2.88	-10.6	25	C	μ=0	68Ha	
T59 1,2,3-TRIAZOLE-4-carboxylic acid, 5-methyl-1-phenyl-			$C_{10}H_9O_2N_3$		see T54		
0.17	3.73	-16.5	25	C	μ=0	68Ha	
T60 1,2,3-TRIAZOLE-4,5-dicarboxylic acid			$C_4HO_4N_3$		see T54		
0.11	1.86	-8.2	25	C	μ=0	68Ha	
0.01	5.90	-27.0	25	C	μ=0	68Ha	
2.26	9.30	-35.0	25	C	μ=0	68Ha	
T61 1,2,3-TRIAZOLE-4,5-dicarboxylic acid, 1-phenyl-			$C_{10}H_7O_4N_3$		see T54		
-0.67	2.13	-11.1	25	C	μ=0	68Ha	
0.06	4.93	-22.3	25	C	μ=0	68Ha	

ΔH, kcal /mole	pK	ΔS, cal /mole°K	T, °C	M	Conditions	Ref.	Remarks

T62 bis-(TRICYANOVINYL)-amine

ΔH, kcal /mole	pK	ΔS, cal /mole°K	T, °C	M	Conditions	Ref.	Remarks
-7.5	-5.8	1	25	T	μ=0	65Bb	

T63 TRIDECANE, 1,5,9,13-tetraaza- $C_9H_{24}N_4$ $H_2N(CH_2)_3NH(CH_2)_3NH(CH_2)_3NH_2$

ΔH, kcal /mole	pK	ΔS, cal /mole°K	T, °C	M	Conditions	Ref.	Remarks
10.88	7.22(+4)	3.5	25	C	μ=0.1($NaNO_3$)	72Ba	
11.65	8.54(+3)	0.0	25	C	μ=0.1($NaNO_3$)	72Ba	
12.47	9.82(+2)	-3.2	25	C	μ=0.1($NaNO_3$)	72Ba	
12.20	10.45(+1)	-6.9	25	C	μ=0.1($NaNO_3$)	72Ba	

T64 TRYPSIN

ΔH, kcal /mole	pK	ΔS, cal /mole°K	T, °C	M	Conditions	Ref.	Remarks
7.0	6.25(25°)	-5.12	25&35	T	μ=0.1M(NaCl)	55Ga	q

T65 TRYPTOPHAN $C_{11}H_{12}O_2N_2$

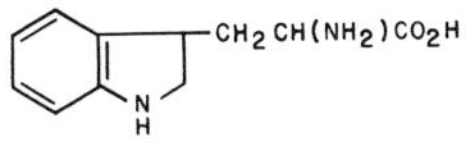

ΔH, kcal /mole	pK	ΔS, cal /mole°K	T, °C	M	Conditions	Ref.	Remarks
0	2.46(+1)	-11.3	25	T	μ=0.1	60Ha	
0.80	2.753(+1)	-9.9	25	C	μ=3.00M($NaClO_4$)	70Wb	e:CO_2H, q:±0.15
10.6	(0)	-----	20	C	μ≃0.01	71Ma	e:α-NH_3^+
10.5	9.41(0)	-7.8	25	T	μ=0.1	60Ha	
10.65	9.28(0)	-6.7	25	C	μ=0.16(KNO_3)	70Md	
9.19	9.923(0)	-14.6	25	C	μ=3.00M($NaClO_4$)	70Wb	e:NH_3^+

T66 TUNGSLIC ACID $H_6O_{21}W_6$ $H_6W_6O_{21}$

ΔH, kcal /mole	pK	ΔS, cal /mole°K	T, °C	M	Conditions	Ref.	Remarks
5.4	8.28(-2)	-20	25	C	μ=3M($NaClO_4$)	70Aa 69Aa	q:±1.3

T67 TUNGSTOTELLURIC HETEROPOLYACIDS

ΔH, kcal /mole	pK	ΔS, cal /mole°K	T, °C	M	Conditions	Ref.	Remarks
9.55	2.86(K_3)	19	25	C	μ≃0 HA=[Composition of acid 1Te/8W]	71Gb	
7.80	2.20(K_5)	16	25	C	μ≃0 HA=[Composition of acid 1Te/8W]	71Gb	
9.10	2.49(K_6)	19	25	C	μ≃0 HA=[Composition of acid 1Te/8W]	71Gb	

T68 o-TYROSINE $C_9H_{11}O_3N$ $HOC_6H_4CH_2CH(NH_2)CO_2H$

ΔH, kcal /mole	pK	ΔS, cal /mole°K	T, °C	M	Conditions	Ref.	Remarks
9.61	8.60(0)	-7.1	25	C	μ=0.16	70Lc	e:NH_3^+
6.84	10.66(-1)	-25.8	25	C	μ=0.16	70Lc	e: Phenol group

T69 m-TYROSINE $C_9H_{11}O_3N$ $HOC_6H_4CH_2CH(NH_2)CO_2H$

ΔH, kcal /mole	pK	ΔS, cal /mole°K	T, °C	M	Conditions	Ref.	Remarks
10.10	9.09(0)	-7.7	25	C	μ=0.16	70Lc	e:NH_3^+
6.39	10.11(-1)	-24.8	25	C	μ=0.16	70Lc	e: Phenol group

ΔH, kcal /mole	pK	ΔS, cal /mole°K	T, °C	M	Conditions	Ref.	Remarks
T70 p-TYROSINE			$C_9H_{11}O_3N$		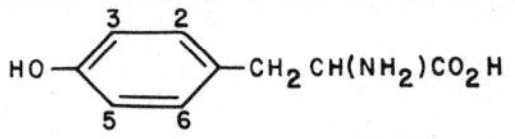		
8.65	(0)	-----	20	C	μ≃0.01	71Ma	e:α-NH_3^+
10.14	9.21(0)	-8.1	25	C	μ=0.16	70Lc	e:NH_3^+
8.65	(-1)	-----	20	C	μ≃0.01	71Ma	e: Phenol group
6.0	10.05(-1)	-26	25	T	μ=0.15	52Ta	
5.83	10.13(-1)	-26.8	25	C	μ=0.16	70Lc	e: Phenol group
T71 p-TYROSINE, N-acetyl-, ethyl ester			$C_{13}H_{17}O_4N$		see T70		
5.65	-----	-----	20	C	μ≃0.01	71Ma	e: Phenol group
T72 p-TYROSINE, N-aspartyl-			$C_{13}H_{16}O_6N_2$		see T70		
0.750	2.13(25°)	-7.23	0&25	T	μ=0.01	33Ga	b,c,e: CO_2H
0	3.57(25°)	-16.3	0&25	T	μ=0.01	33Ga	b,c,e: CO_2H
10.16	8.92(25°)	-6.8	0&25	T	μ=0.01	33Ga	b,c,e: CO_2H
6.20	10.23(25°)	-26.0	0&25	T	μ=0.01	33Ga	b,c,e: oxy-phenol
T73 p-TYROSINE, 3,5-dibromo(L)-			$C_9H_9O_3NBr_2$		see T70		
1.700	2.17(25°)	-4.2	25&40	T	--	35Wa	c
0.860	6.45(25°)	-26.6	25&40	T	--	35Wa	c
9.120	7.60(25°)	-4.2	25&40	T	--	35Wa	c
T74 p-TYROSINE, 3,5-dichloro-			$C_9H_9O_3NCl_2$		see T70		
1.140	2.12(25°)	-5.8	25-40	T	--	35Wa	c
1.420	6.47(25°)	-24.8	25-40	T	--	35Wa	c
8.830	7.62(25°)	-5.3	25-40	T	--	35Wa	c
T75 p-TYROSINE, 3,5-diiodo-			$C_9H_9O_3NI_2$		see T70		
0.980	2.12(25°)	-6.4	0-40	T	--	35Wa	c
0.810	6.48(25°)	-26.9	0-40	T	--	30Ma 35Wa 30Da	c
8.790	7.82(25°)	-6.3	0-40	T	--	30Ma 35Wa 30Da	c
12.77	11.87	-11.5	25	T	μ=0.02	30Ma	

ΔH, kcal /mole	pK	ΔS, cal /mole°K	T, °C	M	Conditions	Ref.	Remarks
			U				
U1 URACIL			$C_4H_4O_2N_2$		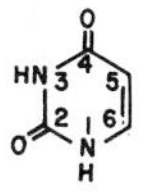		
0.34	0.6(+1)	-1.5	25	C	μ=0	67Ia	
8.30	9.74(0)	-15.2	10	C	μ=0	70Cc	e: N_3-C_4O
7.85	9.46(0)	-16.9	25	C	μ=0	70Cc	e: N_3-C_4O
8.23	9.46(0)	-15.7	25	C	μ=0	67Ia	
7.49	9.14(0)	-17.9	40	C	μ=0	70Cc	e: N_3-C_4O
-----	(0)	-----	10-40	C	μ=0 ΔCp=-27	70Cc	
7.2	9.28(30°,0)	-20	20-50	T	μ=0	64Ld	c,q: ±0.5
U2 URACIL, 5-methyl-			$C_5H_6O_2N_2$		see U1		
8.50	10.18(0)	-16.6	10	C	μ=0	70Cc	e: N_3-C_4O
8.15	9.90(0)	-18.0	25	C	μ=0	70Cc	e: N_3-C_4O
8.83	9.90(0)	-15.7	25	C	μ=0	67Ia	
7.79	9.52(0)	-18.7	40	C	μ=0	70Cc	e: N_3-C_4O
-----	(0)	-----	10-40	C	μ=0 ΔCp=-24	70Cc	
8.0	9.70(30°,0)	-18.0	20-40	T	μ=0	66Lb	c
U3 6-URACILCARBOXYLIC ACID			$C_5H_4O_4N_2$		see U1		
0.5	1.8(+1)	-6.4	25	C	μ=0	70Wd	
8.7	9.55(0)	-14.4	25	C	μ=0	70Wd	
U4 UREA, 1-benzyl-3(2-pyridyl)-2-thio-			$C_{13}H_{13}N_3S$		$C_5H_4NNHC(S)NHCH_2C_6H_5$		
6.752	11.66(20°)	-30.28	10-40	T	μ=0.10 S: 50 v % Acetonitrile	74Pb	c
U5 UREA, 1-methyl-3(2-pyridyl)-2-thio-			$C_7H_9N_3S$		$C_5H_4NNHC(S)NHCH_3$		
5.322	11.75(20°)	-35.61	10-40	T	μ=0.10 S: 50 v % Acetonitrile	74Pb	c
U6 UREA, 1-phenyl-3(2-pyridyl)-2-thio-			$C_{12}H_{11}N_3S$		$C_5H_4NNHC(S)NHC_6H_5$		
5.314	11.32(20°)	-33.68	10-40	T	μ=0.10 S: 50 v % Acetonitrile	74Pb	c
U7 UREA, 3(2-pyridyl)-2-thio-1(2-tolyl)-			$C_{13}H_{13}N_3S$		$C_5H_4NNHC(S)NHC_6H_4CH_3$		
3.647	11.62(20°)	-40.87	10-40	T	μ=0.10 S: 50 v % Acetonitrile	74Pb	c

ΔH, kcal /mole	pK	ΔS, cal /mole°K	T, °C	M	Conditions	Ref.	Remarks
U8 UREA, 3(2-pyridyl)-2-thio-1(4-tolyl)			$C_{13}H_{13}N_3S$		$C_5H_4NNHC(S)NHC_6H_4CH_3$		
5.153	11.45(20°)	-34.82	10-40	T	μ=0.10 S: 50 v % Acetonitrile	74Pb	c
U9 UREA, 2-seleno-			$C_4H_2N_2Se$		$H_2NC(Se)NH_2$		
0.0	0.58	-1.6	25	C	μ=0.01	69Ga	
U10 UREA, 1,1,3,3-tetramethyl-			$C_5H_{12}ON_2$		$(H_3C)_2NC(O)N(CH_3)_2$		
37.6	-----	-----	25	C	C∿10^{-3}M S: Fluorosulfuric acid	74Ab	
U11 URIC ACID			$C_5H_4O_3N_4$		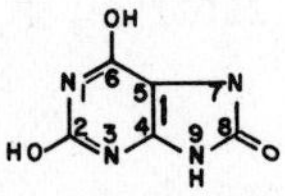		
-5.22	5.468(30°,$\underline{0}$)	$\underline{-42.2}$	20-45	T	μ=0.1M(NaCl)	74Fa	c
U12 URIDINE			$C_9H_{12}O_6N_2$		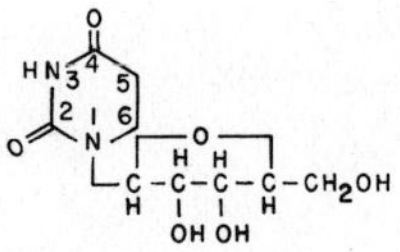		
7.67	9.61($\underline{0}$)	-16.9	10	C	μ=0	70Cc	e: N_3-C_4O
7.24	9.30($\underline{0}$)	-18.3	25	C	μ=0	70Cc	e: N_3-C_4O
7.6	9.30($\underline{0}$)	-17.1	25	C	μ=0	67Ia	
6.8	9.07($\underline{0}$)	-19.8	40	C	μ=0	70Cc	e: N_3-C_4O
-----	($\underline{0}$)	-----	10-40	C	μ=0 ΔCp=-29	70Cc	
8.0	9.51($\underline{0}$)	-16.8	25	C	μ=0.1	64Sd	
11.6	13.03($\underline{-1}$)	-18.8	10	C	μ=0	70Cc	e: Ribose OH
10.9	12.59(-$\underline{1}$)	-21.0	25	C	μ=0	67Ia 70Cc	e: Ribose OH
U13 URIDINE-5'-diphosphoric acid			$C_9H_{14}O_{11}N_2P_2$		see U12		
-1.08	7.16	-36.4	25	T	μ=0	65Pe	
U14 URIDINE-5'-monophosphoric acid			$C_9H_{13}O_9N_2P$		see U12		
-1.12	6.63	-34.0	25	T	μ=0	65Pe	
6.6	9.71(20°)	-21.30	20-50	T	μ=0.015	67Ae	c

ΔH, kcal /mole	pK	ΔS, cal /mole°K	T, °C	M	Conditions	Ref.	Remarks
U15 URIDINE-5-triphosphoric acid			$C_9H_{15}O_{13}N_2P_3$		see U12		
-2.02	7.58	-41.4	25	T	μ=0	65Pe	

ΔH, kcal /mole	pK	ΔS, cal /mole°K	T,°C	M	Conditions	Ref.	Remarks

V

V1 VALINE (DL)			$C_5H_{11}O_2N$		$(CH_3)_2CHCH(NH_2)CO_2H$		
0.890	2.320(+1)	-7.3	1	T	μ=0	37Sa	
0.540	2.297(+1)	-8.6	12.5	T	μ=0	37Sa	
0.080	2.286(+1)	-10.2	25	T	μ=0	37Sa	
						39Hb	
-0.460	2.292(+1)	-12.0	37.5	T	μ=0	37Sa	
-1.100	2.310(+1)	-13.9	50	T	μ=0	37Sa	
0.06	2.23(+1)	-10.3	25	T	--	60Pc	
0.1	3.12(+1)	-10.7	25	T	S: 44.6 % Dioxane	60Pc	
0.5	3.72(+1)	-15.4	25	T	S: 59.7 % Dioxane	60Pc	
0.5	4.04(+1)	-16.8	25	T	S: 69 % Dioxane	60Pc	
10.370	10.413(0)	-8.5	1	T	μ=0	37Sa	
10.800	10.064(0)	-8.2	12.5	T	μ=0	37Sa	
10.73	9.719(0)	-8.5	25	C	μ=0	69Cc	
10.740	9.719(0)	-8.4	25	T	μ=0	37Sa	
10.560	9.405(0)	-9.0	37.5	T	μ=0	37Sa	
10.230	9.124(0)	-10.1	50	T	μ=0	37Sa	
10.37	(0)	-----	20	C	μ≈0.01	71Ma	e:α-NH_3^+
10.99	9.50(0)	-6.61	25	C	μ=0.1M(KNO_3)	72Ia	e:NH_3^+
10.69	(0)	-----	25	C	μ=0.10M	71Ba	e:NH_3^+
10.97	9.44(0)	-6.4	25	C	μ=0.16(KNO_3)	70Md	
10.640	9.744(0)	-8.72	25	C	--	67Ac	
10.65	9.62(0)	-8.8	25	T	--	60Pc	
8.9	10.18(0)	-16.7	25	T	S: 44.6 % Dioxane	60Pc	
11.25	10.57(0)	-10.6	25	T	S: 59.7 % Dioxane	60Pc	
10.00	11.18(0)	-17.6	25	T	S: 69 % Dioxane	60Pc	
V2 VALINE, methyl ester			$C_6H_{13}O_2N$		$(CH_3)_2CHCH(NH_2)CO_2CH_3$		
11.29	7.49(25°,+1)	3.6	25-50	T	μ=0.1M	67Hb	c,e:NH_3^+
V3 VITAMIN B_{12}							
4.6	7.64(25°,+1)	-19	10-25	T	μ=0	67Ha	c,q: ±0.5

ΔH, kcal /mole	pK	ΔS, cal /mole°K	T, °C	M	Conditions		Ref.	Remarks

W

W1 WATER H_2O H_2O

ΔH, kcal /mole	pK	ΔS, cal /mole°K	T, °C	M	Conditions		Ref.	Remarks
15.141	14.955(0)	-13.07	0	C	μ=0	ΔCp=-65.5	58Aa	
14.998	(0)	-----	0	C	μ=0		70Ld	
14.620	(0)	-----	0	C	μ=0		67Vd	
14.513	14.939(0)	-15.23	0	T	μ=0		33Hb	
14.950	15.056(0)	-14.2	0	T	μ=0		10Na	
15.006	14.948(0)	-13.45	0	T	μ=0	ΔCp=-83.9	75Ob	
15.001	14.947(0)	-13.469	0	-	μ=0		72Ca	1
-----	-----	-----	2.5	C	μ=0	ΔCp=-88.6	70Ld	
14.62	(0)	-----	5	C	μ=0		70Gb	
14.555	(0)	-----	5	C	μ=0		70Ld	
14.312	14.730(0)	-15.95	5	T	μ=0		33Hb	
14.564	14.665(0)	-15.060	5	-	μ=0		72Ca	1
14.261	14.536	-16.388	10	C	μ=0		72Ca	
14.437	14.534(0)	-15.59	10	C	μ=0		58Aa	
14.140	(0)	-----	10	C	μ=0		67Vd	
14.109	14.533(0)	-16.67	10	T	μ=0		33Hb	
14.191	14.536(0)	-16.388	10	-	μ=0		72Ca	1
-----	-----	-----	10	C	μ=0	ΔCp=-68.9	70Ld	
13.83	(0)	-----	15	C	μ=0		74Va	
13.866	(0)	-----	15	C	μ=0		70Ld	
13.901	14.345(0)	-17.40	15	T	μ=0		33Hb	
13.872	14.346(0)	-17.497	15	-	μ=0		72Ca	1
13.721	(0)	-----	18	C	μ=0		31Ra	
13.780	(0)	-----	18	C	μ=0		67Vd	
14.055	14.337(0)	-17.3	18	T	μ=0		10Na	
13.702	14.242(0)	-18.092	18	-	μ=0		72Ca	1
13.811	14.161(0)	-17.74	20	C	μ=0		58Aa	
13.650	(0)	-----	20	C	μ=0		31La	1
13.65	(0)	-----	20	C	μ=0		29Ra	
13.66	(0)	-----	20	C	μ=0		22Ra	
13.59	(0)	-----	20	C	μ=0		72Sa	
13.63	(0)	-----	20	C	μ=0		70Gb	
13.692	14.167(0)	-18.12	20	T	μ=0		33Hb	
13.7	15.81(0)	-25.5	20	T	μ=0		49Ea	
13.596	14.168(0)	-18.446	20	-	μ=0		72Ca	1
-----	(0)	-----	20	C	μ=0	ΔCp=-53.2	70Ld	
13.336	14.003(0)	-19.263	25	C	μ=0		72Ca	
13.340	14.001(0)	-19.32	25	C	μ=0		68Gb	
13.522	13.999(0)	-18.70	25	C	μ=0		58Aa	
13.332	(0)	-----	25	C	μ=0		63Gb	
13.320	(0)	-----	25	C	μ=0		52Ba	
13.325	(0)	-----	25	C	μ=0		75Ea	
13.36	(0)	-----	25	C	μ=0	ΔCp=-49.5	73Oa	1
13.50	(0)	-----	25	C	μ=0		56Pa	f
13.331	(0)	-----	25	C	μ=0		71Ha	
							71Hb	
13.336	(0)	-----	25	C	μ=0		63Va	
13.335	(0)	-----	25	C	μ=0		63Hb	
							63Ha	
13.334	(0)	-----	25	C	μ=0		70Ld	
13.53	(0)	-----	25	C	μ=0		71Ad	
							71Ac	
13.35	(0)	-----	25	C	μ=0		70Gb	
13.37	(0)	-----	25	C	μ=0		73Rb	
13.336	(0)	-----	25	C	μ=0		59Sa	
13.45	(0)	-----	25	C	μ=0		61Ta	
13.3	(0)	-----	25	C	μ=0		66Bb	
13.345	(0)	-----	25	C	μ=0		72Oa	
13.46	(0)	-----	25	C	μ=0		72Na	
13.470	(0)	-----	25	C	μ=0		67Vd	
13.363	(0)	-----	25	C	μ=0		37Pa	

ΔH, kcal /mole	pK	ΔS, cal /mole°K	T, °C	M	Conditions	Ref.	Remarks
W1, cont.							
13.463	13.987(0)	-18.9	25	T	μ=0	45Ja	
13.512	13.993(0)	-18.7	25	T	μ=0 ΔCp=-46.8	40Hb	1
13.13	13.571(0)	-19.9	25	T	μ=0 ΔCp=-36	73Ma	
					$\Delta H=18.513 - 6.051 \times 10^{-5}$		
					Log K=-967.06/T + 0.8454 - 0.00316T		
					ΔS=16.18 - 0.1210T		
					ΔCp=-0.1210T		
					298<T °K<473		
13.481	13.996(0)	-18.8	25	T	μ=0 ΔCp=-42.5	33Hb 39Hb	
13.710	14.092(0)	-18.5	25	T	μ=0	10Na	
13.340	13.999(0)	-19.31	25	-	μ=0 ΔCp=-53.7	75Ob	1
13.352	13.999(0)	-19.263	25	-	μ=0	72Ca	1
13.247	13.833(0)	-19.61	30	C	μ=0 ΔCp=-52.7	58Aa	
13.12	(0)	-----	30	C	μ=0 ΔCp=-48.3	73Oa	1
13.267	13.832(0)	-19.53	30	T	μ=0	33Hb	
13.129	13.836(0)	-19.998	30	-	μ=0	72Ca	1
-----	(0)	-----	30	C	μ=0 ΔCp=-40.6	70Ld	
12.80	(0)	-----	35	C	μ=0	70Gb	
12.928	(0)	-----	35	C	μ=0	70Ld	
13.00	(0)	-----	35	C	μ=0	74Va	
13.051	13.680(0)	-20.24	35	T	μ=0	33Hb	
12.916	13.682(0)	-20.687	35	-	μ=0	72Ca	1
12.695	13.537(0)	-21.369	40	C	μ=0	72Ca	
12.734	13.533(0)	-21.24	40	C	μ=0	58Aa	
12.64	(0)	-----	40	C	μ=0 ΔCp=-46.1	73Oa	1
12.810	(0)	-----	40	C	μ=0	67Vd	
12.833	13.535(0)	-20.96	40	T	μ=0	33Hb	
12.703	13.537(0)	-21.369	40	-	μ=0	72Ca	1
-----	(0)	-----	40	C	μ=0 ΔCp=-46.1	70Ld	
12.467	(0)	-----	45	C	μ=0	70Ld	
12.612	13.396(0)	-21.66	45	T	μ=0	33Hb	
12.478	13.398(0)	-22.081	45	-	μ=0	72Ca	1
12.258	13.263(0)	-22.72	50	C	μ=0 ΔCp=-45.2	58Aa	
12.19	(0)	-----	50	C	μ=0 ΔCp=-44.2	73Oa	1
12.10	(0)	-----	50	C	μ=0	70Gb	
12.470	13.347(0)	-22.5	50	T	μ=0	10Na	
12.390	13.262(0)	-22.35	50	T	μ=0	33Hb	
12.230	13.258(0)	-22.844	50	-	μ=0	72Ca	1
12.160	13.276(0)	-23.11	50	-	μ=0 ΔCp=-43.0	75Ob	1
-----	-----	-----	50	C	μ=0 ΔCp=-51.7	70Ld	
12.17	(0)	-----	50.25	C	μ=0	73Oa	
11.950	(0)	-----	55	C	μ=0	70Ld	
12.240	(0)	-----	55	C	μ=0	67Vd	
12.164	13.137(0)	-23.05	55	T	μ=0	33Hb	
11.950	13.147(0)	-23.651	55	-	μ=0	72Ca	1
11.806	13.015(0)	-24.06	60	C	μ=0	58Aa	
11.936	13.017(0)	-23.74	60	T	μ=0	33Hb	
11.364	12.800(0)	-25.34	70	C	μ=0 ΔCp=-42.8	58Aa	
11.33	(0)	-----	70	C	μ=0 ΔCp=-42.3	73Oa	1
11.600	(0)	-----	70	C	μ=0	67Vd	
11.15	(0)	-----	74.40	C	μ=0	73Oa	
11.230	12.772(0)	-26.2	75	T	μ=0	10Na	
11.111	12.711(0)	-26.24	75	-	μ=0 ΔCp=-41.8	75Ob	1
10.921	12.598(0)	-26.59	80	C	μ=0	58Aa	
10.462	12.422(0)	-27.84	90	C	μ=0 ΔCp=-45.7	58Aa	
10.49	(0)	-----	90	C	μ=0 ΔCp=-42.1	73Oa	1
10.06	(0)	-----	100	C	μ=0	73Oa	1
9.975	12.259(0)	-29.13	100	C	μ=0	58Aa	
9.995	12.319(0)	-29.6	100	T	μ=0	10Na	
10.044	12.266(0)	-29.20	100	-	μ=0 ΔCp=-43.7	75Ob	1
9.947	12.126(0)	-30.48	110	C	μ=0 ΔCp=-53.9	58Aa	
9.63	(0)	-----	110	C	μ=0 ΔCp=-44.0	73Oa	1

ΔH, kcal /mole	pK	ΔS, cal /mole°K	T,°C	M	Conditions	Ref.	Re- marks
W1, cont.							
10.05	(0)	-----	110.40	C	μ=0	73Oa	
8.864	12.002(0)	-31.95	120	C	μ=0	58Aa	
8.95	(0)	-----	125	C	μ=0	73Oa	1
8.917	11.916(0)	-32.12	125	-	μ=0 ΔCp=-46.6	75Ob	1
8.93	(0)	-----	125.30	C	μ=0	73Oa	1
8.610	11.943(0)	-33.2	128	T	μ=0	10Na	
8.214	11.907(0)	-33.54	130	C	μ=0 ΔCp=-67.2	58Aa	
8.72	(0)	-----	130	C	μ=0 ΔCp=-47.6	73Oa	1
7.99	(0)	-----	144.60	C	μ=0	73Oa	
7.71	(0)	-----	150	C	μ=0 ΔCp=-53.3	73Oa	1
7.693	11.646(0)	-35.10	150	-	μ=0 ΔCp=-55.6	75Ob	1
7.225	11.664(0)	-36.5	156	T	μ=0	10Na	
4.28	11.302(0)	-42.7	200	-	μ=0 ΔCp=-81.4	75Ob	1
4.155	11.291(0)	-43.2	218	T	μ=0	10Na	
-0.71	11.196(0)	-52.6	250	-	μ=0 ΔCp=-122	75Ob	
-8.90	11.301(0)	-67.2	300	-	μ=0 ΔCp=-230	75Ob	
13.653	(0)	-----	20	C	μ=0.001	72Sa	
13.43	(0)	-----	25	C	μ=0.0036	72Na	
13.692	(0)	-----	20	C	μ=0.004	72Sa	
13.620	(0)	-----	20	C	μ=0.011	64Eb	
13.790	(0)	-----	20	C	μ=0.016	72Sa	
13.43	(0)	-----	25	C	μ=0.017	72Na	
13.931	(0)	-----	20	C	μ=0.025	72Sa	
13.40	(0)	-----	25	C	μ=0.032	72Na	
13.982	(0)	-----	20	C	μ=0.036	72Sa	
14.010	(0)	-----	20	C	μ=0.050	72Sa	
13.35	(0)	-----	25	C	μ=0.05	61Ba	
14.033	(0)	-----	20	C	μ=0.064	72Sa	
13.617	(0)	-----	20	C	μ=0.1	56Ab	
13.7	15.9(0)	-26	20	C	μ=0.1(KNO_3)	65Sb	
13.66	12.90(0)	-17.3	20	C	μ=0.11	73Fa	
13.620	(0)	-----	20	C	μ=0.11	56Ab	
13.70	13.93(0)	-17.0	20	C	μ=0.21	73Fa	
13.75	(0)	-----	--	-	0.4955M Molality	40Ka	
13.96	14.07(0)	-16	15	C	μ=0.5(NaCl) ΔCp=-28	74Va	
13.94	14.08(0)	-16.0	15	C	μ=0.5($NaClO_4$) ΔCp=-43	74Va	
13.90	14.08(0)	-16.2	15	C	μ=0.5(KNO_3) ΔCp=-32	74Va	
14.00	(0)	-----	15	C	μ=0.5($LiNO_3$) ΔCp=-38	74Va	
13.08	13.42(0)	-19.0	35	C	μ=0.5($NaClO_4$) ΔCp=-43	74Va	
13.27	13.41(0)	-18.3	35	C	μ=0.5(KNO_3) ΔCp=-32	74Va	
13.24	(0)	-----	35	C	μ=0.5($LiNO_3$) ΔCp=-38	74Va	
13.39	13.39(0)	-17.8	35	C	μ=0.5(NaCl) ΔCp=-28	74Va	
14.177	(0)	-----	10	C	μ=0.51($NaNO_3$)	67Vd	
13.848	(0)	-----	18	C	μ=0.51($NaNO_3$)	67Vd	
13.80	13.90(0)	-16.5	20	C	μ=0.51	73Fa	
13.599	(0)	-----	25	C	μ=0.51($NaNO_3$)	67Vd	
13.018	(0)	-----	40	C	μ=0.51($NaNO_3$)	67Vd	
12.531	(0)	-----	55	C	μ=0.51($NaNO_3$)	67Vd	
14.516	(0)	-----	0	C	μ=1.0($NaNO_3$)	67Vd	
14.121	(0)	-----	10	C	μ=1.0($NaNO_3$)	67Vd	
14.04	14.09(0)	-15.8	15	C	μ=1.0(NaCl) ΔCp=-27	74Va	
14.20	(0)	-----	15	C	μ=1.0($LiNO_3$) ΔCp=-38	74Va	
13.83	14.11(0)	-16.5	15	C	μ=1.0($NaClO_4$) ΔCp=-43	74Va	

ΔH, kcal /mole	pK	ΔS, cal /mole°K	T, °C	M	Conditions	Ref.	Remarks
W1, cont.							
13.92	14.08(0)	-16.1	15	C	μ=1.0(KNO_3) ΔCp=-31	74Va	
13.857	(0)	-----	18	C	μ=1.0($NaNO_3$)	67Vd	
13.92	13.94(0)	-16.3	20	C	μ=1.0	73Fa	
13.608	(0)	-----	25	C	μ=1.0($NaNO_3$)	67Vd	
13.30	13.41(0)	-18.2	35	C	μ=1.0(KNO_3) ΔCp=-31	74Va	
12.98	13.46(0)	-19.5	35	C	μ=1.0($NaClO_4$) ΔCp=-43	74Va	
13.50	13.41(0)	-17.6	35	C	μ=1.0(NaCl) ΔCp=-27	74Va	
13.43	(0)	-----	35	C	μ=1.0($LiNO_3$) ΔCp=-38	74Va	
13.132	(0)	-----	40	C	μ=1.0($NaNO_3$)	67Vd	
12.645	(0)	-----	55	C	μ=1.0($NaNO_3$)	67Vd	
12.157	(0)	-----	70	C	μ=1.0($NaNO_3$)	67Vd	
12.92	13.010(0)	-19.53	50	T	μ=1M(KCl) ΔCp=-32.9	70Mc	
11.36	11.915(0)	-24.03	100	T	μ=1M(KCl) ΔCp=-29.6	70Mc	
9.96	11.168(0)	-27.55	150	T	μ=1M(KCl) ΔCp=-26.3	70Mc	
8.73	10.662(0)	-30.30	200	T	μ=1M(KCl) ΔCp=-22.9	70Mc	
7.67	10.292(0)	-32.43	250	T	μ=1M(KCl) ΔCp=-19.6	70Mc	
6.78	10.043(0)	-34.07	300	T	μ=1M(KCl) ΔCp=-16.3	70Mc	
14.08	14.19(0)	-16.1	15	C	μ=2.0(NaCl) ΔCp=-24	74Va	
13.69	14.29(0)	-17.9	15	C	μ=2.0($NaClO_4$) ΔCp=-42	74Va	
13.95	14.24(0)	-16.8	15	C	μ=2.0(KNO_3) ΔCp=-29	74Va	
14.55	(0)	-----	15	C	μ=2.0($LiNO_3$) ΔCp=-38	74Va	
14.08	14.07(0)	-16.3	20	C	μ=2.0	73Fa	
12.85	13.63(0)	-20.7	35	C	μ=2.0($NaClO_4$) ΔCp=-42	74Va	
13.60	13.50(0)	-17.6	35	C	μ=2.0(NaCl) ΔCp=-24	74Va	
13.37	13.55(0)	-18.6	35	C	μ=2.0(KNO_3) ΔCp=-29	74Va	
13.79	(0)	-----	35	C	μ=2.0($LiNO_3$) ΔCp=-38	74Va	
14.15	14.14(0)	-16.4	20	C	μ=2.51	73Fa	
15.39	(0)	-----	25	C	2.8462M	40Ka	g:37Pa
14.160	(0)	-----	0	C	μ=3.0($NaNO_3$)	67Vd	
13.800	(0)	-----	10	C	μ=3.0($NaNO_3$)	67Vd	
13.763	(0)	-----	18	C	μ=3.0($NaNO_3$)	67Vd	
14.20	14.20(0)	-16.5	20	C	μ=3.0	73Fa	
13.05	14.22(0)	-21.3	25	C	3M($NaClO_4$)	68Aa 67Ab	
14.245	(0)	-----	25	C	μ=3(NaCl)	52Ba	
13.615	(0)	-----	25	C	μ=3.0($NaNO_3$)	67Vd	
13.350	(0)	-----	40	C	μ=3.0($NaNO_3$)	67Vd	
13.052	(0)	-----	55	C	μ=3.0($NaNO_3$)	67Vd	
12.796	(0)	-----	70	C	μ=3.0($NaNO_3$)	67Vd	
15.83	(0)	-----	25	C	3.2924M	40Ka	g:37Pa
16.27	(0)	-----	25	C	3.7311M	40Ka	g:37Pa
14.555	(0)	-----	25	C	μ=4(NaCl)	52Ba	
16.73	(0)	-----	25	C	4.1625M	40Ka	g:37Pa
17.20	(0)	-----	25	C	4.5868M	40Ka	g:37Pa
14.915	(0)	-----	25	C	μ=5(NaCl)	52Ba	

W1, cont.

ΔH, kcal /mole	pK	ΔS, cal /mole°K	T,°C	M	Conditions	Ref.	Remarks
17.69	(0)	-----	25	C	5.0042M	40Ka	g:37Pa
18.19	(0)	-----	25	C	5.4147M	40Ka	g:37Pa
18.73	(0)	-----	25	C	5.8186M	40Ka	g:37Pa
15.300	(0)	-----	25	C	μ=6(NaCl)	52Ba	
19.15	(0)	-----	25	C	6.156M	40Ka	g:37Pa
19.23	(0)	-----	25	C	6.2161M	40Ka	g:37Pa
19.76	(0)	-----	25	C	6.6072M	40Ka	g:37Pa
20.28	(0)	-----	25	C	6.9922M	40Ka	g:37Pa
15.715	(0)	-----	25	C	μ=7(NaCl)	52Ba	
16.155	(0)	-----	25	C	μ=8(NaCl)	52Ba	
16.630	(0)	-----	25	C	μ=9(NaCl)	52Ba	
17.130	(0)	-----	25	C	μ=10(NaCl)	52Ba	
17.655	(0)	-----	25	C	μ=11(NaCl)	52Ba	
18.195	(0)	-----	25	C	μ=12(NaCl)	52Ba	
18.735	(0)	-----	25	C	μ=13(NaCl)	52Ba	
19.260	(0)	-----	25	C	μ=14(NaCl)	52Ba	
19.825	(0)	-----	25	C	μ=15(NaCl)	52Ba	
20.415	(0)	-----	25	C	μ=16(NaCl)	52Ba	
13.57	(0)	-----	25	C	--	61Ka	
13.519	13.998(0)	-18.7	25	T	--	41Ha	
11.584	13.85(25°,0)	-24.53	25-125	T	--	67Ma	c
7.075	13.12(0)	-36.38	120-200	T	--	67Ma	c
0.0	11.56(0)	-52.8	220-260	T	--	67Ma	c
-6.391	9.08(0)	-62.9	260-300	T	--	67Ma	c
11.869	15.65(25°,0)	-31.8	20-80	T	--	67Ma	c
8.308	14.10(0)	-44.7	80-180	T	--	67Ma	c
5.022	14.26(0)	-48.3	180-240	T	--	67Ma	c
0.0	12.77(0)	-58.4	240-250	T	--	67Ma	c
-5.022	11.18(0)	-67.9	250-300	T	--	67Ma	c
13.45	(0)	-----	25	C	μ=0 S: 5 w % Acetone	75Gb	
13.52	(0)	-----	25	C	μ=0 S: 10 w % Acetone	75Gb	
13.45	(0)	-----	25	C	S: 10 w % Acetone	71Ac	
13.64	(0)	-----	25	C	μ=0 S: 15 w % Acetone	75Gb	
13.79	(0)	-----	25	C	μ=0 S: 20 w % Acetone	75Gb	
13.58	(0)	-----	25	C	S: 20 w % Acetone	71Ac	
13.90	(0)	-----	25	C	μ=0 S: 25 w % Acetone	75Gb	
13.96	(0)	-----	25	C	μ=0 S: 30 w % Acetone	75Gb	
13.63	(0)	-----	25	C	S: 30 w % Acetone	71Ac	
13.95	(0)	-----	25	C	μ=0 S: 35 w % Acetone	71Gb	
13.91	(0)	-----	25	C	μ=0 S: 40 w % Acetone	71Gb	
13.28	(0)	-----	25	C	S: 40 w % Acetone	71Ac	
12.73	(0)	-----	25	C	S: 50 w % Acetone	71Ac	
12.00	(0)	-----	25	C	S: 60 w % Acetone	71Ac	
11.40	(0)	-----	25	C	S: 70 w % Acetone	71Ac	
13.70	(0)	-----	25	C	μ=0 S: 5 w % tert-Butanol	75Gb	
14.09	(0)	-----	25	C	μ=0 S: 10 w % tert-Butanol	75Gb	
14.9	(0)	-----	25	C	S: 10 w % tert-Butanol	71Ad	
14.57	(0)	-----	25	C	μ=0 S: 15 w % tert-Butanol	75Gb	
15.4	(0)	-----	25	C	S: 15 w % tert-Butanol	71Ad	
15.07	(0)	-----	25	C	μ=0 S: 20 w % tert-Butanol	75Gb	
15.6	(0)	-----	25	C	S: 20 w % tert-Butanol	71Ad	
15.7	(0)	-----	25	C	S: 22 w % tert-Butanol	71Ad	

ΔH, kcal /mole	pK	ΔS, cal /mole°K	T,°C	M	Conditions	Ref.	Remarks
W1, cont.							
15.13	(0)	-----	25	C	μ=0 S: 25 w % tert-Butanol	75Gb	
15.8	(0)	-----	25	C	S: 25 w % tert-Butanol	71Ad	
14.95	(0)	-----	25	C	μ=0 S: 30 w % tert-Butanol	75Gb	
15.8	(0)	-----	25	C	S: 30 w % tert-Butanol	71Ad	
14.78	(0)	-----	25	C	μ=0 S: 35 w % tert-Butanol	75Gb	
15.6	(0)	-----	25	C	S: 38 w % tert-Butanol	71Ad	
14.61	(0)	-----	25	C	μ=0 S: 40 w % tert-Butanol	75Gb	
15.5	(0)	-----	25	C	S: 40 w % tert-Butanol	71Ad	
14.9	(0)	-----	25	C	S: 48 w % tert-Butanol	71Ad	
14.7	(0)	-----	25	C	S: 50 w % tert-Butanol	71Ad	
14.1	(0)	-----	25	C	S: 60 w % tert-Butanol	71Ad	
14.78	(0)	-----	25	C	μ=0 S: 0.1 Mole fraction DMSO	73Rb	
14.54	(0)	-----	25	C	μ=0 S: 0.1 Mole fraction DMSO. Standard state of H_2O is unit activity	73Rb	
12.46	(0)	-----	25	C	μ=0 S: 0.2 Mole fraction DMSO	73Rb	
11.88	(0)	-----	25	C	μ=0 S: 0.2 Mole fraction DMSO. Standard state of H_2O is unit activity	73Rb	
11.33	(0)	-----	25	C	μ=0 S: 0.3 Mole fraction DMSO	73Rb	
10.36	(0)	-----	25	C	μ=0 S: 0.3 Mole fraction DMSO. Standard state of H_2O is unit activity	73Rb	
11.04	(0)	-----	25	C	μ=0 S: 0.4 Mole fraction DMSO	73Rb	
9.31	(0)	-----	25	C	μ=0 S: 0.4 Mole fraction DMSO. Standard state of H_2O is unit activity	73Rb	
9.09	(0)	-----	25	C	μ=0 S: 0.5 Mole fraction DMSO	73Rb	
6.86	(0)	-----	25	C	μ=0 S: 0.5 Mole fraction DMSO. Standard state of H_2O is unit activity	73Rb	
11.09	(0)	-----	25	C	μ=0 S: 0.6 Mole fraction DMSO	73Rb	
8.55	(0)	-----	25	C	μ=0 S: 0.6 Mole fraction DMSO. Standard state of H_2O in unit activity	73Rb	
15.45	(0)	-----	25	C	μ=0 S: 0.7 Mole fraction DMSO	73Rb	
12.75	(0)	-----	25	C	μ=0 S: 0.7 Mole fraction DMSO. Standard state of H_2O in unit activity	73Rb	
18.68	(0)	-----	25	C	μ=0 S: 0.8 Mole fraction DMSO	73Rb	
15.96	(0)	-----	25	C	μ=0 S: 0.8 Mole fraction DMSO. Standard state of H_2O in unit activity	73Rb	

ΔH, kcal /mole	pK	ΔS, cal /mole°K	T,°C	M	Conditions	Ref.	Remarks
W1, cont.							
13.60	(0)	-----	25	C	μ=0 S: 5 w % DMSO	75Gb	
13.73	(0)	-----	25	C	μ=0 S: 10 w % DMSO	75Gb	
13.88	(0)	-----	25	C	μ=0 S: 15 w % DMSO	75Gb	
14.04	(0)	-----	25	C	μ=0 S: 20 w % DMSO	75Gb	
14.21	(0)	-----	25	C	μ=0 S: 25 w % DMSO	75Gb	
14.41	(0)	-----	25	C	μ=0 S: 30 w % DMSO	75Gb	
14.65	(0)	-----	25	C	μ=0 S: 35 w % DMSO	75Gb	
15.06	(0)	-----	25	C	μ=0 S: 40 w % DMSO	75Gb	
13.515	14.622(0)	-21.5	25	T	S: 20 w % Dioxane	41Ha 62Aa	
13.65	(0)	-----	25	C	S: 25.68 w % Dioxane	61Ka	
13.190	15.744(0)	-27.8	25	T	S: 45 w % Dioxane	41Ha 62Aa	
13.70	(0)	-----	25	C	S: 50.75 w % Dioxane	61Ka	
12.662	17.859(0)	-39.2	25	T	S: 70 w % Dioxane	41Ha 62Aa	
13.89	(0)	-----	25	C	S: 75.61 w % Dioxane	61Ka	
13.6	(0)	-----	25	C	μ=0 S: 0.0336 Mole fraction Ethanol	66Bb	
13.6	(0)	-----	25	C	μ=0 S: 0.0727 Mole fraction Ethanol	66Bb	
13.6	(0)	-----	25	C	μ=0 S: 0.0948 Mole fraction Ethanol	66Bb	
13.3	(0)	-----	25	C	μ=0 S: 0.145 Mole fraction Ethanol	66Bb	
12.8	(0)	-----	25	C	μ=0 S: 0.174 Mole fraction Ethanol	66Bb	
12.0	(0)	-----	25	C	μ=0 S: 0.240 Mole fraction Ethanol	66Bb	
10.3	(0)	-----	25	C	μ=0 S: 0.372 Mole fraction Ethanol	66Bb	
8.0	(0)	-----	25	C	μ=0 S: 0.566 Mole fraction Ethanol	66Bb	
5.3	(0)	-----	25	C	μ=0 S: 0.757 Mole fraction Ethanol	66Bb	
5.2	(0)	-----	25	C	μ=0 S: 0.881 Mole fraction Ethanol	66Bb	
5.1	(0)	-----	25	C	μ=0 S: 0.938 Mole fraction Ethanol	66Bb	
13.55	(0)	-----	25	C	μ=0 S: 5 w % Ethanol	75Gb	
13.6	(0)	-----	25	C	S: 8.16 w % Ethanol	66Bc	
13.64	(0)	-----	25	C	μ=0 S: 10 w % Ethanol	75Gb	
13.75	(0)	-----	25	C	S: 10 w % Ethanol	67Ad	
13.67	(0)	-----	25	C	μ=0 S: 15 w % Ethanol	75Gb	
13.6	(0)	-----	25	C	S: 16.70 w % Ethanol	66Bc	
13.65	(0)	-----	25	C	μ=0 S: 20 w % Ethanol	75Gb	
13.81	(0)	-----	25	C	S: 20 w % Ethanol	67Ad	
13.6	(0)	-----	25	C	S: 21.12 w % Ethanol	66Bc	
13.54	(0)	-----	25	C	μ=0 S: 25 w % Ethanol	75Gb	
13.34	(0)	-----	25	C	μ=0 S: 30 w % Ethanol	75Gb	

ΔH, kcal /mole	pK	ΔS, cal /mole$°K$	T,$°C$	M	Conditions	Ref.	Re-marks
W1, cont.							
13.52	(0)	-----	25	C	S: 30 w % Ethanol	67Ad	
13.3	(0)	-----	25	C	S: 30.25 w % Ethanol	66Bc	
13.02	(0)	-----	25	C	$\mu=0$ S: 35 w % Ethanol	75Gb	
12.8	(0)	-----	25	C	S: 35.01 w % Ethanol	66Bc	
12.57	(0)	-----	25	C	$\mu=0$	75Gb	
					S: 40 w % Ethanol		
12.76	(0)	-----	25	C	S: 40 w % Ethanol	67Ad	
12.0	(0)	-----	25	C	S: 44.68 w % Ethanol	66Bc	
11.86	(0)	-----	25	C	S: 50 w % Ethanol	67Ad	
11.13	(0)	-----	25	C	S: 60 w % Ethanol	67Ad	
10.3	(0)	-----	25	C	S: 60.24 w % Ethanol	66Bc	
10.13	(0)	-----	25	C	S: 70 w % Ethanol	67Ad	
8.0	(0)	-----	25	C	S: 76.93 w % Ethanol	66Bc	
9.10	(0)	-----	25	C	S: 80 w % Ethanol	67Ad	
5.3	(0)	-----	25	C	S: 88.85 w % Ethanol	66Bc	
5.2	(0)	-----	25	C	S: 94.98 w % Ethanol	66Bc	
5.1	(0)	-----	25	C	S: 97.48 w % Ethanol	66Bc	
14.51	(0)	-----	25	C	S: 10 w % Isopropanol	71Ad	
14.91	(0)	-----	25	C	S: 15 w % Isopropanol	71Ad	
15.20	(0)	-----	25	C	S: 20 w % Isopropanol	71Ad	
15.34	(0)	-----	25	C	S: 25 w % Isopropanol	71Ad	
15.06	(0)	-----	25	C	S: 30 w % Isopropanol	71Ad	
14.91	(0)	-----	25	C	S: 35 w % Isopropanol	71Ad	
14.46	(0)	-----	25	C	S: 40 w % Isopropanol	71Ad	
13.86	(0)	-----	25	C	S: 50 w % Isopropanol	71Ad	
13.41	(0)	-----	25	C	S: 60 w % Isopropanol	71Ad	
13.15	(0)	-----	25	C	S: 5 w % Methanol	71Ac	
12.85	(0)	-----	25	C	S: 10 w % Methanol	71Ac	
11.85	(0)	-----	25	C	S: 20 w % Methanol	71Ac	
11.10	(0)	-----	25	C	S: 30 w % Methanol	71Ac	
10.20	(0)	-----	25	C	S: 40 w % Methanol	71Ac	
9.35	(0)	-----	25	C	S: 50 w % Methanol	71Ac	
8.60	(0)	-----	25	C	S: 60 w % Methanol	71Ac	
7.75	(0)	-----	25	C	S: 70 w % Methanol	71Ac	
7.35	(0)	-----	25	C	S: 80 w % Methanol	71Ac	
13.19	(0)	-----	25	C	$\mu=0$	75Gb	m
					S: 5 w % Urea		
13.01	(0)	-----	25	C	$\mu=0$	75Gb	m
					S: 10 w % Urea		
12.83	(0)	-----	25	C	$\mu=0$	75Gb	m
					S: 15 w % Urea		
12.66	(0)	-----	25	C	$\mu=0$	75Gb	m
					S: 20 w % Urea		
12.52	(0)	-----	25	C	$\mu=0$	75Gb	m
					S: 25 w % Urea		
12.35	(0)	-----	25	C	$\mu=0$	75Gb	m
					S: 30 w % Urea		
12.15	(0)	-----	25	C	$\mu=0$	75Gb	m
					S: 35 w % Urea		
11.91	(0)	-----	25	C	$\mu=0$	75Gb	m
					S: 40 w % Urea		
9.4	3.81(25°,+3)	14.1	46.2-94.6	T	HA=$[Cr(III)(H_2O)_6]^{3+}$	55Pa	c
W2 WATER, deuterated							
15.371	15.740(0)	-16.76	5	T	$\mu=0$	66Cd	
14.488	14.869(0)	-19.45	25	C	$\mu=0$	68Gb	
14.420	14.710(0)	-18.95	25	T	$\mu=0$	36Wa	
14.311	14.955(0)	-20.43	25	T	$\mu=0$	66Cd	
12.884	14.182(0)	-25.03	50	T	$\mu=0$	66Cd	
14.488	14.871(0)	-19.45	25	C	$\mu=0$ S: D_2O	68Gb	

ΔH, kcal /mole	pK	ΔS, cal /mole°K	T,°C	M	Conditions	Ref.	Remarks
			X				
X1 XANTHINE			$C_5H_4O_2N_4$		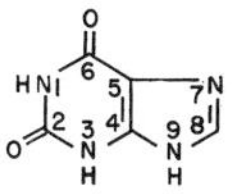		
6.33	7.53(0)	-13.2	25	C	μ=0	70Cb	e: N_1-C_6O
10.18	12.36(-1)	-20.6	10	C	μ=0	70Cb	e:N_9
9.61	11.84(-1)	-22.0	25	C	μ=0	70Cb	e:N_9
9.16	11.51(-1)	-23.4	40	C	μ=0	70Cb	e:N_9
X2 XANTHOSINE			$C_{10}H_{12}O_6N_4$		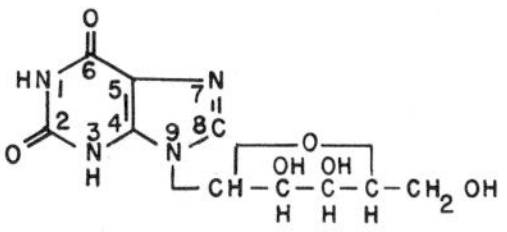		
3.74	5.67(0)	-13.4	25	C	μ=0	70Cb	e:N_1
11.02	12.85(-1)	-19.9	10	C	μ=0	70Cb	e: Ribose OH
10.86	12.00(-1)	-18.9	25	C	μ=0	70Cb	e: Ribose OH
10.75	11.76(-1)	-19.5	40	C	μ=0	70Cb	e: Ribose OH
X3 XYLOSE			$C_5H_{10}O_5$		$HOCH_2CH(OH)CH(OH)CH(OH)CHO$		
9.37	12.61(0)	-24.7	10	C	μ=0	70Cd	
8.2	12.29(0)	-28.7	25	C	μ=0	66Ib	
9.0	12.15(0)	-25.4	25	C	μ=0	70Cd	

EMPIRICAL FORMULA INDEX

$\underline{C_1}$

$CHON$ C21
CHO_6N_3 M23
CHN H35
$CHNS$ T28
CH_2ON_4 T12
CH_2O_2 F6
CH_2O_3 C3
CD_2O_3 C4
$CH_2O_4N_2$ M21
CH_2N_4 T10
CH_2S_3 C6
CH_3O_2N M22
CH_4O M26
CH_4N_2Se U9
CH_4S_2 M25
CH_5ON_3 S5
CH_5N M19
CH_5N_3Se S8
$CH_6O_6P_2$ M24
CH_6N_3S S9

$\underline{C_2}$

$C_2HO_2F_3$ A38
$C_2HO_2F_2Cl$ A9
$C_2HO_2Cl_3$ A37
$C_2HO_2Br_3$ A36
$C_2HN_3Br_2$ T56
$C_2HN_4F_3$ T17
$C_2H_2O_2F_2$ A13
$C_2H_2O_2Cl_2$ A12
$C_2H_2O_2Br_2$ A11
$C_2H_2O_3$ A28
$C_2H_2O_4$ O10
$C_2H_3O_2Cl$ A7
$C_2H_3O_2Br$ A6
$C_2H_3O_2I$ A24
$C_2H_3O_2F$ A15
C_2H_3N A4
$C_2H_3N_3$ T54,T55
C_2H_4OS A35
$C_2H_4O_2$ A1
$C_2H_3DO_2$ A2
$C_2HD_3O_2$ A2
$C_2D_4O_2$ A2
$C_2H_4O_2S$ A25
$C_2H_4O_3$ A16
$C_2H_4O_4N_2$ E29
$C_2H_4N_4$ T13
$C_2H_4N_4S$ T14
C_2H_5ON F8
$C_2H_5O_2N$ E30,G15
C_2H_5N A156
C_2H_5ClS S25
$C_2H_6ON_2$ G16
C_2H_6OS E44,S31
$C_2H_6O_2$ E31
$C_2H_6O_3S$ S37
$C_2H_6O_4S$ D5
C_2H_6S E36
C_2H_7ON E41
$C_2H_7O_3NS$ T2
C_2H_7N A83,E3
C_2H_7NS E38
$C_2H_8O_3NP$ E34
$C_2H_8O_5NP$ E42
$C_2H_8N_2$ E6
$C_2H_8N_4S$ C2
$C_2H_{12}O_4B_{12}$ D11

$\underline{C_3}$

C_3HO_2Cl P187
C_3HO_2Br P186
$C_3H_2OF_6$ P182
$C_3H_2O_2F_6$ P148
$C_3H_2N_2$ M7
$C_3H_3O_2N$ M8
$C_3H_3O_2N_3$ T57
C_3H_3NS T18
$C_3H_4O_2$ P185
$(C_3H_4O_2)_n$ P125
$C_3H_4O_2Cl_2$ P163,P164
$C_3H_4O_2Br_2$ P162
$C_3H_4O_3$ P175
$C_3H_4O_4$ M6
$C_3H_4N_2$ I1,P193
$C_3H_5O_2Cl$ P157,P158
$C_3H_5O_2Br$ P155,P156
$C_3H_5O_2I$ P171
$C_3H_6O_2$ P151
$C_3H_6O_2S$ P172,P173
$C_3H_6O_3$ A27,P168
$C_3H_6O_3N_2$ G21

$C_3H_6O_4N_2$ P141
C_3H_7ON F7
$C_3H_7O_2N$ A53,A54,G18, P142,P143,S1
$C_3H_7O_2NS$ C47
$C_3H_7O_3N$ I24,S10
C_3H_7N A155,C42,P184
$C_3H_8O_2$ P144
$C_3H_8O_2N_2$ P160
$C_3H_8O_2S$ P149
C_3H_8S E40
C_3H_9ON E4,E45,P177, P178,P179
C_3H_9OP P87
$C_3H_9O_4P$ P96
$C_3H_9O_6P$ G13,G14
C_3H_9N A92,P130,P131
C_3H_9NS B190,E5
$C_3H_{10}ON_2$ P181
$C_3H_{10}N_2$ E22,P134,P135

$\underline{C_4}$

$C_4HO_4N_3$ T60
$C_4H_2O_4$ C25
$C_4H_4O_2$ B236
$C_4H_4O_2N_2$ P167,U1
$C_4H_4O_3N_2$ B1
$C_4H_4O_4$ F10,M1
$C_4H_5ON_3$ C60
$C_4H_5O_2N$ S17
C_4H_5NS T22,T23,T24
$C_4H_6ONF_3$ A39
$C_4H_6ON_2$ I6,I7
$C_4H_6O_2$ B234,C43
$C_4H_6O_3$ B224,D7
$C_4H_6O_4$ S16
$C_4H_6O_4S$ T29
$C_4H_6O_5$ A29,S20
$C_4H_6O_6$ T1
$C_4H_6N_2$ I8,I9,I10
$C_4H_6N_4$ T11
$C_4H_7O_2Br$ B215
$C_4H_7O_3N$ G19
$C_4H_7O_4N$ A18,A151,M10
$C_4H_7O_5N$ M9
$C_4H_7N_3$ I3
C_4H_8ONCl A8
C_4H_8OS S33
$C_4H_8O_2$ B201,D6
$C_4H_8O_2$ P174
$C_4H_8O_2N_2$ G31
$C_4H_8O_2S$ S29
$C_4H_8O_3$ B218,P169
$C_4H_8O_3N_2$ A61,A62,A149, G24
$C_4H_8O_4N_2$ B199
C_4H_8S T32
C_4H_9ON A3,M33
C_4H_9ONS E37
$C_4H_9O_2N$ A55,A56,B203, B204,B205, G23,G17,P153, P154
$C_4H_9O_2NS$ C49
$C_4H_9O_3N$ A64,B211,B212, S12,T34,T35
C_4H_9N P305
$C_4H_{10}ON_2$ M34
$C_4H_{10}O_2$ E33
$C_4H_{10}O_2N_2$ P161
$C_4H_{10}O_3ClP$ C9
$C_4H_{10}N_2$ P101
$C_4H_{10}S$ E39,S23
$C_4H_{11}ON$ A77,B225,E43, P180
$C_4H_{11}ONS$ P24
$C_4H_{11}O_2N$ A76,P147
$C_4H_{11}O_3N$ P146
$C_4H_{11}N$ A75,B187,B188, P132,P133
$C_4H_{12}ON_2$ E19,E46
$C_4H_{12}N_2$ B192,B193,B194, B195,E12,E13, P137
$C_4H_{13}N_3$ D4

$\underline{C_5}$

C_5HNCl_4 P291
$C_5H_3ONCl_2$ P229,P230
C_5H_3NClF P224
$C_5H_3NCl_2$ P226,P227,P228
$C_5H_4ON_4$ H43
$C_5H_4O_2N_2$ P284,P285,P286
$C_5H_4O_2N_4$ X1
$C_5H_4O_2S$ T33
$C_5H_4O_3$ F11
$C_5H_4O_3N_4$ U11
$C_5H_4O_4N_2$ U3
C_5H_4NCl P221,P222,P223
C_5H_4NBr P217,P218,P219
$C_5H_4N_4$ P190

$\mathrm{C_5H_5ON}$ P262,P263,P264,P287
$\mathrm{C_5H_5ON_5}$ G32
$\mathrm{C_5H_5O_2N}$ P304
$\mathrm{C_5H_5N}$ P195,P225
$\mathrm{C_5D_5N}$ P225
$\mathrm{C_5H_5NS}$ P268
$\mathrm{C_5H_5N_2Br}$ P204
$\mathrm{C_5H_5N_5}$ P191
$\mathrm{C_5H_6O_2N_2}$ U2
$\mathrm{C_5H_6O_3N_2}$ B14
$\mathrm{C_5H_6O_4}$ C44
$\mathrm{C_5H_6N_2}$ P201,P202,P203
$\mathrm{C_5H_7O_2N}$ P159
$\mathrm{C_5H_7NS}$ T19,T20,T21
$\mathrm{C_5H_7N_3I_2}$ H23
$\mathrm{C_5H_8O_2}$ C23,P13
$\mathrm{C_5H_8O_2N_2}$ C39
$\mathrm{C_5H_8O_3}$ P23
$\mathrm{C_5H_8O_4}$ P8
$\mathrm{C_5H_8N_2}$ I4
$\mathrm{C_5H_8N_3I}$ H24
$\mathrm{C_5H_9ON}$ P307
$\mathrm{C_5H_9O_2N}$ P127
$\mathrm{C_5H_9O_3N}$ A57,G30,P129
$\mathrm{C_5H_9O_4N}$ A23,G9
$\mathrm{C_5H_9O_4N_3S}$ S7
$\mathrm{C_5H_9O_5N_3}$ S6
$\mathrm{C_5H_9N_3}$ H22,I2,I20
$\mathrm{C_5H_{10}O_2}$ B222,B223,P15,P166
$\mathrm{C_5H_{10}O_3}$ C5
$\mathrm{C_5H_{10}O_3N_2}$ A63,B207,B208,G12,G20,P152
$\mathrm{C_5H_{10}O_4}$ R3
$\mathrm{C_5H_{10}O_4N_2}$ P6,S11
$\mathrm{C_5H_{10}O_5}$ A143,L9,R2,X3
$\mathrm{C_5H_{11}ON}$ M39
$\mathrm{C_5H_{11}O_2N}$ B213,B214,M38,P16,P17,V1
$\mathrm{C_5H_{11}O_2NS}$ C50,M28
$\mathrm{C_5H_{11}O_3N_2}$ S2
$\mathrm{C_5H_{11}O_8P}$ R4
$\mathrm{C_5H_{11}N}$ C37,P111,P306
$\mathrm{C_5H_{12}ON_2}$ B232,U10
$\mathrm{C_5H_{12}O_2N}$ B166
$\mathrm{C_5H_{12}O_5}$ P165
$\mathrm{C_5H_{12}N_2}$ P106,P107
$\mathrm{C_5H_{13}O_2N}$ A79,A87
$\mathrm{C_5H_{13}N}$ A68,B191,P3
$\mathrm{C_5H_{14}N_2}$ E28,P5
$\mathrm{C_5H_{19}N}$ A80
$\mathrm{C_5ONCl_5}$ P289
$\mathrm{C_5NCl_4F}$ P261
$\mathrm{C_5NCl_5}$ P288

$\mathrm{C_6}$

$\mathrm{C_6H_3OCl_3}$ P81,P82
$\mathrm{C_6H_3O_6N_4Br}$ A103
$\mathrm{C_6H_3O_7N_3}$ P83
$\mathrm{C_6H_4OCl_2}$ P55,P56
$\mathrm{C_6H_4OBr_2}$ P54
$\mathrm{C_6H_4OI_2}$ P58
$\mathrm{C_6H_4O_2N_2Cl_2}$ A113,A114,A115
$\mathrm{C_6H_4O_2Cl_2}$ B49
$\mathrm{C_6H_4O_4N_3Br}$ A102
$\mathrm{C_6H_4O_5N_2}$ P60,P61,P62,P63
$\mathrm{C_6H_4O_6N_4}$ A137
$\mathrm{C_6H_4N_2}$ P302,P303
$\mathrm{C_6H_4Cl_2}$ B48
$\mathrm{C_6H_5ON}$ P296,P297
$\mathrm{C_6H_5OCl}$ P50,P51,P52
$\mathrm{C_6H_5OCl_2P}$ P86
$\mathrm{C_6H_5OBr}$ P46,P47,P48
$\mathrm{C_6H_5OI}$ P69,P70,P71
$\mathrm{C_6H_5OF}$ P66,P67,P68
$\mathrm{C_6H_5O_2N}$ P298,P299,P300
$\mathrm{C_6H_5O_2N_2Cl}$ A107,A108
$\mathrm{C_6H_5O_2Cl}$ B38
$\mathrm{C_6H_5O_3N}$ P76,P77,P78
$\mathrm{C_6H_5O_4N}$ B53
$\mathrm{C_6H_5O_4N_3}$ A119,A120,A121
$\mathrm{C_6H_5NCl_2}$ A112
$\mathrm{C_6H_5NBr_2}$ A111
$\mathrm{C_6H_5NI_2}$ A116
$\mathrm{C_6H_5N_3}$ B164
$\mathrm{C_6H_6O}$ P44
$\mathrm{C_6H_6ON_2}$ P200
$\mathrm{C_6H_6OS}$ T30
$\mathrm{C_6H_6O_2}$ B50,B51
$\mathrm{C_6H_6O_2N_2}$ A134,A135,A136
$\mathrm{C_6H_6O_4S}$ B92
$\mathrm{C_6H_6O_8S_2}$ B85
$\mathrm{C_6H_6NCl}$ A104,A105,A106
$\mathrm{C_6H_6NBr}$ A99,A100,A101
$\mathrm{C_6H_6NI}$ A128,A129,A130
$\mathrm{C_6H_6NF}$ A125,A126,A127
$\mathrm{C_6H_6N_4}$ B174

C_7

C_7H_9ON A131,A132,A133, P247,P248,P249, P250,P251,P259, P260

$C_7H_9O_2N_3$ S15

$C_7H_9O_3NS$ B90,B91

C_7H_9N T40,P234,P235, P236,P237,P238, P239,P252,P253, P254,T37,T38, T39

$C_7H_9N_2Br$ P220

$C_7H_9N_3S$ U5

$C_7H_{10}O_3N_2$ B13

$C_7H_{10}O_4$ C41,C45,C46

$[C_7H_{10}O_4]_n$ M5

$C_7H_{10}N_2$ B42,P206,P207, P208,P214,P240, P276,P278

$C_7H_{11}O_2N_3$ H26,H28

$C_7H_{11}O_6N$ A19

$C_7H_{11}NS$ T25,T26

$C_7H_{12}O_2$ C32,H6,H7,H18

$C_7H_{12}O_3N_2$ P128

$C_7H_{12}O_4$ H5,M12,P9

$C_7H_{13}N$ Q21

$C_7H_{14}O_2$ H8,H9,P19

$C_7H_{14}O_4N_2S_2$ D10

$C_7H_{15}O_2N$ H10

$C_7H_{15}N$ C36,P114,P115, P116

$C_7H_{16}ON_2$ B229

$C_7H_{16}O_2N_2$ L5

$C_7H_{16}N_2$ P112

$C_7H_{19}N_3$ D9

$C_7H_{20}N_4$ P136

<u>C_8</u>

$C_8H_5O_2N$ B83,B84

$C_8H_6O_4$ B82,P100

$C_8H_7N_3$ I11,I12

C_8H_8O A40

C_8H_8ON P295

$C_8H_8O_2$ A31,A41,B133, B150,B151,B152

$C_8H_8O_2S$ A26

$C_8H_8O_2Se$ A34

$C_8H_8O_3$ A17,A30,B22,B23, B24,B135,B147, B148,B149

$C_8H_8O_3S$ A33

$C_8H_8O_4$ B139

$C_8H_8O_4S$ A32

$C_8H_8N_2$ B98

C_8H_9OCl B39

$C_8H_9O_2N$ B109,B110,B111

$C_8H_9O_7N_3$ B4

$C_8H_{10}O$ B60,B61,B62,B63, B64,B65,B66

$C_8H_{10}O_2$ P64,P65

$C_8H_{10}O_3$ B67,P59

$C_8H_{10}O_8$ B200

$C_8H_{11}ON$ A122,A123,A124, P257,P258

$C_8H_{11}O_2N$ A117

$C_8H_{11}O_3N_3$ H27

$C_8H_{11}N$ A118,B26,B27,B28, B29,B30,B31,B32, B33,B34,P255, P266,P290,P293, P294

$C_8H_{12}O_2N_2$ P192

$C_8H_{12}O_3N$ G27

$C_8H_{12}O_3N_2$ B7

$C_8H_{12}O_4$ C33,C34,C35

$[C_8H_{12}O_4]_n$ M4

$C_8H_{12}O_6N_2S_2$ C52

$C_8H_{12}N$ C28

$C_8H_{12}N_2$ E23,P209,P245, P256,P275,P277

$C_8H_{14}O_2$ A10,H17

$C_8H_{14}O_4$ O6,P10

$C_8H_{14}O_4S_2$ D3

$C_8H_{14}O_5N_4$ G26

$C_8H_{14}O_7N_2$ E20

$C_8H_{16}ON_2$ B226

$C_8H_{16}O_2$ O9

$C_8H_{16}O_2N_2$ O8

$C_8H_{16}O_3N_2$ L3,L6

$C_8H_{16}O_4N_2$ E14

$C_8H_{16}O_4N_2S_2$ H29

$C_8H_{17}O_2N$ L2

$C_8H_{17}N$ A153,C18

$C_8H_{18}ON_2$ B230,B233

$C_8H_{18}OS$ S30

$C_8H_{18}O_4N_2S$ P104

$C_8H_{18}O_6N_2S_2$ P103

$C_8H_{18}S$ S22

$C_8H_{19}O_5N$ P145

$C_8H_{19}N$ A71,A72

$C_8H_{20}N_2$ E17,E18

$C_8H_{22}N_4$ D1

$C_8H_{23}N_5$ T9

C_9

$C_9H_4O_2$	B169
$C_9H_6ON_3ClS$	P53
$C_9H_6O_2$	P188
$C_9H_6O_4NIS$	Q17
$C_9H_6O_6$	B94,B95,B96
$C_9H_6O_6N_2S$	Q18
C_9H_7ON	Q9
$C_9H_7O_2N_3$	T58
$C_9H_7O_2Br$	C15
$C_9H_7O_4NS$	Q16
C_9H_7N	I21,Q2
C_9H_7NS	Q12
$C_9H_8ON_2$	Q1
$C_9H_8O_2$	C14
$C_9H_8O_3$	C16
$C_9H_8N_2$	I22,I23,Q3,Q4,Q5,Q6,Q7,Q8
$C_9H_9O_3NCl_2$	T74
$C_9H_9O_3NBr_2$	T73
$C_9H_9O_3NI_2$	T75
$C_9H_{10}O$	B25
$C_9H_{10}O_2$	B124,B125,B126,B127,B128,P176
$C_9H_{10}O_3$	B130,B138
$C_9H_{10}O_4$	B122,B123
$C_9H_{10}N_2$	B97
$C_9H_{11}ON$	B17,B102
$C_9H_{11}O_2N$	B106,P84
$C_9H_{11}O_2N_3$	P232
$C_9H_{11}O_3N$	P85,T68,T69,T70
$C_9H_{11}O_4NS$	C48
$C_9H_{12}O$	B71,B73,B74,B75,B76,B77
$C_9H_{12}O_4$	B78
$C_9H_{12}O_6N_2$	U12
$C_9H_{13}O_2Br$	B167
$C_9H_{13}O_4N_3$	C54
$C_9H_{13}O_5N_3$	C53
$C_9H_{13}O_9N_2P$	U14
$C_9H_{13}N$	P292
$C_9H_{14}ON$	P79,P80
$C_9H_{14}O_3$	B171
$C_9H_{14}O_7N_3P$	C57
$C_9H_{14}O_8N_3P$	C56,C59
$C_9H_{14}O_{11}N_2P_2$	U13
$C_9H_{14}N_2$	P215,P241,P242,P243,P267
$C_9H_{15}O_2N$	M14
$C_9H_{15}O_{11}N_3P_2$	C55
$C_9H_{15}O_{13}N_2P_3$	U15
$C_9H_{15}N$	A88
$C_9H_{16}O_4$	M13
$C_9H_{16}O_{14}N_3P_3$	C58
$C_9H_{18}O_2N_2$	M20
$C_9H_{19}N$	P121
$C_9H_{20}ON_2$	B227
$C_9H_{21}O_3N$	A94
$C_9H_{21}N$	A93
$C_9H_{24}N_4$	T63

C_{10}

$C_{10}H_6O_8$	B93
$C_{10}H_7O_4N_3$	T61
$C_{10}H_7O_5NS$	N30
$C_{10}H_8O_2$	N3
$C_{10}H_8O_3N_2$	B16
$C_{10}H_8O_8S_2$	N28
$C_{10}H_8N_2$	B178
$C_{10}H_9ON$	Q10,Q11,Q14,Q15
$C_{10}H_9ON_3S$	T51
$C_{10}H_9O_2N_3$	T59
$C_{10}H_9O_2N_3S$	P75
$C_{10}H_9O_3NS$	N29
$C_{10}H_9N$	N1,N2
$C_{10}H_9NS$	Q13
$C_{10}H_{11}O_5N_4$	I13
$C_{10}H_{12}O_2$	B158
$C_{10}H_{12}O_2N_2$	B202
$C_{10}H_{12}O_3N_2$	B6
$C_{10}H_{12}O_6N_4$	X2
$C_{10}H_{12}O_8N_4P$	I15
$C_{10}H_{13}O_2N$	A66,B170
$C_{10}H_{13}O_3N_5$	A48
$C_{10}H_{13}O_4N_5$	A47,G34
$C_{10}H_{13}O_5N_5$	G33
$C_{10}H_{13}O_{11}N_4P_2$	I14
$C_{10}H_{13}N_3S_2$	B181
$C_{10}H_{14}O$	B35,B36,B37
$C_{10}H_{14}O_2$	B58
$C_{10}H_{14}O_2Si$	B159
$C_{10}H_{14}O_2Si$	B160
$C_{10}H_{14}O_3$	P57
$C_{10}H_{14}O_4$	B173
$[C_{10}H_{14}O_5]_n$	M3
$C_{10}H_{14}O_5N_2$	T36

$C_{10}H_{14}O_6N_5P$ A49
$C_{10}H_{14}O_7N_5P$ A51,G37
$C_{10}H_{14}O_8N_5P$ G36
$C_{10}H_{14}O_{14}N_4P_3$ I16
$C_{10}H_{14}N_2$ P110
$C_{10}H_{15}ON$ E1
$C_{10}H_{15}O_{10}N_5P_2$ A50
$C_{10}H_{15}O_{11}N_5P_2$ G35
$C_{10}H_{16}O_3N_4$ A138
$C_{10}H_{16}O_4$ C1,P12
$C_{10}H_{16}O_6N_2S_2$ C51
$C_{10}H_{16}O_8N_2$ E24
$C_{10}H_{16}O_{13}N_5P_3$ A52
$C_{10}H_{16}O_{14}N_5P_3$ G38
$C_{10}H_{16}N_2$ P244,P279
$C_{10}H_{18}O_4$ M15
$C_{10}H_{18}O_7N_2$ E21
$C_{10}H_{19}N$ A74
$C_{10}H_{20}O_4N_2$ D2
$C_{10}H_{23}N$ A81
$C_{10}H_{24}O_4N_2$ E26
$C_{10}H_{24}N_2$ E9
$C_{10}H_{26}N_4$ D12
$C_{10}H_{28}N_6$ E25

C_{11}

$C_{11}H_8O_3$ N31
$C_{11}H_{10}O_3N_2$ B15
$C_{11}H_{11}ON_3S$ P45
$C_{11}H_{11}O_6N$ A20
$C_{11}H_{12}ON_4S$ B59
$C_{11}H_{12}O_2N_2$ T65
$C_{11}H_{13}ON_3$ A140
$C_{11}H_{14}O_2$ B115,B116,B156
$C_{11}H_{14}O_3N_2$ G29
$C_{11}H_{17}N$ T46
$C_{11}H_{18}O_3N_2$ B5,B10,B11
$C_{11}H_{18}O_8N_2$ P138,P139
$C_{11}H_{21}O_5N_3$ G11
$C_{11}H_{30}N_6$ P140

C_{12}

$C_{12}H_5O_{12}N_7$ A86
$C_{12}H_6O_{12}$ B86
$C_{12}H_7O_2N_3$ P38,P39,P40
$C_{12}H_8ON_2$ B165,P42,P43
$C_{12}H_8O_6N_4$ A84
$C_{12}H_8N_2$ P31,P32,P33
$C_{12}H_9O_4N_3$ B55,B56
$C_{12}H_9N_3$ B99
$C_{12}H_{10}O_2NCl$ S28
$C_{12}H_{10}O_2N_2$ B54
$C_{12}H_{10}O_5N_2S$ B57
$C_{12}H_{10}S$ S26
$C_{12}H_{11}ON$ P216
$C_{12}H_{11}O_8N$ A21
$C_{12}H_{11}N$ B175,B176, B177,P283
$C_{12}H_{11}N_3S$ U6
$C_{12}H_{12}O_3N_2$ B12
$C_{12}H_{12}O_4N_2S$ B70
$C_{12}H_{18}ON_2$ P105
$C_{12}H_{18}O_3N_2$ B3
$C_{12}H_{18}O_4$ B168
$C_{12}H_{18}N_2$ P108,P109
$C_{12}H_{20}O_8N_2$ B197,B198
$C_{12}H_{20}O_8N_2S$ S24
$C_{12}H_{20}O_9N_2$ E47
$C_{12}H_{22}O_2N$ B172
$C_{12}H_{23}N$ A73
$C_{12}H_{24}O_4$ S18
$C_{12}H_{24}O_4N_4$ A60
$C_{12}H_{26}O_3N_4$ L7
$C_{12}H_{26}O_4N_2$ C19

C_{13}

$C_{13}H_9O_5N_3$ B140,B141
$C_{13}H_9N$ B162,B163
$C_{13}H_{10}N_2$ P37
$C_{13}H_{11}ON$ B161
$C_{13}H_{11}N$ F5
$C_{13}H_{13}N_3S$ U4,U7,U8
$C_{13}H_{14}ON$ A43,A44,A45
$C_{13}H_{14}N$ A42
$C_{13}H_{16}O_6N_2$ T72
$C_{13}H_{17}ON_3$ A141
$C_{13}H_{17}O_4N$ T71
$C_{13}H_{20}O_8N_2$ C38
$C_{13}H_{22}O_8N_2$ P7

C_{14}

$C_{14}H_5N_5$ M11
$C_{14}H_{11}N$ A139,P27,P28, P29,P30
$C_{14}H_{12}O_3$ A14
$C_{14}H_{12}N_2$ P35
$C_{14}H_{13}O_2N$ B100
$C_{14}H_{13}N_3S_2$ B179
$C_{14}H_{14}O_3N_3SNa$ M29

$C_{14}H_{15}N$ A70
$C_{14}H_{16}N_2$ E15,E16
$C_{14}H_{21}OBr$ P49
$C_{14}H_{21}O_3N$ B47
$C_{14}H_{22}O$ B44,B45,B46
$C_{14}H_{22}O_8N_2$ C31
$C_{14}H_{23}O_{10}N_2$ P4
$C_{14}H_{24}O_8N_2$ H13
$C_{14}H_{24}O_{10}N_2$ O1
$C_{14}H_{30}O_2N_4$ D14
$C_{14}H_{34}N_4$ D13

C_{15}

$C_{15}H_{10}O_6N_4$ Q19
$C_{15}H_{11}O_4N_3S$ Q20
$C_{15}H_{14}OS$ P183
$C_{15}H_{15}N_3S_2$ B182,B183
$C_{15}H_{22}O_2$ B43
$C_{15}H_{23}O_3N_5$ A145
$C_{15}H_{23}O_4N_5$ A146

C_{16}

$C_{16}H_{11}N$ P194
$C_{16}H_{12}O_5N_2S$ N4
$C_{16}H_{12}O_6N_3ClS_2$ N12,N13, N14
$C_{16}H_{12}O_8N_4S_2$ N22,N23
$C_{16}H_{13}O_6N_3S_2$ N24
$C_{16}H_{13}O_7N_3S_2$ N15
$C_{16}H_{13}O_9N_3S_3$ N25,N26,N27
$C_{16}H_{16}O_2N_2$ P34
$C_{16}H_{16}N_2$ P41
$C_{16}H_{17}N_3S_2$ B180
$C_{16}H_{28}O_8N_2$ O4

C_{17}

$C_{17}H_{13}O_7N_3S_2$ N8
$C_{17}H_{13}O_8N_3S_2$ N9,N10,N11
$C_{17}H_{15}O_6N_3S_2$ N19,N20,N21
$C_{17}H_{15}O_7N_3S_2$ N16,N17,N18
$C_{17}H_{20}O_6N_4$ R5
$C_{17}H_{35}N$ A154

C_{18}

$C_{18}H_{15}OP$ P88
$C_{18}H_{17}O_6N_3S_2$ N5,N6,N7
$C_{18}H_{20}N_2$ P36
$C_{18}H_{30}O$ B72
$C_{18}H_{36}O_6N_2$ C20

C_{22}

$C_{22}H_{24}O_8N_2Cl_2$ T5,T8
$C_{22}H_{25}O_8N_2Cl$ T4,T7

C_{23}

$C_{23}H_{26}N_2$ P189
$C_{24}H_{28}O_8N_2Cl_2$ T6

O

OH H39
OH_2 W1
OD_2 W2
O_2H P26
O_2H_2 H37
O_2Ge G2
O_3H_2Se S4
O_3H_3As A148
O_3H_3B B184
O_4H_2Cr C11
O_4H_2Mo M31,M32
O_4H_2Se S3
O_4H_3As A147
O_6H_6Te T3
$O_{21}HW$ W3
$O_{21}H_6W_6$ T66

N

NHO_2 N33
NHO_3 N32
NH_3 A95
NH_3O H40
NH_3O_3S S21
$N_2H_2O_2$ N34
N_2H_4 H31
N_3H H32

Cl

ClH H34
$ClHO$ H42
$ClHO_2$ C10
$ClHO_4$ C8
ClH_4N A96
$ClHO_3Cr$ C12

Br

BrH H33
$BrHO$ H41
$BrHO_3$ B186

SYNONYM INDEX

REFERENCE INDEX

1975

75Aa G. Anderegg, *Helv. Chim. Acta*, *58*, 1218 (1975).

75Ba R. Barbucci, *Inorg. Chim. Acta*, *12*, 113 (1975).

75Bb S. Bergstrom and G. Olofsson, *J. Solution Chem.*, *4*, 535 (1975).

75Bc M. J. Blais, O. Enea and G. Berthon, *Thermochimica Acta*, *12*, 25 (1975).

75Bd A. Braibanti, G. Mori, F. Dallavalle and E. Leporati, *J. C. S. Dalton*, 1319 (1975).

75Ca R. Cali, S. Gurrieri, E. Rizzarelli and S. Sammartano, *Thermochimica Acta*, *12*, 19 (1975).

75Da U. N. Dash, *Thermochimica Acta*, *11*, 25 (1975).

75Db U. N. Dash and B. Nayak, *Aust. J. Chem.*, *28*, 1377 (1975).

75Ea D. J. Eatough, J. J. Christensen and R. M. Izatt, *J. Chem. Thermo.*, *7*, 417 (1975).

75Ga A. Gergely, I. Nagypal and E. Farkas, *J. Inorg. Nucl. Chem.*, *37*, 551 (1975).

75Gb H. Gillet, L. Avedikian and J. P. Morel, *Can. J. Chem.*, *53*, 455 (1975).

75Ka F. Kai, Y. Sadakane, N. Tanaka and T. Matsuda, *J. Inorg. Nucl. Chem.*, *37*, 1311 (1975).

75Kb E. J. King, *J. Amer. Chem. Soc.*, *97*, 88 (1975).

75La P. Lumme and E. Kari, *Acta. Chem. Scand.*, *29*, 117 (1975).

75Ma W. L. Marshall and R. Slusher, *J. Inorg. Nucl. Chem.*, *37*, 1191 (1975).

75Mb W. L. Marshall and R. Slusher, *J. Inorg. Nucl. Chem.*, *37*, 2165 (1975).

75Oa G. Olofsson, *J. Chem. Thermo.*, *7*, 507 (1975).

75Ob G. Olofsson and L. G. Hepler, *J. Solution Chem.*, *4*, 127 (1975).

75Pa G. H. Parsons and C. H. Rochester, *Faraday Trans.*, *71*, 1069 (1975).

75Sa T. P. Sharma and Y. K. Agrawal, *J. Inorg. Nucl. Chem.*, *37*, 1830 (1975).

75Ta J. G. Travers, K. G. McCurdy, D. Dolman and L. G. Hepler, *J. Solution Chem.*, *4*, 267 (1975).

1974

74Aa N. K. Agarwal, A. D. Taneja and K. P. Srivastava, *Indian J. Chem.*, *12*, 874 (1974).

74Ab E. M. Arnett, E. J. Mitchell and T. S. S. R. Murty, *J. Amer. Chem. Soc.*, *96*, 3875 (1974).

74Ac S. A. Attiga and C. H. Rochester, *J. C. S. Perkin II*, 1624 (1974).

74Ba J. L. Banyasz and J. E. Stuehr, *J. Amer. Chem. Soc.*, *96*, 6481 (1974).

74Bb R. G. Bates, R. N. Roy and R. A. Robinson, *J. Solution Chem.*, *3*, 905 (1974).

74Bc A. C. Baxter and D. R. Williams, *J. C. S. Dalton*, 1117 (1974).

74Bd M. A. Bernard, N. Bois and M. Daireaux, *Bull. Soc. Chim. France*, 27 (1974).

74Be M. J. Blais, O. Enea and G. Berthon, *Thermochimica Acta*, *8*, 433 (1974).

74Ca R. D. Cannon and J. Gardiner, *Inorg. Chem.*, *13*, 390 (1974).

74Cb M. J. Cook, N. L. Dassanyake, C. D. Johnson, A. R. Katritzky and T. W. Toone, *J. C. S. Perkin II*, 1069 (1974).

74Cc V. Crescenzi, F. Delben, S. Paoletti and J. Skerjanc, *J. Phys. Chem.*, *78*, 607 (1974).

74Da H. F. DeBrabander, G. G. Herman and L. C. Van Poucke, *Thermochimica Acta*, *10*, 385 (1974).

74Fa B. Finlayson and A. Smith, *J. Chem. Eng. Data*, *19*, 94 (1974).

74Ga N. N. Ghosh and G. Mukhopadhyay, *Indian J. Chem.*, *12*, 636 (1974).

74Gb I. Grenthe and G. Gardhammar, *Acta Chem. Scand.*, *28*, 125 (1974).

74Ha G. R. Hedwig and H. K. J. Powell, *J. C. S. Dalton*, 47 (1974).

74Ja L. Juliano and A. C. M. Paiva, *Biochemistry*, *13*, 2445 (1974).

74Ka Kabir-ud-Din, *Z. Physik. Chem. Neue Folge*, *88*, 316 (1974).

74Kb F. Kai and Y. Sadakane, *J. Inorg. Nucl. Chem.*, *36*, 1404 (1974).

74Kc N. Kiba and T. Takeuchi, *J. Inorg. Nucl. Chem.*, *36*, 847 (1974).

74Kd H. Koffer, *J. C. S. Perkin II*, 1428 (1974).

74La R. J. Lemire and M. W. Lister, *Thermochimica Acta*, *8*, 291 (1974).

74Lb C. L. Liotta, E. M. Perdue and H. P. Hopkins, *J. Amer. Chem. Soc.*, *96*, 7308 (1974).

74Lc C. L. Liotta, E. M. Perdue and H. P. Hopkins, *J. Amer. Chem. Soc.*, *96*, 7981 (1974).

74Ld P. Lumme and I. Pilkanen, *Acta Chem. Scand.*, *28*, 1106 (1974).

74Ma T. Matsui, L. G. Hepler and E. M. Woolley, *Can J. Chem.*, *52*, 1910 (1974).

74Mb T. Matsui, H. C. Ko and L. G. Hepler, *Can J. Chem.*, *52*, 2912 (1974).

74Mc T. Matsui, H. C. Ko and L. G. Hepler, *Can J. Chem.*, *52*, 2906 (1974).

74Md J. P. Morel, J. Fauve, L. Avedikian and J. Juillard, *J. Solution Chem.*, *3*, 403 (1974).

74Na I. Nagypal, A. Gergely and E. Farkas, *J. Inorg. Nucl. Chem.*, *36*, 699 (1974).

74Pa B. N. Palmer and H. K. J. Powell, *J. C. S. Dalton*, 2089 (1974).

74Pb C. S. G. Prasad and S. K. Banerji, *J. Indian Chem.*, *51*, 358 (1974).

1974 (continued)

74Ra F. Rodante, F. Rallo and P. Fiordiponti, *Thermochimica Acta, 9,* 269 (1974).

74Sa A. K. Sinha, J. C. Ghosh and B. Prasad, *J. Indian Chem. Soc., 51,* 586 (1974).

74Sb S. K. Srivastava and H. B. Mathur, *Indian J. Chem., 12,* 736 (1974).

74Va V. P. Vasilev and L. D. Shekhanova, *Russ. J. Inorg. Chem. (English Trans.), 19,* 1623 (1974). (Russian Page No. 2969.)

74Vb V. P. Vasilev, L. A. Kochergina and T. D. Yastrebova, *Zhurnal Obshchei Khimii, 44,* 1371 (1974).

1973

73Aa J. L. Ault, H. J. Harries and J. Burgess, *J. C. S. Dalton,* 1095 (1973).

73Ba J. L. Banyasz and J. E. Stuehr, *J. Amer. Chem. Soc., 95,* 7226 (1973).

73Bb R. Barbucci, L. Fabbrizzi, P. Paoletti and A. Vacca, *J. C. S. Dalton,* 1763 (1973).

73Bc M. A. Beg, Kabir-ud-Din and R. A. Khan, *Aust. J. Chem., 26,* 671 (1973).

73Bd I. R. Bellobono and M. A. Monetti, *J. C. S. Perkin II,* 790 (1973).

73Be G. Berthon, O. Enea and E. M'Foundou, *Bull. Soc. Chim., France,* 2967 (1973).

73Bf P. D. Bolton, J. Ellis, K. A. Fleming and I. R. Lantzke, *Aust. J. Chem., 26,* 1005 (1973).

73Bg A. Braibanti, F. Dallavalle, E. Leporati and G. Mori, *J. C. S. Dalton,* 323 (1973).

73Ca M. R. Chakrabarty, C. S. Handloser and M. W. Mosher, *J. C. S. Perkin II,* 938 (1973).

73Cb V. Crescenzi, F. Delben, F. Quadrifoglio and D. Dolar, *J. Phys. Chem., 77,* 539 (1973).

73Da I. Dellien, *Acta Chem. Scand., 27,* 733 (1973).

73Db I. Dellien, I. Grenthe and G. Hessler, *Acta Chem. Scand., 27,* 2431 (1973).

73Dc I. Dellien and L. A. Malmsten, *Acta Chem. Scand., 27,* 2877 (1973).

73Ea E. S. Etz, R. A. Robinson and R. G. Bates, *J. Solution Chem., 2,* 407 (1973).

73Fa L. Fabbrizzi, P. Paoletti, M. C. Zobrist and G. Schwarzenbach, *Helv. Chim. Acta, 56,* 670 (1973).

73Ga A. Gergely and I. Sovago, *J. Inorg. Nucl. Chem., 35,* 4355 (1973).

73Ha L. D. Hansen and E. A. Lewis, *J. Phys. Chem., 77,* 286 (1973).

73Hb G. R. Hedwig and H. K. J. Powell, *J. C. S. Dalton,* 793 (1973).

73Hc G. R. Hedwig and H. K. J. Powell, *J. C. S. Dalton,* 1942 (1973).

73Ia T. P. I, E. J. Burke, J. L. Meyer and G. H. Nancollas, *Thermochimica Acta, 5,* 463 (1973).

73Ka C. Klofutar, S. Paljk and D. Kremser, *Spectrochimica Acta, 29A,* 139 (1973).

73Kb C. Klofutar, S. Paljk and B. Barlic, *Spectrochimica Acta, 29A,* 1069 (1973).

73Kc Ts. B. Konunova and L. S. Kachkar, *Russ. J. Inorg. Chem. (English Trans.), 18,* 805 (1973). (Russian Page No. 1527.)

73Kd Yu. A. Kozlov, V. V. Blokhin, V. V. Shurukhin and V. E. Mironov, *Russ. J. Phys. Chem. (English Trans.), 47,* 1343 (1973). (Russian Page No. 2386.)

73Ke L. A. Kutulya, L. P. Pivovarevich, P. A. Grigorov, N. N. Magdesieva, S. V. Tsukerman and V. F. Lavrushin, *J. Gen. Chem. U.S.S.R., (English Trans.), 43,* 1745 (1973). (Russian Page No. 1762.)

73La P. Lingaiah and E. V. Sundaram, *J. Indian Chem. Soc., L,* 479 (1973).

73Lb C. L. Liotta, E. M. Perdue and H. P. Hopkins, *J. Amer. Chem. Soc., 95,* 2439 (1973).

73Ma D. D. Macdonald, P. Butler and D. Owen, *Can. J. Chem., 51,* 2590 (1973).

73Mb T. H. Marshall and V. Chen, *J. Amer. Chem. Soc., 95,* 5400 (1973).

73Mc T. Matsui and L. G. Hepler, *Can. J. Chem., 51,* 1941 (1973).

73Oa G. Olofsson and I. Olofsson, *J. Chem. Thermodynamics, 5,* 533 (1973).

73Pa B. C. Patel, A. H. Gandhi and S. R. Patel, *Indian J. Chem., 11,* 468 (1973).

73Pb N. Purdie and M. B. Tomson, *J. Amer. Chem. Soc., 95,* 48 (1973).

73Ra G. Reinhard, R. Dreyer and R. Munze, *Z. Phys. Chemie, Leipzig, 254,* 226 (1973).

73Rb F. Rodante, F. Rallo and P. Fiordiponti, *Thermochimica Acta, 6,* 369 (1973).

73Rc R. N. Roy, R. A. Robinson and R. G. Bates, *J. Amer. Chem. Soc., 95,* 8231 (1973).

73Sa P. Stenius, *Acta Chem. Scand., 27,* 3452 (1973).

73Va V. P. Vasilev, S. A. Aleksandrova and L. A. Kochergina, *Russ. J. Inorg. Chem. (English Trans.) 18,* 1549 (1973). (Russian Page No. 2912.)

73Vb V. P. Vasilev, L. A. Kochergina and T. D. Yastrebova, *J. Gen. Chem. U.S.S.R., (English Trans.) 43,* 971 (1973). (Russian Page No. 975.)

73Vc V. P. Vasilev and E. V. Kozlovskii, *Russ. J. Inorg. Chem. (English Trans.) 18,* 1544 (1973). (Russian Page No. 2902.)

73Vd V. P. Vasilev, L. D. Shekhanova and L. A. Kochergina, *J. Gen. Chem., U.S.S.R., (English Trans.) 43,* 967 (1973). (Russian Page No. 971.)

73Za J. Zsako, Z. Finta and Cs. Varhelyi, *J. Inorg. Nucl. Chem., 35,* 2819 (1973).

1972

72Aa R. Arnek and L. Barcza, *Acta Chem. Scand., 26,* 213 (1972).

72Ab R. Arnek and S. R. Johansson, *Acta Chem. Scand., 26,* 2903 (1972).

72Ba R. Barbucci, L. Fabbrizzi and P. Paoletti, *J. C. S. Dalton,* 745 (1972).

72Bb I. R. Bellobono and E. Diani, *J. C. S. Perkin II,* 1707 (1972).

72Bc C. Bjurulf, *Eur. J. Biochem.*, *30*, 33 (1972).

72Bd P. D. Bolton, K. A. Fleming and F. M. Hall, *J. Amer. Chem. Soc.*, *94*, 1033 (1972).

72Be P. D. Bolton, F. M. Hall and J. Kudrynski, *Aust. J. Chem.*, *25*, 75 (1972).

72Bf F. Bolza and F. E. Treloar, *J. Chem. Eng. Data*, *17*, 197 (1972).

72Bg A. Braibanti, E. Leporati, F. Dallavalle and G. Mori, *Inorg. Chimica Acta*, *6*, 395 (1972).

72Bh A. Braibanti, G. Mori, F. Dallavalle and E. Leporati, *Inorg. Chim. Acta*, *6*, 106 (1972).

72Bi A. P. Brunetti, E. J. Burke, M. C. Lim and G. H. Nancollas, *J. Solution Chem.*, *1*, 153 (1972).

72Ca J. J. Christensen, G. L. Kimball, H. D. Johnston and R. M. Izatt, *Thermochimica Acta*, *4*, 141 (1972).

72Cb J. J. Christensen, D. E. Smith, M. D. Slade and R. M. Izatt, *Thermochimica Acta*, *4*, 17 (1972).

72Cc J. J. Christensen, D. E. Smith, M. D. Slade and R. M. Izatt, *Thermochimica Acta*, *5*, 35 (1972).

72Cd H. S. Creyf and L. C. VanPoucke, *Thermochimica Acta*, *4*, 485 (1972).

72Da I. P. Dasgupta and M. L. Tobe, *Inorg. Chem.* *11*, 1011 (1972).

72Db G. Degischer and G. R. Choppin, *J. Inorg. Nucl. Chem.*, *34*, 2823 (1972).

72Dc H. S. Dunsmore and D. Midgley, *J. C. S. Dalton*, 64 (1972).

72Dd P. J. Dynes, G. S. Chapman, E. Kebede and F. W. Schneider, *J. Amer. Chem. Soc.*, *94*, 6356 (1972).

72Ea O. Enea, K. Houngbossa and G. Berthon, *Electrochimica Acta*, *17*, 1585 (1972).

72Eb E. S. Etz, R. A. Robinson and R. G. Bates, *J. Solution Chem.*, *1*, 507 (1972).

72Fa L. Fabbrizzi, R. Barbucci and P. Paoletti, *J. C. S. Dalton*, 1529 (1972).

72Ga R. D. Graham, D. R. Williams and P. A. Yeo, *J. C. S. Perkin II*, 1876 (1972).

72Gb A. Gergely and I. Sovago, *Acta Chim. Acad. Sci. Hung.*, *74*, 273 (1972).

72Gc A. Gergely, J. Mojzes and Zs. Kassai-Bazsa, *J. Inorg. Nucl. Chem.*, *34*, 1277 (1972).

72Gd I. Grenthe and G. Gardhammar, *Acta Chem. Scand.*, *26*, 3207 (1972).

72Ge I. Grenthe and H. Ots, *Acta Chem. Scand.*, *26*, 1229 (1972).

72Ha R. W. Hay and P. J. Morris, *J. C. S. Perkin II*, 1021 (1972).

72Hb E. Hayon, A. Treinin and J. Wilf, *J. Amer. Chem. Soc.*, *94*, 47 (1972).

72Ia T. P. I and G. H. Nancollas, *Inorg. Chem.*, *11*, 2414 (1972).

1972 (continued)

72Ib R. M. Izatt, H. D. Johnson and J. J. Christensen, *J. C. S. Dalton,* 1152 (1972).

72Ja R. F. Jameson and M. F. Wilson, *J. C. S. Dalton,* 2610 (1972).

72Jb R. F. Jameson and M. F. Wilson, *J. C. S. Dalton,* 2617 (1972).

72Ka R. G. Kallen, R. O. Viale and L. K. Smith, *J. Amer. Chem. Soc., 94,* 576 (1972).

72Kb V. A. Kogan, N. I. Dorokhova, O. A. Osipov and S. G. Kochin, *Russ. J. Phys. Chem. (English Trans.) 46,* 120 (1972). (Russian Page No. 205.)

72La S. C. Lahiri, U. C. Bhattacharyya and S. Aditya, *Z. Phys. Chemie, Leipzig, 249,* 49 (1972).

72Lb P. Lingaiah, G. Punnaiah and E. V. Sundaram, *Indian J. Chem., 10,* 521 (1972).

72Lc C. L. Liotta, D. F. Smith, H. P. Hopkins and K. A. Rhodes, *J. Phys. Chem., 76,* 1909 (1972).

72Ma J. N. Mathur, *Indian J. Chem., 10,*299 (1972).

72Mb M. W. Mosher, C. B. Sharma and M. R. Chakrabarty, *J. Magnetic Resonance, 7,* 247 (1972).

72Na M. Nakanishi and S. Fujieda, *Anal. Chem., 44,* 574 (1972).

72Oa H. Ots, *Acta Chem. Scand., 26,* 3810 (1972).

72Pa A. N. Pant, R. N. Soni and S. L. Gupta, *Indian J. Chem., 10,* 90 (1972).

72Pb A. N. Pant, R. N. Soni and S. L. Gupta, *Indian J. Chem., 10,* 632 (1972).

72Pc A. N. Pant, R. N. Soni and S. L. Gupta, *Indian J. Chem., 10,* 724 (1972).

72Pd G. H. Parsons and C. H. Rochester, *Trans. Faraday Soc., 68,* 523 (1972).

72Pe S. M. Parsons and M. A. Raftery, *Biochemistry, 11,* 1630 (1972).

72Pf N. Purdie, M. B. Tomson and N. Riemann, *J. Solution Chem., 1,* 465 (1972).

72Ra H. S. Randhawa and W. U. Malik, *J. Indian Chem. Soc., 49,* 7 (1972).

72Rb M. C. Rose and J. E. Stuehr, *J. Amer. Chem. Soc., 94,* 5532 (1972).

72Sa Vl. Simeon, N. Ivicic and M. Tkalcec, *Z. Physik. Chem. Neue Folge, 78,* 1 (1972).

72Va C. E. Vanderzee, D. L. King and I. Wadso, *J. Chem. Thermo., 4,* 685 (1972).

72Vb Cs. Varhelyi, J. Zsako and Z. Finta, *J. Inorg. Nucl. Chem., 34,* 2583 (1972).

72Wa D. R. Williams, *J. C. S. Dalton,* 790 (1972).

72Za S. Zimmer and R. Biltonen, *J. Solution Chem., 1,* 291 (1972).

72Zb J. Zsako, Z. Finta and Cs. Varhelyi, *J. Inorg. Nucl. Chem., 34,* 2887 (1972).

1971

71Aa S. Ahrland and L. Kullberg, *Acta Chem. Scand.*, *25*, 3471 (1971).

71Ab S. Ahrland and L. Kullberg, *Acta Chem. Scand.*, *25*, 3677 (1971).

71Ac L. Avedikian, *Bull. Soc. Chim. (France)*, *No. 8*, 2832 (1971).

71Ad L. Avedikian and J. P. Morel, *J. Chim. Nucl. Chem.*, *33*, 2177 (1971).

71Ba D. S. Barnes and L. D. Pettit, *J. Inorg. Phys.*, *68*, 2177 (1971).

71Bb J. H. Baxendale, M. D. Ward and P. Wardman, *Trans. Farad. Soc.*, *67*, 2532 (1971).

71Bc J. L. Bear and M. E. Clark, *J. Inorg. Nucl. Chem.*, *33*, 3805 (1971).

71Bd J. T. Bell, R. D. Baybarz and D. M. Helton, *J. Inorg. Nucl. Chem.*, *33*, 3077 (1971).

71Be L. Beres and J. M. Sturtevant, *Biochem.*, *10*, 2120 (1971).

71Bf G. Berthon, O. Enea and K. Houngrossa, *C. R. Acad. Sc. (Paris)*, *273*, 1140 (1971).

71Bg A. Braibanti, F. Dallavalle, E. Leporati and G. Mori, *Inorg. Chim. Acta*, *5*, 449 (1971).

71Ca S. Cabani, G. Conti and L. Lepori, *Trans. Faraday Soc.*, *67*, 1933 (1971).

71Cb S. Chakrabarti and S. Aditya, *J. Indian Chem. Soc.*, *48*, 493 (1971).

71Da R. Davies and R. B. Jordan, *Inorg. Chem.*, *10*, 2132 (1971).

71Ea K. G. Everett and D. A. Skoog, *Anal.Chem.*, *43*, 1541 (1971).

71Fa B. J. Felber and Neil Purdie, *J. Phys. Chem.*, *75*, 1136 (1971).

71Fb P. U. Fruh, J. T. Clerc and W. Simon, *Helv. Chim. Acta*, *54*, 1445 (1971).

71Ga A. H. Gandhi and S. R. Patel, *Bull. Chem. Soc. Japan*, *44*, 455 (1971).

71Gb E. Sh. Ganelina and V. A. Borgoyakov, *Russ. J. Inorg. Chem. (English Trans.)*, *16*, 111 (1971). (Russian Page No. 214.)

71Gc Von G. Gattow and M. Drager, *Z. Anorg. Allg. Chem.*, *384*, 235 (1971).

71Gd G. Geier and U. Karlen, *Helv. Chim. Acta*, *54*, 135 (1971).

71Ge A. M. Goeminne and Z. Eeckhaut, *Bull. Soc. Chim. Belges*, *80*, 605 (1971).

71Gf S. Goldman, P. Sagner and R. G. Bates, *J. Phys. Chem.*, *75*, 826 (1971).

71Ha L. D. Hansen and E. A. Lewis, *Anal. Chem.*, *43*, 1393 (1971).

71Hb L. D. Hansen and E. A. Lewis, *J. Chem. Thermodynamics*, *3*, 35 (1971).

71Hc L. D. Hansen, E. A. Lewis, J. J. Christensen, R. M. Izatt and D. P. Wrathall, *J. Amer. Chem. Soc.*, *93*, 1099 (1971).

71Hd R. W. Hay and P. J. Morris, *J. Chem. Soc. (A)*, 1518 (1971).

71He R. W. Hay and P. J. Morris, *J. Chem. Soc. (A)*, 3562 (1971).

71Hf H. J. Hinz, D. D. F. Shiao and J. M. Sturtevant, *Biochemistry*, *10*, 1347 (1971).

1971 (continued)

71Ia Y. J. Israeli and R. Volpe, *Bull. Soc. Chim. France,* 3119 (1971).

71Ja A. D. Jones and D. R. Williams, *Inorg. Nucl. Chem. Letters, 7,* 369 (1971).

71Jb A. D. Jones and D. R. Williams, *J. Chem. Soc. (A),* 3159 (1971).

71Ka Y. D. Kim and R. Lumry, *J. Amer. Chem. Soc., 93,* 1003 (1971).

71Kb Y. Kim and R. Lumry, *J. Amer. Chem. Soc., 93,* 5882 (1971).

71La M. C. Lim and G. H. Nancollas, *Inorg. Chem., 10,* 1957 (1971).

71Ma M. A. Marini and R. L. Berger, *Anal. Biochem., 43,* 188 (1971).

71Mb J. N. Mathur, *Indian J. Chem., 9,* 567 (1971).

71Mc A. J. Murphy, *Biochem., 10,* 3723 (1971).

71Oa G. Olofsson, *J. Chem. Thermo., 3,* 217 (1971).

71Pa P. Paoletti, R. Barbucci, A. Vacca and A. Dei, *J. Chem. Soc. (A),* 310 (1971).

71Pb P. Paoletti, A. Dei and A. Vacca, *J. Chem. Soc. (A)* 2656 (1971).

71Pc P. Paoletti, R. Walser, A. Vacca and G. Schwarzenbach, *Helv. Chim. Acta, 54,* 243 (1971).

71Sa R. S. Saxena and P. Singh, *Z. Phys. Chemie, Leipzig, 247,* 250 (1971).

71Sb L. M. Schwartz and L. O. Howard, *J. Phys. Chem., 75,* 1798 (1971).

71Va W. Van de Poel, *Bull. Soc. Chim. Belges, 80,* 401 (1971).

71Vb V. P. Vasilev, L. A. Kochergina and V. I. Eremenko, *Russ. J. Phys. Chem. (English Trans.) 45,* 1196 (1971). (Russian Page No. 1197.)

71Wa E. M. Woolley and L. G. Hepler, *Can. J. Chem., 49,* 3054 (1971).

71Wb E. M. Woolley, L. G. Hepler and R. S. Roche, *Can. J. Chem., 49,* 3054 (1971).

1970

70Aa R. Arnek, *Arkiv for Kemi., 32,* 55 (1970).

70Ab L. Avedikian, J. Morin and J. P. Morel, *C. R. Acad. Sci., Paris, Series C, 271,* 988 (1970).

70Ba R. Barbucci, P. Paoletti and A. Vacca, *J. Chem. Soc. (A),* 2202 (1970).

70Bb E. W. Baumann, *J. Inorg. Nucl. Chem., 32,* 3823 (1970).

70Bc P. D. Bolton, J. Ellis and F. M. Hall, *J. Chem. Soc. (B),* 1252 (1970).

70Bd P. D. Bolton and F. M. Hall, *J. Chem. Soc. (B),* 1247 (1970).

70Be P. D. Bolton, C. D. Johnson, A. R. Katritzky and S. A. Shapiro, *J. Amer. Chem. Soc., 92,* 1567 (1970).

70Bf P. D. Bolton, C. H. Rochester and B. Rossall, *Trans. Faraday Soc., 66,* 1348 (1970).

1970 (continued)

70Ca J. J. Christensen, H. D. Johnston and R. M. Izatt, *J. Chem. Soc. (A)*, 454 (1970).

70Cb J. J. Christensen, J. H. Rytting and R. M. Izatt, *Biochemistry*, *9*, 4907 (1970).

70Cc J. J. Christensen, J. H. Rytting and R. M. Izatt, *J. Chem. Soc. (B)*, 1643 (1970).

70Cd J. J. Christensen, J. H. Rytting and R. M. Izatt, *J. Chem. Soc. (B)*, 1646 (1970).

70Ce J. J. Christensen, M. D. Slade, D. E. Smith, R. M. Izatt and J. Tsang, *J. Amer. Chem. Soc.*, *92*, 4164 (1970).

70Cf A. J. Cunningham, D. A. House and H. K. J. Powell, *Aust. J. Chem.*, *23*, 2375 (1970).

70Da G. Degischer and G. H. Nancollas, *Inorg. Chem.*, *9*, 1259 (1970).

70Ea D. J. Eatough, *Anal. Chem.*, *42*, 635 (1970).

70Eb W. J. Eilbeck, F. Holmes and G. Phillips, *J. Chem. Soc. (A)*, 689 (1970).

70Ec W. J. Eilbeck, F. Holmes and T. W. Thomas, *J. Chem. Soc. (A)*, 2062 (1970).

70Ga R. Ghosh and V. S. K. Nair, *J. Inorg. Nucl. Chem.*, *32*, 3041 (1970).

70Gb I. Grenthe, H. Ots and O. Ginstrup, *Acta Chem. Scand.*, *24*, 1067 (1970).

70Ha R. W. Hay and P. J. Morris, *J. Chem. Soc. (B)*, 1577 (1970).

70Hb L. D. Hansen, E. J. Baca and P. Scheiner, *J. Heterocyclic Chem.*, *7*, 991 (1970).

70Ia D. J. G. Ives and P. G. N. Moseley, *J. Chem. Soc. (B)*, 1655 (1970).

70Ib D. J. G. Ives and D. Prasad, *J. Chem. Soc. (B)*, 1652 (1970).

70Ja A. D. Jones and D. R. Williams, *J. Chem. Soc. (A)*, 1154 (1970).

70Ka F. Kopecky, M. Pesak and J. Celechovsky, *Coll. Czechoslov Chem. Commun.*, *35*, 576 (1970).

70Kb G. C. Kugler and G. H. Carey, *Talanta*, *17*, 907 (1970).

70Kc K. K. Kundu, P. K. Chattopadhyay and M. N. Das, *J. Chem. Soc. (A)*, 2034 (1970).

70Kd K. K. Kundu, P. K. Chattopadhyay, D. Jana and M. N. Das, *J. Phys. Chem.*, *74*, 2633 (1970).

70La A. N. Lazarev, Yu. A. Makashev and V. E. Mironov, *Russ. J. Inorg. Chem. (English Trans.)*, *15*, 237 (1970). (Russian Page No. 459.)

70Lb J. E. Letter and J. E. Bauman, *J. Amer. Chem. Soc.*, *92*, 437 (1970).

70Lc J. E. Letter and J. E. Bauman, *J. Amer. Chem. Soc.*, *92*, 443 (1970).

70Ld C. S. Leung and E. Grunwald, *J. Phys. Chem.*, *74*, 687 (1970).

70Le C. S. Leung and E. Grunwald, *J. Phys. Chem.*, *74*, 696 (1970).

1970 (continued)

70Ma A. Martin and E. Uhlig, *Z. Anorg. and Allg. Chem.*, *375*, 166 (1970).

70Mb A. Martin and E. Uhlig, *Z. Anorg. and Allg. Chem.*, *376*, 282 (1970).

70Mc R. E. Mesmer, C. F. Baes and F. H. Sweeton, *J. Phys. Chem.*, *74*, 1937 (1970).

70Md J. L. Meyer and J. E. Bauman, *J. Chem. Eng. Data*, *15*, 404 (1970).

70Me J. L. Meyer and J. E. Bauman, *J. Amer. Chem. Soc.*, *92*, 4210 (1970).

70Pa M. Paabo and R. G. Bates, *J. Research N.B.S.*, *74A*, 667 (1970).

70Pb M. Paabo and R. G. Bates, *J. Phys. Chem.*, *74*, 702 (1970).

70Pc M. Paabo and R. G. Bates, *J. Phys. Chem.*, *74*, 706 (1970).

70Pd T. B. Paiva, M. Tominaga and A. C. M. Paiva, *J. Medicinal Chem.*, *13*, 689 (1970).

70Sa L. M. Schwartz and L. O. Howard, *J. Phys. Chem.*, *74*, 4374 (1970).

70Ta A. Treinin and J. Wilf, *J. Phys. Chem.*, *74*, 4131 (1970).

70Va W. Van de Poel and P. J. Slootmaekers, *Bull. Soc. Chim. Belges*, *79*, 223 (1970).

70Wa J. B. Walker, *J. Inorg. Nucl. Chem.*, *32*, 2793 (1970).

70Wb D. R. Williams, *J. Chem. Soc. (A)*, 1550 (1970).

70Wc J. M. Wilson, A. G. Briggs, J. E. Sawbridge, P. Tickle and J. J. Zuckerman, *J. Chem. Soc. (A)*, 1024 (1970).

70Wd E. M. Woolley, R. W. Wilton and L. G. Hepler, *Can. J. Chem.*, *48*, 3249 (1970).

70We A. Wrobel, A. Rabczenko and D. Shugar, *Acta Biochimica Polonica*, *17*, 339 (1970).

70Ya T. Yamaoka, H. Hosoya and S. Nagakura, *Tetrahedron*, *26*, 4125 (1970).

1969

69Aa R. Arnek, *Acta Chem. Scand.*, *23*, 1986 (1969).

69Ab L. Avedikian and N. Dollet, *Bull Soc. Chim. France*, 4551 (1968).

69Ba D. Barnes, P. G. Laye and L. D. Pettit, *J. Chem. Soc. (A)*, 2073 (1969).

69Bb I. R. Bellobono and P. Beltrame, *J. Chem. Soc. (B)*, 620 (1969).

69Bc U. C. Bhattacharyya, S. C. Lahiri and S. Aditya, *J. Indian Chem. Soc.*, *46*, 247 (1969).

69Bd P. D. Bolton and F. M. Hall, *J. Chem. Soc. (B)*, 259 (1969).

69Be P. D. Bolton and F. M. Hall, *J. Chem. Soc. (B)*, 1047 (1969).

69Bf A. G. Briggs, J. E. Sawbridge, P. Tickle and J. M. Wilson, *J. Chem. Soc. (B)*, 802 (1969).

69Bg A. P. Brunetti, G. H. Nancollas and P. N. Smith, *J. Amer. Chem. Soc.*, *91*, 4680 (1969).

1969 (continued)

69Ca S. Cabani, G. Conti and P. Gianni, *J. Chem. Soc. (A)*, 1363 (1969).

69Cb A. S. Carson, P. G. Laye and W. V. Steele, *J. Chem. Thermo. 1*, 515 (1969).

69Cc J. J. Christensen, R. M. Izatt, D. P. Wrathall and L. D. Hansen, *J. Chem. Soc. (A)*, 1212 (1969).

69Cd E. Coates, C. G. Marsden and B. Rigg, *Trans. Faraday Soc., 65*, 863 (1969).

69Ce E. Coates, C. G. Marsden and B. Rigg, *Trans. Faraday Soc., 65*, 3032 (1969).

69Da A. G. Desai and R. M. Milburn, *J. Am. Chem. Soc., 91*, 1958 (1969).

69Ea W. D. Ellis and H. B. Dunford, *Archives of Biochem. Biophys., 133*, 313 (1969).

69Ga D. R. Boodard, B. D. Lodham, J. O. Ajayi and M. J. Cambell, *J. Chem. Soc. (A)*, 506 (1969).

69Gb R. W. Green, *Aust. J. Chem., 22*, 721 (1969).

69Gc J. Guilleme and B. Wojtkowiak, *Bull. Soc. Chim. France, No. 9*, 3007 (1969).

69Ha H. J. Harries and G. Wright, *J. Inorg. Nucl. Chem., 31*, 3149 (1969).

69Hb J. O. Hill and R. J. Irving, *J. Chem. Soc. (A)*, 2759 (1969).

69Hc J. O. Hill, G. Ojelund and I. Wadso, *J. Chem. Thermo. 1*, 111 (1969).

69Ia R. M. Izatt, D. Eatough, J. J. Christensen and C. H. Bartholomew, *J. Chem. Soc. (A)*, 45 (1969).

69Ja C. D. Johnson, A. R. Katritzky and S. A. Shapiro, 6654 (1969).

69Ka I. V. Kolosov and Z. F. Andreeva, *Russ. J. Inorg. Chem. (English Trans.) 14*, 346 (1969). (Russian Page No. 664.)

69Kb J. L. Kurz and J. M. Farrar, *J. Amer. Chem. Soc., 91*, 6057 (1969).

69La J. W. Larson and L. G. Hepler, Chapter in *Solute-Solvent Interactions*, Coetzee and Ritchie, Ed., Marcel Dekker, New York (1969).

69Lb P. Lumme and P. Virtanen, *Suomen Kemistilehti B, 42*, 333 (1969).

69Ma R. E. Mesmer and C. F. Baez, *Inorg. Chem., 8*, 618 (1969).

69Na G. H. Nancollas and D. J. Poulton, *Inorg. Chem., 8*, 680 (1969).

69Nb A. Neuberger and A. P. Fletcher, *J. Chem. Soc. (B)*, 178 (1969).

69Pa M. Paabo and R. G. Bates, *J. Phys. Chem., 73*, 3014 (1969).

69Pb S. K. Pal, U. C. Bhattacharyya, S. C. Lahiri and S. Aditya, *J. Indian Chem. Soc., 46*, 497 (1969).

69Pc K. H. Pearson, J. R. Baker and J. D. Goodrich, *Anal. Letters, 2*, 577 (1969).

69Rb J. M. Readnour and J. W. Cobble, *Inorg. Chem., 8*, 2174 (1969).

69Rc G. C. K. Roberts, D. H. Meadows and O. Jardetzky, *Biochemistry, 8*, 2053 (1969).

1969 (continued)

69Rd C. H. Rochester and B. Rossall, *Trans. Faraday Soc.*, *65*, 1004 (1969).

69Sa L. Sharma and B. Prasad, *J. Indian Chem. Soc.*, *46*, 241 (1969).

69Sb J. J. R. F. Silva and M. L. S. Simoes, *Rev. Port. Quim.*, *11*, 54 (1969).

69Sc Vl. Simeon, B. Svigir and N. Paulic, *J. Inorg. Nucl. Chem.*, *31*, 2085 (1969).

69Ta A. C. R. Thornton and H. A. Skinner, *Trans. Faraday Soc.*, *65*, 2044 (1969).

69Ua E. Uhlig and A. Martin, *J. Inorg. Nucl. Chem.*, *31*, 2781 (1969).

69Wa D. R. Williams, *J. Chem. Soc. (A)*, 2695 (1969).

69Wb A. C. M. Wanders and Th. N. Zwietering, *J. Phys. Chem.*, *73*, 2076 (1969).

69Wc E. W. Wilson and D. F. Smith, *Anal. Chem.*, *41*, 1903 (1969).

1968

68Aa R. Arnek and C. C. Patel, *Acta Chem. Scand.*, *22*, 1097 (1968).

68Ab R. Arnek and I. Szilard, *Acta Chem. Scand.*, *22*, 1334 (1968).

68Ba S. K. Banerjee, K. K. Kundu and M. N. Das, *J. Chem. Soc. (A)*, 139 (1968).

68Bb P. D. Bolton and F. M. Hall, *Aust. J. Chem.*, *21*, 939 (1968).

68Bc P. D. Bolton, F. M. Hall and J. Kudrynski, *Aust. J. Chem.*, *21*, 1541 (1968).

68Bd A. P. Brunetti, M. C. Lim and G. H. Nancollas, *J. Am. Chem. Soc.*, *90*, 5120 (1968).

68Ca S. Cabani, G. Conti and L. Lepori, *La Ricerca Scientifica*, *38*, 1039 (1968).

68Cb J. P. Calmon, Y. Cazaux-Maraval and P. Maroni, *Bull. Soc. Chim. France*, 3779 (1968).

68Cc A. S. Carson, P. G. Laye, and P. N. Smith, *J. Chem. Soc. (A)*, 141 (1968).

68Cd J. J. Christensen, J. L. Oscarson and R. M. Izatt, *J. Amer. Chem. Soc.*, *90*, 5949 (1968).

68Ce J. J. Christensen, D. P. Wrathall, and R. M. Izatt, *Anal. Chem.*, *40*, 175 (1968).

68Cf M. C. Cox, D. H. Everett, D. A. Landsman and R. J. Munn, *J. Chem. Soc. (B)*, 1373 (1968).

68Fa J. Fuger and E. Merciny, *Bull. Soc. Chim. Belges*, *77*, 59 (1968).

68Ga P. S. Gentile and A. Dadgar, *J. Chem. Engr. Data*, *13*, 367 (1968).

68Gb R. N. Goldberg and L. G. Hepler, *J. Phys. Chem.*, *72*, 4654 (1968).

68Gc R. W. Green and W. G. Goodwin, *Aust. J. Chem.*, *21*, 1165 (1968).

68Gd G. Gutnikov and H. Freiser, *Anal. Chem.*, *40*, 39 (1968).

68Ha L. D. Hansen, B. D. West, E. J. Baca and C. L. Blank, *J. Amer. Chem. Soc.*, *90*, 6588 (1968).

68Hb H. B. Hetzer, R. A. Robinson and R. G. Bates, *J. Phys. Chem.*, *72*, 2081 (1968).

68La D. Levine and I. B. Wilson, *Inorg. Chem.*, *7*, 818 (1968).

68Ma E. E. Mercer and D. T. Farrar, *Can. J. Chem.*, *46*, 2679 (1968).

68Oa W. F. O'Hara, *Can J. Chem.*, *46*, 1965 (1968).

68Ob G. Ojelund and I. Wadso, *Acta Chem. Scand.*, *22*, 2691 (1968).

68Pa H. H. Perkampus and G. Prescher, *Ber. Bunsenges. Physik. Chem.*, *72*, 429 (1968).

68Pb S. M. Petrov and Yu. I. Umanskii, *Russ. J. Phys. Chem.*, *(English Trans.)*, *42*, 178 (1968). (Russian Page No. 331.)

68Pc S. M. Petrov and Yu. I. Umanskii, *Russ. J. Phys. Chem.*, *(English Trans.)*, *42*, 1627 (1968). (Russian Page No. 3052.)

68Ra R. J. Raffa, M. J. Stern and L. Malspeis, *Anal. Chem.*, *40*, 70 (1968).

68Ta J. Tummavuroi and P. Lumme, *Acta Chem. Scand.*, *22*, 2003 (1968).

68Tb B. A. Timimi and D. H. Everett, *J. Chem. Soc. (B)*, 1380 (1968).

68Va V. P. Vasil'ev and L. A. Kockergina, *Russ. J. Phys. Chem.*, *42*, 199 (1968).

1967

67Aa S. Ahrland, *Helv. Chim. Acta*, *50*, 306 (1967).

67Ab R. Arnek and K. Wladyslaw, *Acta Chem. Scand.*, *21*, 1449 (1967).

67Ac L. Avedikian, *Bull. Soc. Chim. France*, 254 (1967).

67Ad L. Avedikian and N. Dollet, *Bull. Soc. Chim. France*, *12*, 4551 (1967).

67Ae N. N. Aylward, *J. Chem. Soc. (B)*, 401 (1967).

67Ba P. D. Bolton and F. M. Hall, *Aust. J. Chem.*, *20*, 1797 (1967).

67Bb P. D. Bolton, F. M. Hall and I. H. Reece, *J. Chem. Soc. (B)*, 709 (1967).

67Ca J. J. Christensen, R. M. Izatt and L. D. Hansen, *J. Amer. Chem. Soc.*, *89*, 213 (1967).

67Cb J. J. Christensen, D. P. Wrathall, R. M. Izatt and D. O. Tolman, *J. Phys. Chem.*, *71*, 3001 (1967).

67Ga P. Gerding, *Acta Chem. Scand.*, *21*, 2007 (1967).

67Gb D. R. Goddard, S. I. Nwankwo and L. A. K. Stavley, *J. Chem. Soc. (A)*, 1376 (1967).

67Ha G. I. H. Hanania, D. H. Irvine, W. A. Eaton and P. George, *J. Phys. Chem.*, *71*, 2022 (1967).

67Hb R. W. Hay and L. J. Porter, *J. Chem. Soc. (B)*, 1261 (1967).

67Hc F. Holmes and D. R. Williams, *J. Chem. Soc. (A)*, 1256 (1967).

67Ia R. M. Izatt, J. H. Rytting, and J. J. Christensen, *J. Phys. Chem.*, *71*, 2700 (1967).

1967 (continued)

67La S. C. Lahiri and S. Aditya, *Z. Phys. Chime, Neue Folge (Frankfurt)*, *55*, 6 (1967).

67Lb C. L. Liotta, K. H. Leavell and D. F. Smith, Jr., *J. Phys. Chem.*, *71*, 3091 (1967).

67Ma I. N. Maksimova, *Russ. J. Phys. Chem.*, *41*, 27 (1967).

67Oa W. F. O'Hara, H. C. Ko, M. N. Ackerman and L. G. Hepler, *J. Phys. Chem.*, *71*, 3107 (1967).

67Ob G. Ojelund and I. Wadso, *Acta Chem. Scand.*, *21*, 1408 (1967).

67Pa A. J. Poe, K. Shaw, and M. J. Wendt, *Inorg. Chim. Acta*, *1*, 371 (1967).

67Pb E. Popper, L. Roman and P. Mareu, *Talanta*, *14*, 1163 (1967).

67Ra C. D. Ritchie and G. H. Megerle, *J. Amer. Chem. Soc.*, *89*, 1447 (1967).

67Rb C. D. Ritchie and G. H. Megerle, *J. Amer. Chem. Soc.*, *89*, 1452 (1967).

67Sa Sallavo and P. Lumme, *Suom. Kemistilehti*, *40*, 155 (1967).

67Va A. Vacca and D. Arenare, *J. Phys. Chem.*, *71*, 1495 (1967).

67Vb V. P. Vasilev and L. A. Kochergina, *Russ. J. Phys. Chem. (English Trans.)* *41*, 681 (1967). (Russian Page No. 1282.)

67Vc V. P. Vasil'ev and L. A. Kochergina, *Russ. J. Phys. Chem.*, *41*, 1149 (1967).

67Vd V. P. Vasil'ev and G. A. Lobanov, *Russ. J. Phys. Chem.*, *41*, 434 (1967).

67Wa J. M. Wilson, N. E. Gore, J. E. Sawbridge and F. Cardenas-Cruz, *J. Chem. Soc. (B)*, 852 (1967).

67Wb M. R. Wright, *J. Chem. Soc. (B)*, 1265 (1967).

67Za A. A. Zavitsas, *J. Chem. Eng. Data*, *12*, 94 (1967).

1966

66Aa A. Aboul-Seoud and M. Doheim, *Can. J. Chem.*, *44*, 521 (1966).

66Ab J. C. Ahluwalia, F. J. Millero, R. N. Goldberg and L. G. Hepler, *J. Phys. Chem.*, *70*, 319 (1966).

66Ac K. P. Anderson, W. O. Greenhalgh and R. M. Izatt, *Inorg. Chem.*, *5*, 2106 (1966).

66Ad K. P. Anderson, D. A. Newell and R. M. Izatt, *Inorg. Chem.*, *5*, 62 (1966).

66Ae L. Avedikian, *Bull. Soc. Chim. France*, 2570 (1966).

66Ba L. Z. Benet and J. E. Goyan, *J. Pharm. Sci.*, *55*, 1184 (1966).

66Bb G. L. Bertrand, F. J. Millero, C. Wu, and L. G. Hepler, *J. Phys. Chem.*, *70*, 699 (1966).

66Bd P. D. Bolton, F. M. Hall and I. H. Reece, *J. Chem. Soc. (B)*, 717 (1966).

66Bf P. D. Bolton, F. M. Hall and I. H. Reece, *Spectrochem. Acta*, *22*, 1149 (1966).

66Bg P. D. Bolton, F. M. Hall and L. H. Reece, *Spectrochem. Acta, 22,* 1825 (1966).

66Bh J. H. Boughton and R. N. Keller, *J. Inorg. Nucl. Chem., 28,* 2851 (1966).

66Bi L. G. Bunville and S. J. Schwalbe, *Biochemistry, 5,* 3521 (1966).

66Ca J. J. Christensen, R. M. Izatt, L. D. Hansen and J. A. Partridge, *J. Phys. Chem., 70,* 2003 (1966).

66Cb J. J. Christensen, J. H. Rytting and R. M. Izatt, *J. Amer. Chem. Soc., 88,* 5105 (1966).

66Cc E. C. W. Clarke and D. N. Glew, *Trans. Faraday Society, 62,* 539 (1966).

66Cd A. K. Covington, R. A. Robinson and R. G. Bates, *J. Phys. Chem., 70,* 3820 (1966).

66Ce R. A. Cox, *Biochem. J., 100,* 146 (1966).

66Cf F. T. Crimmins, C. Dymek, M. Flood and W. F. O'Hara, *J. Phys. Chem., 70,* 931 (1966).

66Da S. P. Datta and A. K. Grzybowski, *J. Chem. Soc. (B),* 136 (1966).

66Fa G. Faita, T. Mussini and R. Oggioni, *J. Chem. and Eng. Data, 11,* 162 (1966).

66Ga P. Goursot and I. Wadso, *Acta Chem. Scand., 20,* 1314 (1966).

66Ha G. I. H. Hanania, D. H. Irvine and M. V. Irvine, *J. Chem. Soc. (A),* 296 (1966).

66Hb L. H. Hansen, J. A. Partridge, R. M. Izatt and J. J. Christensen, *Inorg. Chem., 5,* 569 (1966).

66Hc H. B. Hetzer, R. G. Bates and R. A. Robinson, *J. Phys. Chem., 70,* 2869 (1966).

66Ia R. R. Irani and T. A. Taulii, *J. Inorg. Nucl. Chem., 28,* 1011 (1966).

66Ib R. M. Izatt, J. H. Rytting, L. D. Hansen and J. J. Christensen, *J. Amer. Chem. Soc., 88,* 2641 (1966).

66La J. W. Larson, G. L. Bertrand and L. G. Hepler, *J. Chem. Eng. Data, 11,* 595 (1966).

66Lb S. Lewin and M. A. Barnes, *J. Chem. Soc. (B),* 478 (1966).

66Lc S. Lewin and D. A. Humphreys, *J. Chem. Soc. (B),* 210 (1966).

66Ma W. L. Marshall and E. V.Jones, *J. Phys. Chem., 70,* 4028 (1966).

66Mb R. P. Mitra, H. C. Malhotra and D. V. S. Jain, *Trans. Faraday Soc., 62,* 167 (1966).

66Mc T. Moeller and S. Chu, *J. Inorg. Nucl. Chem., 28,* 153 (1966).

66Md J. C. Morris, *J. Phys. Chem., 70,* 3798 (1966).

66Pa M. Paabo, R. G. Bates and R. A. Robinson, *J. Phys. Chem., 70,* 540 (1966).

66Pb M. Paabo, R. G. Bates and R. A. Robinson, *J. Phys. Chem., 70,* 2073 (1966).

1966 (continued)

66Pc M. Paabo, R. A. Robinson and R. G. Bates, *Anal. Chem.*, *38*, 1573 (1966).

66Pd P. Paoletti, F. Nuzzi and A. Vacca, *J. Chem. Soc. (A)*, 1385 (1966).

66Pe J. A. Partridge, J. J. Christensen and R. M. Izatt, *J. Amer. Chem. Soc.*, *88*, 1649 (1966).

66Pf R. C. Phillips, S. J. P. Eisenberg, P. George and R. J. Rutman, *J. Amer. Chem. Soc.*, *88*, 2631 (1966).

66Pg H. K. J. Powell and N. F. Curtis, *J. Chem. Soc. (B)*, 1205 (1966).

66Ta M. M. Taqui Khan and A. E. Martell, *J. Amer. Chem. Soc.*, *88*, 668 (1966).

66Tb J. Y. Tong and R. L. Johnson, *Inorg. Chem.*, *5*, 1902 (1966).

66Va A. Vacca and D. Arenare, *Ricerca Scientifica*, *36*, 1361 (1966).

66Vb J. Vaissermann, *Compt. Rend.*, *262 C*, 692 (1966).

66Vc V. P. Vasil'ev and L. A. Kochergina, *Russ. J. Phys. Chem.*, *40*, 1622 (1966).

1965

65Aa G. Anderegg, *Helv. Chim. Acta*, *48*, 1718 (1965).

65Ab G. Anderegg, *Helv. Chim. Acta*, *48*, 1722 (1965).

65Ba J. G. Beetlestone and D. H. Irvine, *J. Chem. Soc.*, 3271 (1965).

65Bb R. H. Boyd and C. H. Wang, *J. Amer. Chem. Soc.*, *87*, 430 (1965).

65Bc S. Boyd, A. Bryson, G. H. Nancollas and K. Torrance, *J. Chem. Soc.*, 7353 (1965).

65Ca S. Cabani and G. Conti, *Gazz. Chim. Ital.*, *45*, 533 (1965).

65Cb P. J. Conn and D. F. Swinehart, *J. Phys. Chem.*, *69*, 2653 (1965).

65Cc W. A. Connor, M. M. Jones and D. L. Tuleen, *Inorg. Chem.*, *4*, 1129 (1965).

65Da R. L. Davies and K. W. Dunning, *J. Chem. Soc.*, 4168 (1965).

65Ga R. Gary, R. G. Bates and R. A. Robinson, *J. Phys. Chem.*, *69*, 2750 (1965).

65Gb S. L. Gupta and R. N. Soni, *J. Indian Chem. Soc.*, *42 (6)*, 377 (1965).

65Ha G. I. H. Hanania, D. H. Irvine and F. Shurayh, *J. Chem. Soc.*, 1149 (1965).

65Hb J. Hermans and G. Rialdi, *Biochemistry*, *4*, 1277 (1965).

65Hc H. P. Hopkins, C. Wu and L. G. Hepler, *J. Phys. Chem.*, *69*, 2244 (1965).

65Ia D. J. G. Ives and P. D. Marsden, *J. Chem. Soc.*, 649 (1965).

65Ib R. M. Izatt, L. D. Hansen, J. H. Rytting and J. J. Christensen, *J. Amer. Chem. Soc.*, *87*, 2760 (1965).

65Ka V. A. Korobova and G. A. Prik, *Russ. J. Inorg. Chem.*, *10*, 456 (1965).

65La P. Lumme, P. Lahermo and J. Tummavuori, *Acta Chem. Scand.*, *19*, 2175 (1965).

65Lb I. Lipchitz, *Diss. Abs.*, *26*, 662 (1965).

65Ma F. J. Millero, J. C. Ahluwalia and L. G. Hepler, *J. Chem. Eng. Data*, *10*, 199 (1965).

65Mb P. N. Milyukov and N. V. Polenova, *Zzv. Vysshikh Uchebn. Zavadenii, Khim. i Khim. Technol.*, *8*, 42 (1965).

65Pa M. Paabo, R. A. Robinson and R. G. Bates, *J. Amer. Chem. Soc.*, *87*, 415 (1965).

65Pb P. Paoletti, J. H. Stern and A. Vacca, *J. Phys. Chem.*, *69*, 3759 (1965).

65Pc P. Papoff and P. G. Zambonin, *La Ricerca Scientifica*, *35*, 93 (1965).

65Pd A. Pekkarinen, *Suom. Kemistilehti*, *38B (3)*, 63 (1965).

65Pe R. Phillips, S. J. P. Eisenberg, P. George and R. J. Rutman, *J. Biol. Chem.*, *240*, 4393 (1965).

65Ra C. D. Ritchie and P. D. Heffley, *J. Amer. Chem. Soc.*, *87*, 5402 (1965).

65Sa H. C. Saraswat and U. D. Tripathi, *Bull. Chem. Soc. Japan*, *38*, 1555 (1965).

65Sb G. Schwarzenbach and M. Schellenberg, *Helv. Chim. Acta*, *48*, 28 (1965).

65Va J. Vaissermann and M. Quintin, *J. Chim. Phys.*, *63*, 731 (1965).

65Wa D. D. Wagman, W. H. Evans, I. Halow, V. B. Parker, S. M. Baily and R. H. Schumm, *Tech. Note 270-1, US Natl. Bur. Standards* (1965).

65Wb M. Woodhead, M. Paabo, R. A. Robinson and R. G. Bates, *J. Res. Natl. Bur. Standards*, *69A*, 263 (1965).

65Wc D. P. Wrathall, R. M. Izatt and J. J. Christensen, *J. Amer. Chem. Soc.*, *87*, 5809 (1965).

1964

64Aa G. Anderegg, *Experientia, Supplementum*, *9*, 75 (1964).

64Ab G. Anderegg, *Helv. Chim. Acta*, *47*, 1801 (1964).

64Ba J. E. Bauman and J. C. Wang, *Inorg. Chem.*, *3*, 368 (1964).

64Bb J. G. Beetlestone and D. H. Irvine, *J. Chem. Soc.*, 5090 (1964).

64Bc J. G. Beetlestone and D. H. Irvine, *Royal Soc. London, Proc.*, *277*, 401 (1964).

64Ca R. K. Chaturvedi, P. Dinkar and B. Biswas, *Proc. Natl. Acad. Sci. India*, *34*, 22 (1964).

64Da S. P. Datta, A. K. Grzybowski and R. G. Bates, *J. Phys. Chem.*, *68*, 275 (1964).

64Ea P. L. Edelin de la Praudiere and L. A. K. Staveley, *J. Inorg. Nucl. Chem.*, *26*, 1713 (1964).

64Eb P. L. Edelin de la Praudiere and L. A. K. Staveley, *J. Inorg. Nucl. Chem.*, *26*, 1713 (1964).

64Ec Z. L. Ernst, R. J. Irving and J. Menashi, *Trans. Faraday Soc.*, *60*, 56 (1964).

64Ed E. Eyal and A. Treinin, *J. Amer. Chem. Soc., 86,* 4287 (1964).

64Fa A. N. Fletcher, *J. Inorg. Nucl. Chem., 26,* 955 (1964).

64Ga R. Gary, R. G. Bates and R. A. Robinson, *J. Phys. Chem., 68,* 3806 (1964).

64Gb P. George, G. I. H. Hanania, D. H. Irvine and I. Abu-Issa, *J. Chem. Soc.,* 5689 (1964).

64Gc I. Grenthe, *Acta Chem. Scand., 18,* 283 (1964).

64Ha J. A. Hull, R. H. Davies and L. A. K. Staveley, *J. Chem. Soc.,* 5422 (1964).

64Ia R. J. Irving, L. Nelander and I. Wadso, *Acta Chem. Scand., 18,* 769 (1964).

64Ib R. J. Irving and I. Wadso, *Acta Chem. Scand., 18,* 195 (1964).

64Ic R. M. Izatt, J. J. Christensen and V. Kolhari, *Inorg. Chem., 3,* 1565 (1964).

64Ka H. C. Ko, W. F. O'Hara, T. Hu and L. G. Hepler, *J. Amer. Chem. Soc., 86,* 1003 (1964).

64La S. C. Lahiri and S. Aditya, *Z. Physik. Chem., 41,* 173 (1964).

64Lb S. C. Lahiri and S. Aditya, *J. Indian Chem. Soc., 41,* 469 (1964).

64Lc S. C. Lahiri and S. Aditya, *Z. Physik. Chemie Neue Folge (Frankfurt), 43,* 282 (1964).

64Ld S. Lewin, *J. Chem. Soc. (London),* 792 (1964).

64Ma F. J. Milleró, J. C. Ahluwalia and L. G. Hepler, *J. Phys. Chem., 68,* 3435 (1964).

64Mb F. J. Millero, J. C. Ahluwalia and L. G. Hepler, *J. Chem. Eng. Data, 9,* 192 (1964).

64Mc F. J. Millero, J. C. Ahluwalia and L. G. Hepler, *J. Chem. Eng. Data, 9,* 319 (1964).

64Na V. S. K. Nair, *J. Inorg. Nucl. Chem., 26,* 1911 (1964).

64Nb L. Nelander, *Acta Chem. Scand., 18,* 973 (1964).

64Pa P. Paoletti and A. Vacca, *J. Chem. Soc.,* 5051 (1964).

64Pb G. E. Perlmann, *J. Biol. Chem., 239,* 3762 (1964).

64Sa L. Sacconi, P. Paoletti and M. Ciampolini, *J. Chem. Soc.,* 5046 (1964).

64Sb P. Sellers, S. Sunner and I. Wadso, *Acta Chem. Scand., 18,* 202 (1964).

64Sc L. G. Sillen and A. E. Martell, *Stability Constants, Publication No. 17, The Chemical Society (London)* (1964).

64Sd B. I. Sukhorukov, V. I. Poltev and L. A. Blyumenfel'd, *Abhandl. Deut. Akad. Wiss. Berlin, KL. Med.,* 381 (1964).

64Wa J. G. Wang, J. E. Bauman and R. K. Murmann, *J. Phys. Chem., 68,* 2296 (1964).

64Wb D. P. Wrathall, R. M. Izatt and J. J. Christensen, *J. Am. Chem. Soc., 86,* 4779 (1964).

1963

63Aa G. Anderegg, *Helv. Chim. Acta, 46*, 2813 (1963).

63Ab G. Anderegg, *Helv. Chim. Acta, 46*, 1833 (1963).

63Ba K. Batzar, A. Chester and D. E. Goldberg, *J. Chem. Eng. Data, 8*, 293 (1963).

63Da S. P. Datta, A. K. Grzybowski and B. A. Weston, *J. Chem. Soc.*, 792 (1963).

63Ea L. Eberson and I. Wadso, *Acta Chem. Scand., 17*, 1552 (1963).

63Ga V. G. Gattow and B. Krebs, *Z. Anorg. Allgem. Chem., 323*, 13 (1963).

63Gb P. Gerding, I. Leden and S. Sunner, *Acta Chem. Scand., 17*, 2190 (1963).

63Gc P. S. Gentile, M. Cefola and A. V. Celiano, *J. Phys. Chem., 67*, 1447 (1963).

63Ha J. D. Hale, R. M. Izatt and J. J. Christensen, *Proc. Chem. Soc. (London)*, 240 (1963).

63Hb J. D. Hale, R. M. Izatt and J. J. Christensen, *J. Phys. Chem., 67*, 2605 (1963).

63Hc R. P. Held and D. E. Goldberg, *Inorg. Chem., 2*, 585 (1963).

63Hd J. Hermans, Jr., S. J. Leach and P. A. Scheraga, *J. Amer. Chem. Soc., 85*, 1390 (1963).

63He H. B. Hetzer, R. G. Bates and R. A. Robinson, *J. Phys. Chem., 67*, 1124 (1963).

63Hf M. N. Hughes and G. Stedman, *J. Chem. Soc.*, 1239 (1963).

63Ia H. Irving and J. J. R. F. da Silva, *J. Chem. Soc. (London)*, 448 (1963).

63La J. Llopis and D. Ordonez, *J. Electroanal Chem., 5*, 129 (1963).

63Ma S. D. Morrett and D. F. Swinehart, *J. Phys. Chem., 67*, 717 (1963).

63Mb R. K. Murmann and D. R. Foerster, *J. Phys. Chem., 67*, 1383 (1963).

63Oa W. F. O'Hara, T. Hu and L. G. Hepler, *J. Phys. Chem., 67*, 1933 (1963).

63Pa P. Paoletti and M. Ciampolini, *Ric. Sci. Rend. Sex., A, 3(4)*, 405 (1963).

63Pb P. Paoletti, M. Ciampolini and A. Vacca, *J. Phys. Chem., 67*, 1065 (1963).

63Pc R. C. Phillips, P. George and R. J. Rutman, *Biochemistry, 2*, 501 (1963).

63Ra R. A. Robinson and A. Peiperl, *J. Phys. Chem., 67*, 1723 (1963).

63Rb R. A. Robinson and A. Peiperl, *J. Phys. Chem., 67*, 2860 (1963).

63Ta E. Thorogood and G. I. H. Hanania, *Biochem. J., 87*, 123 (1963).

63Va C. E. Vanderzee and J. A. Swanson, *J. Phys. Chem., 67*, 2608 (1963).

63Vb F. Vilallonga and M. I. Pouchan, *Biochim. Biophys. Acta, 75*, 449 (1963).

1962

62Aa G. Aksnes, *Acta Chem. Scand.*, *16*, 1967 (1962).

62Ab G. F. Allen, R. A. Robinson and V. E. Bower, *J. Phys. Chem.*, *66*, 171 (1962).

62Ba V. E. Bower, R. A. Robinson and R. G. Bates, *J. Research Natl. Bur. Standards*, *66A*, 71 (1962).

62Bb G. Briere, N. Felici, E. Piot, *Compt. Rend.*, *255*, 107 (1962).

62Ca R. K. Chaturvedi and S. S. Katiyar, *Bull. Chem. Soc. Japan*, *35*, 1416 (1962).

62Cb D. T. Y. Chen and K. J. Laidler, *Trans. Faraday Soc.*, *58*, 480 (1962).

62Cc J. J. Christensen and R. M. Izatt, *J. Phys. Chem.*, *66*, 1030 (1962).

62Cd E. Coates and B. Rigg, *Trans. Faraday Soc.*, *58*, 88 (1962).

62Da S. P. Datta and A. K. Grzybowski, *J. Chem. Soc.*, 3068 (1962).

62Ea H. R. Ellison, J. O. Edwards and E. A. Healy, *J. Amer. Chem. Soc.*, *84*, 1820 (1962).

62Ha G. I. H. Hanania and D. H. Irving, *J. Chem. Soc.*, 2745 (1962).

62Hb G. I. H. Hanania and D. H. Irving, *J. Chem. Soc.*, 2750 (1962).

62Hc H. B. Hetzer and R. G. Bales, *J. Phys. Chem.*, *66*, 308 (1962).

62Hd H. B. Hetzer, R. A. Robinson and R. G. Bates, *J. Phys. Chem.*, *66*, 2696 (1962).

62Ia R. R. Irani and K. Moedritzer, *J. Phys. Chem.*, *66*, 1349 (1962).

62Ib R. M. Izatt, J. J. Christensen, R. T. Pack and R. Bench, *Inorg. Chem.*, *1*, 828 (1962).

62Ja J. Jordan and G. J. Ewing, *Inorg. Chem.*, *1*, 587 (1962).

62Ka F. Ya. Kul'ba and Yu. A. Makashev, *Russ. J. Inorg. Chem.*, *7*, 661 (1962).

62La S. Lewin and N. W. Tann, *J. Chem. Soc.*, 1466 (1962).

62Ma T. Moeller and R. Ferrus, *Inorg. Chem.*, *1*, 49 (1962).

62Mb T. Moeller and T. M. Hseu, *J. Inorg. Nucl. Chem.*, *24*, 1635 (1962).

62Mc T. Moeller and L. C. Thompson, *J. Inorg. Nucl. Chem.*, *24*, 499 (1962).

62Sa V. I. Slovetskii, V. B. Belikov, I. M. Zavilovich and L. V. Epishina, *Izv. Akad. Nauk SSSR, otd. Khim. Nauk.*, 520 (1962).

62Wa I. Wadso, *Acta Chem. Scand.*, *16*, 479 (1962).

62Ya K. B. Yatsimirskii and G. A. Prik, *Russ. J. Inorg. Chem.*, *7*, 30 (1962).

1961

61Ba J. Barthel, F. Becker and N. G. Schmahl, *Z. Physik. Chemie Neue Folge*, *29*, 58 (1961).

61Bb R. G. Bates and H. B. Hetzer, *J. Phys. Chem.*, *65*, 667 (1961).

1961 (continued)

61Bc A. I. Biggs, *J. Chem. Soc.*, 2572 (1961).

61Ca M. Ciampolini and P. Paoletti, *J. Phys. Chem.*, *65*, 1224 (1961).

61Cb E. Coates and B. Rigg, *Trans. Faraday Soc.*, *57*, 1088 (1961).

61Da S. P. Datta and A. K. Grzybowski, *Biochem. J.*, *78*, 289 (1961).

61Db M. M. Davis and H. B. Hetzer, *J. Research Natl. Bur. Standards*, *65A*, 209 (1961).

61Dc S. N. Das and D. J. B. Ives, *Proc. Chem. Soc. (London)*, 373 (1961).

61Ea J. O. Edwards and R. L. Sederstrom, *J. Phys. Chem.*, *65*, 862 (1961).

61Eb J. M. Essery and K. Schofield, *J. Chem. Soc.*, 3939 (1961).

61Ga R. W. Green and I. R. Freer, *J. Phys. Chem.*, *65*, 2211 (1961).

61Ia R. R. Irani, *J. Phys. Chem.*, *65*, 1463 (1961).

61Ib R. M. Izatt, J. W. Wrathall and K. P. Anderson, *J. Phys. Chem.*, *65*, 1914 (1961).

61Ja T. B. Jackson and J. O. Edwards, *J. Amer. Chem. Soc.*, *83*, 355 (1961).

61Ka H. Kido and W. C. Fernelius, *J. Phys. Chem.*, *65*, 574 (1961).

61La L. Laloi and P. Rumpf, *Bull. Soc. Chim. France*, 1645 (1961).

61Lb M. H. Lietzke, R. W. Stoughton and T. F. Young, *J. Phys. Chem.*, *65*, 2247 (1961).

61Ma T. Moeller and R. Ferrus, *J. Inorg. Nucl. Chem.*, *20*, 261 (1961).

61Na W. C. Nicholas and W. C. Fernelius, *J. Phys. Chem.*, *65*, 1047 (1961).

61Oa W. F. O'Hara and L. G. Hepler, *J. Phys. Chem.*, *65*, 2107 (1961).

61Pa J. M. Pagano, D. E. Goldberg and W. C. Fernelius, *J. Phys. Chem.*, *65*, 1062 (1961).

61Ra R. E. Reichard and W. C. Fernelius, *J. Phys. Chem.*, *65*, 380 (1961).

61Sa K. Schlyter and D. L. Martin, *Trans. Royal Inst. Tech., Stockholm*, *175*, 1 (1961).

61Sb V. I. Slovetskii, S. A. Shevelov, A. A. Fainzil'berg and S. S. Novikov, *Zh. Vses Khim Okshchestva im DI Mendeleeva*, *6*, 599 (1961).

61Sc V. I. Slovetskii, S. A. Shevelov, A. A. Fainzil'berg and S. S. Novikov, *Zh. Vses Khim Okshchestva im DI Mendeleeva*, *6*, 707 (1961).

61Ta B. C. Tyson, W. H. McCurdy and C. E. Bricker, *Anal. Chem.*, *33*, 1640 (1961).

1960

60Ba R. G. Bates and G. F. Allen, *J. Res. Natl. Bur. Standards*, *64A*, 343 (1960).

60Bb R. G. Bates and H. B. Hetzer, *J. Res. Natl. Bur. Standards*, *64A*, 427 (1960).

1960 (continued)

60Ha J. Hermans, Jr., J. W. Donovan and H. A. Scheraga, *J. Biol. Chem.*, *235*, 91 (1960).

60Hb G. C. Hood and C. A. Reilly, *J. Chem. Phys.*, *32*, 127 (1960).

60Ka D. M. Kern, *J. Chem. Ed.*, *37*, 14 (1960).

60Kb E. J. King, *J. Amer. Chem. Soc.*, *82*, 3575 (1960).

60La T. J. Lane and K. P. Quinlan, *J. Amer. Chem. Soc.*, *82*, 2994 (1960).

60Lb P. Lumme, *Suomen Kem.*, *33B*, 87 (1960).

60Pa V. V. Pal'chevskii, M. S. Zakhar'evskii and E. A. Malinina, *Yestnik Leningrad Univ.*, *15*, *No. 16*, *Ser. Fiz. i Khim*, *No. 3*, 95 (1960).

60Pb S. Pelletier, *J. Chim. Phys. Physicochim. Biol.*, *57*, 301 (1960).

60Pc S. Pelletier, *J. Chim. Phys. Physicochim. Biol.*, *57*, 311 (1960).

60Pd H. H. Perkampus and H. Kohler, *Z. Elektrochem.*, *64*, 365 (1960).

60Ra M. Rawitscher and J. M. Sturtevant, *J. Amer. Chem. Soc.*, *82*, 3739 (1960).

60Rb R. Riccardi and M. Bresesti, *Annali Chimica*, *50*, 1305 (1960).

60Sa L. Sacconi, P. Paoletti and M. Ciampolini, *J. Amer. Chem. Soc.*, *82*, 3831 (1960).

60Wa H. R. Weimer and W. C. Fernelius, *J. Phys. Chem.*, *64*, 1951 (1960).

1959

59Ba R. P. Bell, *The Proton in Chemistry*, Cornell University Press, Ithaca, N.Y., 1959, p. 64.

59Ca L. V. Coulter, J. R. Sinclair and A. G. Roper, *J. Amer. Chem. Soc.*, *81*, 2986 (1959).

59Ea M. Eden and R. G. Bates, *J. Res. Natl. Bur. Standards*, *62*, 161 (1959).

59Eb J. J. Elliott and S. F. Mason, *J. Chem. Soc.*, 2352 (1959).

59Fa L. P. Fernandez and L. G. Hepler, *J. Amer. Chem. Soc.*, *81*, 1783 (1959).

59Fb F. P. Fernandez and L. G. Hepler, *J. Phys. Chem.*, *63*, 110 (1959).

59Ga D. E. Goldberg and W. C. Fernelius, *J. Phys. Chem.*, *63*, 1246 (1959).

59Gb D. E. Goldberg and W. C. Fernelius, *J. Phys. Chem.*, *63*, 1328 (1959).

59Gc L. Gruen, M. Laskowski and H. A. Scheraga, *J. Amer. Chem. Soc.*, *81*, 3891 (1959).

59Ha T. R. Harkins and H. Freiser, *J. Phys. Chem.*, *63*, 309 (1959).

59Ka C. Kitzinger and R. Hems, *Biochem. J.*, *71*, 395 (1959).

59Kb J. W. Kury, A. D. Paul, L. G. Hepler and R. E. Connick, *J. Amer. Chem. Soc.*, *81*, 4185 (1959).

59La J. R. Lotz, B. P. Block and W. C. Fernelius, *J. Phys. Chem.*, *63*, 541 (1959).

1959 (continued)

59Ma G. H. McIntyre, B. P. Block and W. C. Fernelius, *J. Amer. Chem. Soc.*, *81*, 529 (1959).

59Mb C. T. Mortimer and K. J. Laidler, *Trans. Faraday Soc.*, *55*, 1731 (1959).

59Pa H. M. Papee, W. J. Canady, T. W. Zawidzki and K. J. Laidler, *Trans. Faraday Soc.*, *55*, 1734 (1959).

59Ra T. Randall and L. A. K. Staveley, *J. Chem. Soc.*, 472 (1959).

59Rb W. Rubaszewska and Z. R. Grobowski, *Roczniki Chem.*, *33*, 781 (1959).

59Sa L. Sacconi, P. Paoletti and M. Ciampolini, *Ricerca Scientifica*, *29*, 2412 (1959).

59Sb S. N. R. von Schalien, *Suomen Kemistilehti*, *B32*, 148 (1959).

59Sc K. Schwabe, W. Groichen and D. Spiethoff, *Z. Physik Chem.*, *20*, 68 (1959).

59Za T. W. Zawidzki, H. M. Papee, W. J. Canady and K. J. Laidler, *Trans. Faraday Soc.*, *55*, 1738 (1959).

59Zb T. W. Zawidzki, H. M. Papee and K. J. Laidler, *Trans. Faraday Soc.*, *55*, 1743 (1959).

59Zc A. J. Zielen, *J. Amer. Chem. Soc.*, *81*, 5022 (1959).

1958

58Aa Th. Ackermann, *Z. Elektrochemie*, *62*, 411 (1958).

58Ca W. J. Canady, H. M. Papee and K. J. Laidler, *Trans. Faraday Soc.*, *54*, 502 (1958).

58Cb R. Caramazza, *Gazzetta Chimie Italiana*, *88*, 308 (1958).

58Da S. P. Datta and A. K. Grzybowski, *Trans. Faraday Soc.*, *54*, 1188 (1958).

58Db S. P. Datta and A. K. Grzybowski, *Biochem. J.*, *69*, 218 (1958).

58Ea J. T. Edsall and J. Wyman, *Biophysical Chemistry*, *Vol. 1*, Academic Press, Inc., New York, (1958) pp. 452, 464.

58Eb D. H. Everett and J. P. Hyne, *J. Chem. Soc.*, 1636 (1958).

58Fa B. F. Freasier, A. G. Oberg and W. W. Wendlandt, *J. Phys. Chem.*, *62*, 700 (1958).

58Ga A. K. Grzybowski, *J. Phys. Chem.*, *62*, 555 (1958).

58Ha T. R. Harkins and H. Freiser, *J. Amer. Chem. Soc.*, *80*, 1132 (1958).

58Hb L. G. Hepler, *J. Amer. Chem. Soc.*, *80*, 6181 (1958).

58Ja J. Jordan, *Rec. Chem. Prog.*, *19*, 193 (1958).

58Ma S. Miyamoto and K. Michima, *Kyoritsu Yakka Daigaku Kenkya Nempo*, *4*, 12 (1958).

58Na V. S. K. Nair and G. H. Nancollas, *J. Chem. Soc.*, 4144 (1958).

58Pa H. H. Perkampus and T. Rossel, *Z. Elektrochem.*, *62*, 94 (1958).

58Sa P. G. Scheurer, R. M. Brownell and J. E. LuValle, *J. Phys. Chem.*, *62*, 809 (1958).

1958 (continued)

58Sb E. L. Smith, V. J. Chavre and M. J. Parker, *J. Biol. Chem.*, *230*, 283 (1958).

58Ta M. J. L. Tillotson and L. A. K. Staveley, *J. Chem. Soc.*, 3613 (1958).

1957

57Ba J. J. Banewicz, C. W. Reed and M. E. Levitch, *J. Amer. Chem. Soc.*, *79*, 2693 (1957).

57Da R. N. Diebel and D. F. Swinehart, *J. Phys. Chem.*, *61*, 333 (1957).

57Ka E. J. King, *J. Amer. Chem. Soc.*, *79*, 6151 (1957).

57Kb J. Koskikallio, *Suomen Kemistilehti*, *30*, 155 (1957).

57Ma C.B. Murphy and A. E. Martell, *J. Biol. Chem.*, *226*, 37 (1957).

57Na Y. Nozaki, F. R. N. Gurd, R. F. Chen and J. T. Edsall, *J. Amer. Chem. Soc.*, *79*, 2123 (1957).

57Pa S. Pelletier and M. Quintin, *Comptes Rendus*, *244*, 894 (1957).

57Sa E. E. Sager and F. C. Byers, *J. Res. Natl. Bur. Standards*, (1957).

57Sb K. Y. Salnis, K. P. Mishchenko and I. E. Flis, *Zh. Neorgan. Khim.*, *2*, 1985 (1957).

57Ua E. Uusitalo, *Ann. Acad. Sci. Fennicae*, *AII*, 87 (1957).

1956

56Aa A. Agren, *Svensk Kemisk Tidskrift*, *86*, 181 (1956).

56Ab P. J. Antikainen, *Suom. Kemistilehti*, *B29*, 14 (1956).

56Ba R. G. Bates and V. E. Bower, *J. Res. Natl. Bur. Standards*, *57*, 153 (1956).

56Bb S. A. Bernhard, *J. Biol. Chem.*, *218*, 961 (1956).

56Ea E. Ellenbogen, *J. Amer. Chem. Soc.*, *78*, 369 (1956).

56Fa F. S. Feates and D. J. G. Ives, *J. Chem. Soc.*, 2798 (1956).

56Ha V. L. Hughes and A. E. Martell, *J. Amer. Chem. Soc.*, *78*, 1319 (1956).

56Ka C. M. Kelley and H. V. Tartar, *J. Amer. Chem. Soc.*, *78*, 5752 (1956).

56Kb E. J. King, *J. Amer. Chem. Soc.*, *78*, 6020 (1956).

56Kc E. J. King and G. W. King, *J. Amer. Chem. Soc.*, *78*, 1089 (1956).

56La G. I. Loeb and H. A. Scheraga, *J. Phys. Chem.*, *60*, 1638 (1956).

56Na R. Nasanen and E. Uusitalo, *Suom. Kemistilehti*, *29B*, No. 2, 11 (1956).

56Oa N. E. Ockerbloom and A. E. Martell, *J. Amer. Chem. Soc.*, *78*, 267 (1956).

56Pa H. M. Papee, W. J. Canady and K. J. Laidler, *Can. J. Chem.*, *34*, 1677 (1956).

56Sa S. Searles, M. Tamres, F. Block and L. A. Quarterman, *J. Amer. Chem. Soc.*, *78*, 4917 (1956).

1955

55Aa J. H. Ashby, H. B. Clarke, E. M. Croon and S. P. Datta, *Biochem. J.*, *59*, 203 (1955).

55Ca R. Cecil and J. R. McPhee, *Biochem. J.*, *60*, 496 (1955).

55Cb H. B. Clarke, S. P. Datta and B. R. Rabin, *Biochem. J.*, *59*, 209 (1955).

55Da C. Duculot, *Comptes. Rendes.*, *241*, 1925 (1955).

55Fa W. F. Fyfe, *J. Chem. Soc.*, 1347 (1955).

55Ga H. Gutfreund, *Trans. Faraday. Soc.*, *51*, 441 (1955).

55Ia D. J. G. Ives and J. H. Pryon, *J. Chem. Soc.*, 2104 (1955).

55Ib R. M. Izatt, W. C. Fernelius and B. P. Block, *J. Phys. Chem.*, *59*, 235 (1955).

55Ka A. A. Krawetz, *Ph.D. Thesis*, University of Chicago, Ill., (1955).

55Ma B. L. Mickel and A. C. Andrews, *J. Amer. Chem. Soc.*, *77*, 5291 (1955).

55Ha T. R. Narkins and H. Freiser, *J. Amer. Chem. Soc.*, *77*, 1374 (1955).

55Na R. Nasanen, *Suom. Kemistilehti*, *28B*, 161 (1955).

55Pa C. Postmus and E. L. King, *J. Phys. Chem.*, *59*, 1208 (1955).

55Sa J. M. Sturtevant, *J. Amer. Chem. Soc.*, *77*, 1495 (1955).

55Ta C. lanford, J. D. llavcnstein and D. G. Rands, *J. Amer. Chem. Soc.*, *77*, 6409 (1955).

1954

54Aa R. J. L. Andon, J. D. Cox and E. F. G. Herington, *Trans. Faraday Soc.*, *50*, 918 (1954).

54Ab J. H. Ashby, E. M. Crook and S. P. Datta, *Biochem. J.*, *56*, 198 (1954).

54Da S. P. Datta and A. K. Grzybowski, *J. Chem. Soc.*, 1091 (1954).

54Db T. Davies, S. S. Singer and L. A. K. Staveley, *J. Chem. Soc.*, 2304 (1954).

54Ea J. T. Edsall, G. Felsenfeld, D. S. Goodman and F. R. N. Gurd, *J. Amer. Chem. Soc.*, *76*, 3054 (1954).

54Ga R. J. Gibbs, *Arch. Biochem. Biophys.*, *51*, 277 (1954).

54Ja W. D. Johnston and H. Freiser, *Anal. Chimi. Acta*, *11*, 201, (1954).

54Ka E. J. King, *J. Amer. Chem. Soc.*, *75*, 2204 (1954).

54Kb E. J. King, *J. Amer. Chem. Soc.*, *76*, 1006 (1954).

54Wa K. F. Wissbrun, D. M. French and A. Patterson, *J. Phys. Chem.*, *58*, 693 (1954).

1953

53Ba F. Basolo, R. K. Murman and Y. T. Chen, *J. Amer. Chem. Soc.*, *75*, 1478 (1953).

53Bb J. H. Baxendale and H. R. Hardy, *Trans. Faraday Soc.*, *49*, 1140 (1953).

53Ca F. F. Carini and A. E. Martell, *J. Amer. Chem. Soc.*, *75*, 4810 (1953).

53Ea A. Ellila, *Ann. Acad. Sci. Fennicae. Ser. A II*, nr 51, (1953).

53Ga P. George and G. Hanania, *Biochem. J.*, *55*, 236 (1953).

53Ha L. G. Hepler, W. L. Jolly and W. M. Latimer, *J. Amer. Chem. Soc.*, *75*, 2809 (1953).

53Ka M. Kilpatrick and C. A. Arenberg, *J. Amer. Chem. Soc.*, *75*, 3812 (1953).

53Kb G. Kilde and W. F. K. Wynne-Jones, *Trans. Faraday Soc.*, *49*, 243 (1953).

53Ta C. Tanford and N. L. Wagner, *J. Amer. Chem. Soc.*, *75*, 434 (1953).

1952

52Ba P. Bender and W. J. Biermann, *J. Amer. Chem. Soc.*, *74*, 322 (1952).

52Da C. W. Davies, H. W. Jones and C. B. Monk, *Trans. Faraday Soc.*, *48*, 921 (1952).

52Ea D. H. Everett, D. A. Landsman and B. R. W. Pinsent, *Proc. Royal Soc. (A) 215*, 403 (1952).

52Eb D. H. Everett and B. R. W. Pinsent, *Proc. Royal Soc. (A) 215*, 416 (1952).

52Ec D. H. Everett and W. F. K. Wynne-Jones, *Trans. Faraday Soc.*, *48*, 531 (1952).

52Ga P. George and C. L. Tsou, *Biochem. J.*, *50*, 440 (1952).

52Gb P. George and G. Hanania, *Biochem. J.*, *52*, 517 (1952).

52Ja W. L. Jolly, *J. Amer. Chem. Soc.*, *74*, 6199 (1952).

52Jb V. Jones and H. N. Parton, *Trans. Faraday Soc.*, *48*, 8 (1952).

52Kb E. J. King and G. W. King, *J. Amer. Chem. Soc.*, *74*, 1212 (1952).

52Ma G. G. Manov, K. E. Schuette and F. S. Kirk, *J. Res. Natl. Bur. Standards*, *48*, 84 (1952).

52Ta C. Tanford and G. L. Roberts, *J. Amer. Chem. Soc.*, *74*, 2509 (1952).

52Wa H. F. Walton and A. A. Schilt, *J. Amer. Chem. Soc.*, *74*, 4995 (1952).

1951

51Ba R. G. Bates, *J. Res. Natl. Bur. Standards*, *47*, 127 (1951).

51Bb R. G. Bates and R. G. Carham, *J. Res. Natl. Bur. Standards*, *47*, 343 (1951).

51Bc R. G. Bates and G. D. Pinching, *J. Res. Natl. Bur. Standards*, *46*, 349 (1951).

51Ea A. G. Evans and S. D. Hamann, *Trans. Faraday Soc.*, *47*, 25 (1951).

51Eb A. G. Evans and S. D. Hamann, *Trans. Faraday Soc.*, *47*, 34 (1951).

1951 (continued)

51Ka E. J. King, *J. Amer. Chem. Soc.*, *73*, 155 (1951).

51Ma R. O. MacLoren and D. F. Swinehart, *J. Amer. Chem. Soc.*, *73*, 1822 (1951).

51Mb M. May and W. A. Felsing, *J. Amer. Chem. Soc.*, *73*, 406 (1951).

1950

50Aa T. Ayrapaa, *Suensk. Kem. Tid.*, *62*, 135 (1950).

50Ba C. R. Bertsch, W. C. Fernelius, and B. P. Block, *J. Phys. Chem.*, *62*, 444 (1950).

50Ha H. S. Harned and B. B. Owen, *The Physical Chemistry of Electrolytic Solutions*," *2nd Ed.*, Reinhold Pub. Corp., New York, N.Y. (1950), p. 514.

50Ja H. B. Jonassen, R. Bruce, R. B. LeBlanc, A. W. Meibohm and R. M. Rogan, *J. Amer. Chem. Soc.*, *72*, 2430 (1950).

50Pa G. D. Pinching and R. G. Bates, *J. Res. Natl. Bur. Standards*, *45*, 322 (1950).

50Pb G. D. Pinching and R. G. Bates, *J. Res. Natl. Bur. Standards*, *45*, 444 (1950).

50Sa E. E. Sager and I. J. Sieweis, *J. Res. Natl. Bur. Standards*, *45*, 489 (1950).

1949

49Ba R. G. Bates and G. D. Pinching, *J. Res. Natl. Bur. Standards*, *42*, 419 (1949).

49Bb R. G. Bates and G. D. Pinching, *J. Amer. Chem. Soc.*, *71*, 1274 (1949).

49Ea M. G. Evans and N. Uri, *Trans. Faraday Soc.*, *45*, 224 (1949).

49Ha J. Z. Hearon, D. Burk and A. L. Schade, *J. Natl. Cancer Inst.*, *9*, 337 (1949).

49Ka P. Krumholz, *J. Amer. Chem. Soc.*, *71*, 3654 (1949).

49La D. L. Levi, W. S. McEwan and J. H. Wolfenden, *J. Chem. Soc.*, 760 (1949).

1948

48Ca T. L. Cottrell, G. W. Drake, D. L. Levi, K. J. Tully and J. H. Wolfenden, *J. Chem. Soc.*, 1016 (1948).

48Cb T. L. Cottrell and J. H. Wolfenden, *J. Chem. Soc.*, 1019 (1948).

48Pa G. D. Pinching and R. G. Bates, *J. Res. Natl. Bur. Standards*, *40*, 405 (1948).

1947

47Ma R. L. Moore and W. A. Felsing, *J. Amer. Chem. Soc.*, *69*, 2420 (1947).

1946

46Aa J. G. Aston and C. W. Ziemer, *J. Amer. Chem. Soc.*, *68*, 1405 (1946).

1945

45Ha W. J. Hamer and S. F. Acree, *J. Res. Natl. Bur. Standards*, *35*, 381 (1945).

45Hb W. J. Hamer, G. D. Pinching and S. F. Acree, *J. Res. Natl. Bur. Standards*, *35*, 539 (1945).

45Ja H. O. Jenkins, *Trans. Faraday Soc.*, *41*, 138 (1945).

45Ka E. J. King, *J. Amer. Chem. Soc.*, *67*, 2178 (1945).

1943

43Ba R. G. Bates and S. F. Acree, *J. Res. Natl. Bur. Standards*, *30*, 129 (1943).

43Bb R. G. Bates, G. L. Siegel and S. F. Acree, *J. Res. Natl. Bur. Standards*, *31*, 205 (1943).

43Bc J. S. P. Blumberger, *Rec. Trav. Chim.*, *62*, 753 (1943).

43Ta D. Turnbull and S. H. Maron, *J. Amer. Chem. Soc.*, *65*, 212 (1943).

1942

42Sa P. K. Smith, A. T. Gorham and E. R. B. Smith, *J. Biol. Chem.*, *144*, 737 (1942).

42Sb E. R. B. Smith and P. K. Smith, *J. Biol. Chem.*, *146*, 187 (1942).

42Sc J. M. Sturtevant, *J. Amer. Chem. Soc.*, *64*, 77 (1942).

42Sd J. M. Sturtevant, *J. Amer. Chem. Soc.*, *64*, 762 (1942).

1941

41Ea D. H. Everett and W. F. K. Wynne-Jones, *Proc. Royal Soc. (London)*, *A177*, 499 (1941).

41Ha H. S. Harned and T. R. Dedell, *J. Amer. Chem. Soc.*, *63*, 3308 (1941).

41Hb H. S. Harned and S. R. Scholes, *J. Amer. Chem. Soc.*, *63*, 1706 (1941).

41La N. C. C. Li and Y. Tulo, *J. Amer. Chem. Soc.*, *63*, 394 (1941).

41Ra F. J. W. Roughton, *J. Amer. Chem. Soc.*, *63*, 2930 (1941).

41Sa J. M. Sturtevant, *J. Amer. Chem. Soc.*, *63*, 88 (1941).

1940

40Aa J. E. Ablard, D. S. McKinney and J. C. Warner, *J. Amer. Chem. Soc.*, *62*, 2181 (1940).

40Ha W. J. Hamer, J. O. Burton and S. F. Acree, *J. Res. Natl. Bur. Standards*, *24*, 269 (1940).

40Hb H. S. Harned and R. A. Robinson, *Trans. Faraday Soc.*, *36*, 973 (1940).

40Ka G. Kegeles, *J. Amer. Chem. Soc.*, *62*, 3230 (1940).

1939

39Ea D. H. Everett and Wynne-Jones, *Trans. Faraday Soc.*, *35*, 1380 (1939).

39Ha H. S. Harned and L. D. Fallon, *J. Amer. Chem. Soc.*, *61*, 3111 (1939).

39Hb H. S. Harned and B. S. Owen, *Chem. Reviews*, *25*, 31 (1939).

39La W. M. Latimer and H. W. Zimmerman, *J. Amer. Chem. Soc.*, *61*, 1550 (1939).

39Wa A. W. Walde, *J. Phys. Chem.*, *43*, 431 (1939).

1938

38Wa W. F. K. Wynne-Jones and G. Saloman, *Trans. Faraday Soc.*, *34*, 1321 (1938).

1937

37Ma A. W. Martin and H. V. Tartar, *J. Amer. Chem. Soc.*, *59*, 2672 (1937).

37Pa K. S. Pitzer, *J. Amer. Chem. Soc.*, *59*, 2365 (1937).

37Sa P. K. Smith, A. C. Taylor and E. R. B. Smith, *J. Biol. Chem.*, *122*, 109 (1937).

37Wa H. Wilson and R. K. Cannon, *J. Biol. Chem.*, *119*, 309 (1937).

1936

36Na L. F. Nims, *J. Amer. Chem. Soc.*, *58*, 987 (1936).

36Nb L. F. Nims and P. K. Smith, *J. Biol. Chem.*, *113*, 145 (1936).

36Wa W. F. K. Wynne-Jones, *Trans. Faraday Soc.*, *32*, 1397 (1936).

1935

35Sa T. Shedlovsky and D. A. MacInnes, *J. Amer. Chem. Soc.*, *57*, 1705 (1935).

35Wa P. S. Winnek and C. L. A. Schmidt, *J. Gen. Physiol.*, *18*, 889 (1935).

1934

34Ca H. D. Crockford and T. R. Douglas, *J. Amer. Chem. Soc.*, *56*, 1472 (1934).

34Ha H. S. Harned and R. O. Sutherland, *J. Amer. Chem. Soc.*, *56*, 2039 (1934).

34Hb H. S. Harned and N. D. Embree, *J. Amer. Chem. Soc.*, *56*, 1042 (1934).

34Hc H. S. Harned and N. D. Embree, *J. Amer. Chem. Soc.*, *56*, 1050 (1934).

34Ja H. F. Johnstone and P. W. Leppla, *J. Amer. Chem. Soc.*, *56*, 2233 (1934).

34Na L. F. Nims, *J. Amer. Chem. Soc.*, *56*, 1110 (1934).

34Oa B. B. Owen, *J. Amer. Chem. Soc.*, *56*, 24 (1934).

34Ob B. B. Owen, *J. Amer. Chem. Soc.*, *56*, 1695 (1934).

34Wa D. D. Wright, *J. Amer. Chem. Soc.*, *56*, 314 (1934).

1933

33Ea J. T. Edsall and M. H. Blanchard, *J. Amer. Chem. Soc.*, *55*, 2337 (1933).

33Ga J. P. Greenstein, *J. Bio. Chem.*, *101*, 603 (1933).

33Ha H. S. Harned and R. W. Ehlers, *J. Amer. Chem. Soc.*, *55*, 652 (1933).

33Hb H. S. Harned and W. J. Hamer, *J. Amer. Chem. Soc.*, *55*, 2194 (1933).

33Hc H. S. Harned and R. W. Ehlers, *J. Amer. Chem. Soc.*, *55*, 2379 (1933).

33Na L. F. Nims and P. K. Smith, *J. Biol. Chem.*, *101*, 401 (1933).

1931

31La R. H. Lambert and L. J. Gillespie, *J. Amer. Chem. Soc.*, *53*, 2632 (1931).

31Ra F. D. Rossini, *J. Res. Natl. Bur. Standards*, *6*, 847 (1931).

1930

30Ba G. E. K. Branch and S. Miyamoto, *J. Amer. Chem. Soc.*, *52*, 863 (1930).

30Da J. B. Dalton, P. L. Kirk and C. L. A. Schmidt, *J. Biol. Chem.*, *88*, 589 (1930).

30Ma S. Miyamoto and C. L. A. Schmidt, *J. Biol. Chem.*, *90*, 165 (1930).

30Sa C. L. A. Schmidt, P. L. Kirk and W. R. Appleman, *J. Biol. Chem.*, *88*, 285 (1930).

1929

29Ra T. W. Richards and L. P. Hall, *J. Amer. Chem. Soc.*, *51*, 731 (1929).

29Rb T. W. Richards and B. J. Moir, *J. Amer. Chem. Soc.*, *51*, 737 (1929).

1928

28Sa W. C. Stadie and E. R. Hawes, *J. Biol. Chem.*, *77*, 241 (1928).

1927

27Wa J. H. Wolfenden, W. Jackson and H. B. Hartley, *J. Phys. Chem.*, *31*, 850 (1927).

1925

25Ba F. Bradley and W. C. M. Lewis, *J. Phys. Chem.*, *29*, 782 (1925).

25Bb N. Bjerrum, A. Unmack and L. Zechmeisler, *Kgl. Danske Videnskab. Selskab Math fys Medd.*, *5*, *No. 11*, (1925).

1922

22Ra T. W. Richards and A. W. Rowe, *J. Amer. Chem. Soc.*, *44*, 684 (1922).

1921

21Na S. M. Neale, *Trans. Faraday Soc.*, *17*, 505 (1921).

1910

10Na A. A. Noyes, Y. Kato and R. B. Sosman, *Zeitschr. Physik. Chemie*, *73*, 1 (1910).

1907

07Na A. A. Noyes, *The Electrical Conductivity of Aqueous Solutions*, *Carnegie Institution of Washington, DC*, *Publ. No. 63*, (1907).

1882

82Ta J. Thompson, *Thermochemische Untersuchugen*, *1*, 162 (1882).